AF361738

Sustainable Land Management and Ecosystem Services in Agroforestry Systems

Sustainable Land Management and Ecosystem Services in Agroforestry Systems

Editors

Jay Mar D. Quevedo
Norie Tamura
Yuta Uchiyama
Ryo Kohsaka

MDPI • Basel • Beijing • Wuhan • Barcelona • Belgrade • Manchester • Tokyo • Cluj • Tianjin

Editors
Jay Mar D. Quevedo
The University of Tokyo
Japan

Norie Tamura
Research Institute for
Humanity and Nature
Japan

Yuta Uchiyama
Kobe University
Japan

Ryo Kohsaka
The University of Tokyo
Japan

Editorial Office
MDPI
St. Alban-Anlage 66
4052 Basel, Switzerland

This is a reprint of articles from the Special Issue published online in the open access journal *Land* (ISSN 2073-445X) (available at: https://www.mdpi.com/journal/land/special_issues/sustainable_land_management_ecosystem_services_agroforestry_systems).

For citation purposes, cite each article independently as indicated on the article page online and as indicated below:

LastName, A.A.; LastName, B.B.; LastName, C.C. Article Title. *Journal Name* **Year**, *Volume Number*, Page Range.

ISBN 978-3-0365-5489-1 (Hbk)
ISBN 978-3-0365-5490-7 (PDF)

Contents

land

Article

Evaluating the Impact of Forest Tenure Reform on Farmers' Investment in Public Welfare Forest Areas: A Case Study of Gansu Province, China

Yuge Wang [1,†], Apurbo Sarkar [1,†], Min Li [1], Zehui Chen [1], Ahmed Khairul Hasan [2], Quanxing Meng [1,*], Md. Shakhawat Hossain [3] and Md. Ashfikur Rahman [4]

[1] College of Economics and Management, Northwest A&F University, Xianyang 712100, China; wangyuge@nwafu.edu.cn (Y.W.); apurbo@nwafu.edu.cn (A.S.); limin1983@nwafu.edu.cn (M.L.); chenzh@nwafu.edu.cn (Z.C.)

[2] Department of Agronomy, Bangladesh Agricultural University, Mymensingh 2202, Bangladesh; akhasan@bau.edu.bd

[3] College of Agronomy, Northwest Agriculture and Forestry University, Xianyang 712100, China; shakhawat@nwafu.edu.cn

[4] Development Studies Discipline, Social Science School, Khulna University, Khulna 9208, Bangladesh; ashfikur@ku.ac.bd

[*] Correspondence: qsmeng@nwafu.edu.cn

[†] These authors contributed equally to the study.

Citation: Wang, Y.; Sarkar, A.; Li, M.; Chen, Z.; Hasan, A.K.; Meng, Q.; Hossain, M.S.; Rahman, M.A. Evaluating the Impact of Forest Tenure Reform on Farmers' Investment in Public Welfare Forest Areas: A Case Study of Gansu Province, China. *Land* 2022, 11, 708. https://doi.org/10.3390/land11050708

Academic Editors: Jay Mar D. Quevedo, Norie Tamura, Yuta Uchiyama and Ryo Kohsaka

Received: 9 April 2022
Accepted: 5 May 2022
Published: 9 May 2022

Publisher's Note: MDPI stays neutral with regard to jurisdictional claims in published maps and institutional affiliations.

Abstract: In recent times, forest tenure reform has become one of the most discussed agendas among local and global policymakers. Forest tenure is a contract that specifies who has rights to forestry resources and depicts who should utilize, maintain, and acquire them. It can have a significant impact on whether farmers invest in their forestland. The study's primary purpose is to explore whether and how the reform of forest rights affects farmers' investment in public welfare forestry. More specifically, the study thoroughly analyzes the impact of primary and supplementary reforms on farmers' investment in public welfare forest areas. We have outlined the theoretical framework using the theory of property rights and utilized the fixed-effect model and the Difference in Differences (DID) model to achieve research objectives. However, the empirical setup of the study has comprised time series data of 500 farmers, which was collected via interviews conducted at regular time intervals (2011—before the reform; 2013, 2015, and 2017— after the reform). The collective forest land welfare areas in Gansu Province, China, have been selected as the key data collection area. The study concludes that: (i) although the principle reform of forest tenure can stimulate farmers' investment intensity in the short term, it is insufficient in the long term. (ii) The supplementary reform of forest tenure can significantly promote farmers' long-term effective investment. There is a significant difference in forest land investment between the experimental and control groups, and this difference gradually expands over time. The study suggests that the government should pay more attention to the relevance of additional reforms to encourage the growth of forest rights mortgages and circulation. Moreover, the core themes of sustainable development in forestry should be highlighted.

Keywords: collective forest rights; land tenure; reform; forestland investment; public welfare; forest land

1. Introduction

Nowadays, the present world confronts several interconnected socio-economic and ecological problems such as the climatological issue, the widespread grasp of novel pandemics, increasing social discrimination with pervasive hunger, and the threat of losing global biodiversity [1]. These issues lead to the pressing necessity of restoring and managing land and forests sustainably [2,3]. According to recent studies, land and forest tenure reforms can boost collective forest occupancy [4,5], and it has progressively been cited as an efficient way to decrease deforestation, combat the adverse impact of climate change, retain biodiversity, and restore natural ecosystems [6–8]. Specifically, it may foster a vital boost for facilitating a smooth transition to the sustainable development of forest land within remote Chinese mountainous regions [9]. Since the Brundtland Report was released in the 1970s, also called Our Common Future, the publication presented by the World Commission on Environment and Development (WCED) that familiarized the notion of sustainable development and defined how it could be attained, the importance of collective forest rights has retained much appreciation towards sustainable forest management [10]. The essential element of the tenure reform was to provide farmers user rights on land collectively owned by villages [11]. Various regions have endeavored to expand formal forest ownership structures and use the potential opportunities for involving local communities. In Africa, Asia, and Latin America, around 28% of forests are officially possessed or allocated for usage by native communities [12,13].

In the early 1980s, China adopted agricultural land tenure reform, which also fostered seeds for the availing of rural forest tenure reform, and since then, the government has allowed the privatization of some collective community forestry resources [14]. Aligned with agriculture land tenure reform, China has implemented the reform of collective forest tenure reform, which is divided into two stages: (i) the primary reform stage and (ii) supplementary reform. The primary reform stage began in 2008 and was completed by the end of 2011. The purpose of the primary reform is to clarify the property rights, endow farmers with the right to use and benefit from collective forest land, enable farmers to obtain important means of production, and promote farmers' employment and income [15]. Moreover, it also ensures the active participation of associated farmers, determines the ranges of forest land, and issues forest tenure certificates for farmers to exercise their rights and supervise the completion of the reform. The supporting reform began in 2012 and is still ongoing. The purpose of the supplementary reform is to promote the improvement of forest land mortgage and circulation and so on, to realize the sustainable development of collective forest land [16]. Seemingly, it also monitors the implementation of supporting reforms through follow-up surveys. Interestingly, the collective forests cover around 58% of China's forest territory and have the potential to significantly improve livelihood opportunities [17]. China has adopted several policies to uphold the possibilities of forest land reform, which concentrates on providing local families land-use rights and forest governance in collective forest regions. This enables communities to make earnings and enhance their lifestyles by embracing communal forest lands and forests. The ongoing forestry tenancy reforms will distribute 167 million hectares of forest land to local households, with around 500 million farmers expected to benefit by 2025 [18]. Within the reform mechanism, around 35% of the overall communal forest has already been distributed to individual families [19].

Moreover, with the ever-increasing pressure of mitigating climate change and global warming, China has initiated a new phase of collective reform of the forest property rights system to enhance the poor productivity of communal forest land while restoring and protecting the ecological environment of vital water conservation regions [20]. The core aspect of the changes (such as demarcation, confirmation, and certification) was completed at the end of 2011, and farmers were given the complete rights to manage forestland and the ownership of forest trees [21]. Around 99% of the collective forest land was distributed to the local communities till then [22]. Moreover, several regions have begun additional changes by boosting the development of forest land credit and forest property rights transfers. As a result, farmers' mortgage and transaction rights are strengthened substantially, and the diversification of collective forest land development is encouraged [23,24].

Existing literature highlights that the relationship between forest property rights reform and forest land investment is primarily focused on two dimensions. The first dimension deals with farmers' interpersonal behaviors and feelings about forest tenure reform [25–27]. The second dimension fosters the impact of farmers' subjective evaluations of forest tenure reform on investment [28–30]. However, most of the existing studies focus exclusively on the primary reform of forest tenure and overlook the impact of subsidiary reforms on forest land investment [31,32]. The development of public welfare forests may be a significant challenge for sustaining the ecological environment, as it is a continuous development process and inextricably linked to forest farmers' long-term and effective investment [33,34]. Farmers need to make a long-term and effective investments to ensure the sustainable growth of communal forests, which is one of the main objectives of new rounds of collective forest rights system reforms [35,36]. While, past studies have relied substantially on small, short-term samples and have failed to capture the dynamic influence of the new round of collective forest land usufruct confirmation on farmers' long-term investment in forest land. The research on forest land investment is also disproportionately focused on economic forest areas, while minimal attention has been paid to public welfare forest areas. Specifically, economic wellbeing and communal growth are critical aspects of investments that may foster the farmers' collective forest management and conservation [37–39]. Therefore, studying forest land investment in public welfare forest areas is necessary.

After the forest tenure reform, the trees in public welfare forest areas are protected and cannot be cut down, reducing farmers' timber income [40]. Forestry subsidies and benefits from secondary economics make up the majority of farmers' forestry revenue, causing the forestry development of public welfare forest regions to shift to other sustainable forest usage elements such as non-timber forest products on forest land [41,42]. This context impacts the method of forest rights reforms in public benefit forest regions and farmers' investments. Interestingly, some studies have found that farmers' long-term investment in forest land shows an unstable growing trend [43,44]. This notion leads to the following research questions, which need to be explored for understanding the critical impacts of forest tenure reform: (i) Whether the ongoing incentive of collective forest rights reforms affects forest land investments? (ii) What would be the possible effect of the supplementary forest tenure reforms on farmers' investments? (iii) Is there any deviation between the investment intensity within a certain period? (iv) Is there any combined effects from the primary and supplementary reform policies?

This study tracks the changes in forest farmers' forest land investment during regular intervals (2011, 2013, 2015, and 2017) using property rights theory and monitoring panel data from the forest rights system reform. The impact of the primary reform (forest land certification) and supplementary reform policies (development of forest right mortgages and forest right transfers) on farmers' forestland investment has been assessed using the fixed-effect and DID models. These are the main innovations of this study. This study's findings might help determine if the forest property rights reform's primary and supplementary reforms have long-term consequences. On the other hand, this study demonstrates a more profound knowledge of the relationship between forest property rights reform and long-term investment by farmers and a prospective for a new direction for follow-up reform.

2. Materials and Methods

2.1. Literature Review

In this part, this study discusses the theoretical background, portrays the formulation of the hypothesis, and outlines the adopted methods and methodology.

2.1.1. Background Studies and Hypothesis Development

Small farm sizes and low productivity can be ameliorated by letting farmers transfer farmland to others for agricultural production. However, private land ownership is banned in China, and therefore insecure ownership of farmland may cause a serious burden for the farmers to invest spontaneously. Under China's current Household Responsibility System (HRS), which started in the early 1980s, all rural land is owned by rural collectives, which allocate contract rights for parcels of farmland to eligible households (NPC 2017). Under the contract rights, farmers can decide what to plant and how, keep returns from their agricultural production, and lease their land to others for agricultural production.

Currently, land transfers to firms represent 10.5 percent of all transfers (or 3.8 percent of all arable land), but their growth has been slow in recent years [45]. An important factor is that the property rights of rural land are insecure and unclear. This is manifested in inaccurate land borders and sizes, incomplete land use right certificates, and a limited HRS tenure. Therefore, a property rights system with a clear definition is essential in realizing the effective allocation of production factors [46,47]. The Chinese government just extended HRS tenure to 2057 and is in the process of issuing land use right certificates with more accurate land positions and size information. This effort is expected to boost the land rental market in the future.

The impact of forest rights reform on forest land investment is mainly divided into two aspects: the primary reform's impact and the supplementary reform's impact [48], which are shown in Figure 1. The primary reform improves farmers' expectations of stable property rights and security perceptions through issuing certificates [49,50]. Previous studies have shown that unstable property rights will make farmers lack long-term expectations for the plots they use [51] and have a negative impact on farmers' investment incentives [52–54]. After analyzing the impacts of forestland distributions among local farmers, private producers, and land-poor households of Nicaragua, Deininger and Chamorro [55] identified that the confirmation of forest land ownership can reduce the risk of the random adjustment and expropriation of forest land, increase farmers' investment enthusiasm, improve farmers' sense of security in obtaining income, and impel their investment willingness and behavior. Moreover, Ghebru and Holden [56] explored farm-level data of Ethiopian rural forestland and confirmed that through the accurate mapping of the plot, the ownership can be clarified, the unclear income ownership caused by the ambiguity of property rights can be avoided, the cost of forest land disputes and mediation can be reduced, and farmers' investment in forest land can be increased.

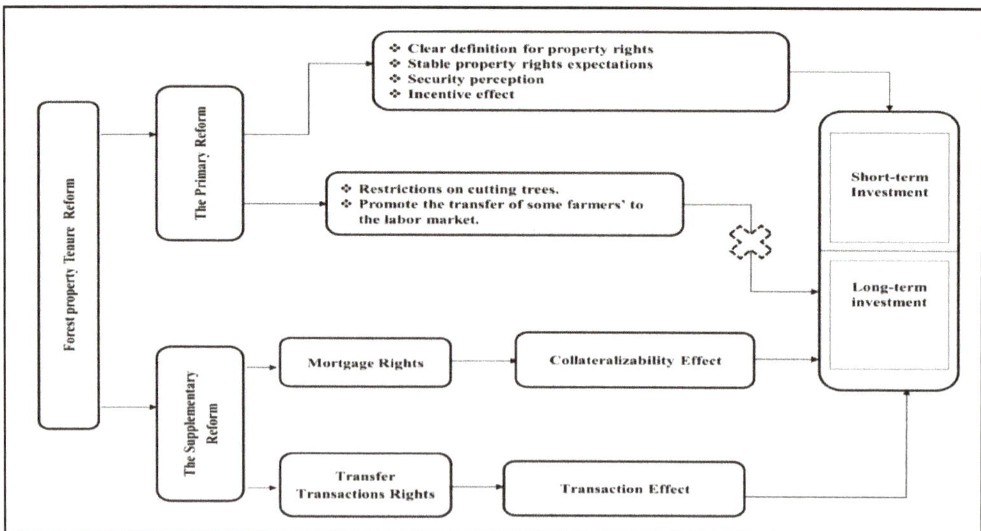

Figure 1. Effect mechanism of forest tenure reform on farmer's investment in the public welfare forest area.

Moreover, existing studies have studied the relationship between property ownership, management rights, mortgages, transaction rights, and different investments. However, there is no consensus on the relationship between property rights and land investment [55]. Some scholars have affirmed the positive incentive between rights confirmation and farmers' investment [57–59] and believe that the stability and duration of property rights play an essential role in encouraging farmers' investment and production decisions [45,60]. Some studies have not observed the relationship between right granting and land investment (for example the study of Haley and Nelson [61] regarding crown forest tenure systems; Holden and Yohannes [62], exploring Southern Ethiopian farm households and Carter and Olinto [63], evaluating Paraguan farmers forest tenure rights systems). Moreover, by exploring existing trends in the literature it can be found that they mostly focus on exploring the relationship between the primary reform of property rights and investment rather than supplementary reform [20,31]. Interestingly, most of the related literature rarely explored the potential impacts of the primary reform and supplementary reform in fostering short term and long-term investment behavior in public welfare forest regions. For example, in a study of African nations, Conigliani et al. [64] explored the relationship between farmers' investment behaviors and institutional contracts, but the study exclusively focused on large-scale farm dimensions with long-term investments. Lönnstedt and Sedjo [65] explored how investment in forestland is changing as per the alteration of forestland ownership changes in the United States and Sweden, and they solely focused on long-terms effects.

However, the nature of forest land in public welfare forest areas makes farmers lose their ownership of trees and only have the usufruct and management rights of forest land. This limits the incentive effect of the primary reform in public welfare forest areas on farmers' investment behavior. Low returns on investment are an important reason why farmers are unwilling to invest [16]. Borras [66] explored the investment behavior of farmers in the public forest of the Philippines and found that farmers' investment behavior has the characteristics of short-term monetization, and the characteristics of forestry management often require long-term investment to obtain corresponding returns. Therefore, when the income cycle does not match farmers' expected cycle, they often choose to reduce forestland investment. At the same time, the reform of forest property rights promotes the transfer of some farmers to the labor market [67,68], which further reduces farmers' enthusiasm for forest investment [69,70]. Therefore, the lack of property rights may make the long-term incentive of the primary reform regarding forest land investment insufficient. The study outlines Hypothesis one (H1) and Hypothesis two (H2) as follows:

H1: *the primary reform of forest tenure doesn't have any significant effects on farmers' short-term investment.*

H2: *the primary reform of forest tenure has no positive incentive for farmers' long-term investment.*

Supplementary reforms liberalize the rights of farmers' mortgage and transfer transactions, which is necessary for forest land investment. Credit shortage is the key factor determining the production performance and development of agroforestry, and budget constraints are an important factor that limits farmers' input [71–73]. The supplementary reform of forest tenure gives farmers the right to forest land mortgages and makes varying degrees of efforts for the implementation of the mortgage (such as actively communicating with relevant banks and promoting the improvement and revision of loan treaties on forest rights). This measure makes it possible for farmers to obtain forest right credits and effectively promotes farmers' resource allocation so that the potential collateralizability effect can appear. In a study of Latin American rural and indigenous women, Bose [74] has shown that farmers' access to credit can increase the dual impact on variable and fixed inputs. Similarly, Ceddia et al. [75] explored the relationship between land rights and agricultural expansion among Latin American indigenous communities and identified that if the continuous improvement of the credit right of forest land can be ensured, farmers' long-term expectations of obtaining forest rights credits is more stable, and they tend to be more willing to make a long-term investment in forest land. Therefore, the success of the supplementary reform of forest rights mortgages has a long-term and positive impact on forest land investment.

The development of the forest land circulation market can liberalize the allocation of forest land means of production and enhance the efficiency of farmers' resource allocation [76]. Households with obtained land certification are more likely to rent out or rent in the land than the not-obtained ones [77,78]. There is evidence that leased land's input use and productivity are higher than self-owned land [79,80]. Moreover, safe trading rights can encourage farmers to invest more and grow long-term trees, which is supported by the study of the Brazilian Amazon [81]. In addition, even if the long-term investment of farmers cannot be recovered temporarily, farmers can also realize the realization effect through circulation to reduce the investment risk [82,83]. Stickler et al. [84] identified similar assumptions among community-driven forest owners of Zambia. Therefore, to maintain the circulation value of forest land, farmers are bound to maintain the management and protection of forest land or fertilizer for a long time. Therefore, this article depicts the third hypothesis as follows:

H3: *There is no association of the supplementary reform of forest tenure to promoting farmers' long-term investment in forest land.*

2.1.2. Methodology

This study uses a combination of the fixed-effect model and DID model to compare the changes in forest land investment of forest households before and after implementing the forest rights reform policy. We mainly considered the different implementations of the primary and supplementary reforms. The state predominantly initiates the primary reform, and all sample farmers have to carry out the reform to ensure that the forest rights certificate is issued in the hands of each farmer. Therefore, using the fixed-effect model can better highlight the relationship between the reform of forest property rights and farmers' long-term investment behavior, as suggested by Lu et al. [85]. The fixed-effect model set in this study is as follows:

$$y = \beta_0 + \beta_i \chi'_{it} + \delta_i z'_i + \lambda_t + \mu_i + \varepsilon_{it} \tag{1}$$

In the formula, y is the farmers' investment, χ'_{it} is the explanatory variable such as the right confirmation period, β_0 is the intercept term, β_i is the coefficient parameter corresponding to the explanatory variable. Seemingly, z'_i is the control variable, δ_i is its corresponding parameter, λ_t represents the time effect that does not change due to the individual, μ_i represents the individual effect that does not change with time, and ε_{it} is the random disturbance term.

However, the supplementary reform is carried out in an orderly manner in combination with the actual conditions [22]. Due to the different development of various regions, it is more suitable to use the DID model for analysis. The DID model is a standard method to identify the effectiveness of policies [86], which can test the average change in forest land investment in the experimental and control groups before and after implementing the forest tenure supplementary reform. Referring to Nunn and Qian [87], the model set in the article is as follows:

$$y_{it} = \alpha_0 + \alpha_1 du + \alpha_2 dt + \alpha_3 du * dt + \varepsilon_{it} \tag{2}$$

In the formula, y_{it} is the investment of farmer i in year t, and du is a grouped dummy variable. If the individual i is affected by the implementation of the policy, individual i belongs to the treatment group, and the corresponding du value is 1. If individual i is not affected by the implementation of the policy, individual i belongs to the control group, and the corresponding du value is 0. Where, dt is the dummy variable of policy implementation. Before policy implementation, dt is 0, and DT is 1 after policy implementation. While, $du*dt$ is the interaction between the grouped dummy variables and policy implementation dummy variables, and its coefficient reflects the net effect of policy implementation.

Farmers' investment has been selected as the dependent variable. The article chooses farmers' forestry production and management inputs to measure farmers' investment. It is mainly divided into the management and protection costs of forest land and inputs in understory planting. So, the investment includes the materials and labor cost for forest management and protection and the inputs of seeding, chemical fertilizers, pesticides, machinery, and the labor force in the development of the understory planting industry and other forestry management. Regarding the primary reform of forest tenure, the study selects the certification duration as the index to verify the long-term impact of the primary reform on forest land investment.

The interaction term between *dt* and *du* has been used as independent variable two. In terms of supplementary reform, the article uses the interaction term between *du* (regional dummy variable) and *dt* (time dummy variable) as an index to measure the effectiveness of the supplementary reform. According to the different development of forest right mortgages and forest right transfers, Jingchuan County, Hui County, and Huining county, with a perfect supplementary reform, are set as the experimental group with a value of 1 (*du* = 1), and other areas are set as the control group with a value of 0 (*du* = 0). Considering the lag effect of policy implementation, the article set the *dt* value before 2013 as 0 and after 2013 as 1.

The study used farmers' individual characteristics, family capital, and forest land characteristics as the control variables. These variables may affect forest farmers' investment and mainly include sex, age, the number of laborers, the number of migrant workers, the forest land area, the number of forest land blocks, and total household income [88–90]. The definition and descriptive statistics of the variables are shown in Table 1.

Table 1. Variable selection and descriptive statistics.

Variable	Implication	Mean	Min	Max
Farmers' Investment	Yuan	2672.969	0	650,000
Confirmation of Tenure for Two Years	Yes = 1, No = 0	0.250	0	1
Confirmation of Tenure for Four Years	Yes = 1, No = 0	0.250	0	1
Confirmation of Tenure for Six Years	Yes = 1, No = 0	0.250	0	1
dt	After 2013 = 1, others = 0	0.750	0	1
du	After policy implement = 1, Before policy implement = 0	0.300	0	1
DID	*du*∗*dt*	0.225	0	1
Age	age of the householder (years)	52.252	21	85
Sex	Male = 1, female = 0	0.949	0	1
Number of labors	The actual number of adult laborers in the family	2.739	0	8
Number of migrant labors	The actual number of migrant workers in the family	0.969	0	5
Forest land area	mu	35.822	0.3	940.66
Number of forest land blocks	blocks	2.892	1	20
Household total income	yuan	43,281.730	149	1,230,000

2.2. Data Resources

The data in the study has been comprised of time series data with regular time intervals of 2 years (2011, 2013, 2015, and 2017) to reflect the impact of forest land investment before and after the forest tenure reform, especially on long-term investment. The data for 2011 denotes before the reform, and 2013, 2015, and 2017 represent after the reform. The survey uses the stratified sampling method to select the peasant who participated in communal forestland. More specifically, we explored the collective public welfare forest areas, which refer to the forest, trees, and forest land owned by the collective, mainly to protect and improve the human living environment, maintain ecological balance, encourage tourism, and foster better livelihood opportunities for local communities [25]. It can be divided into shelter forests and special-purpose forests. First, a set of 10 counties has been chosen among 86 counties using the random sampling technique (which is shown in Figure 2). Next, five sample townships were chosen randomly from those counties, and then one village was chosen from each township. A set of ten peasant farming individuals were chosen randomly from each village, which comprised 500 peasants for final interviews. The main strengths of this sampling technique were that each member of the population had an exactly equal probability of being chosen using this sampling procedure [91]. It also employs randomization; any research conducted on this sample has excellent internal and external validity [92], because randomization is the most effective strategy for reducing the impact of potential confounding variables [93]. The empirical data has been collected through face-to-face interviews, where the responses have been taken based on a structured questionnaire. Finally, the study has obtained a total of 2000 questionnaires (500 peasants multiplied by four subsequent years) which have been further analyzed to fulfill the research objectives.

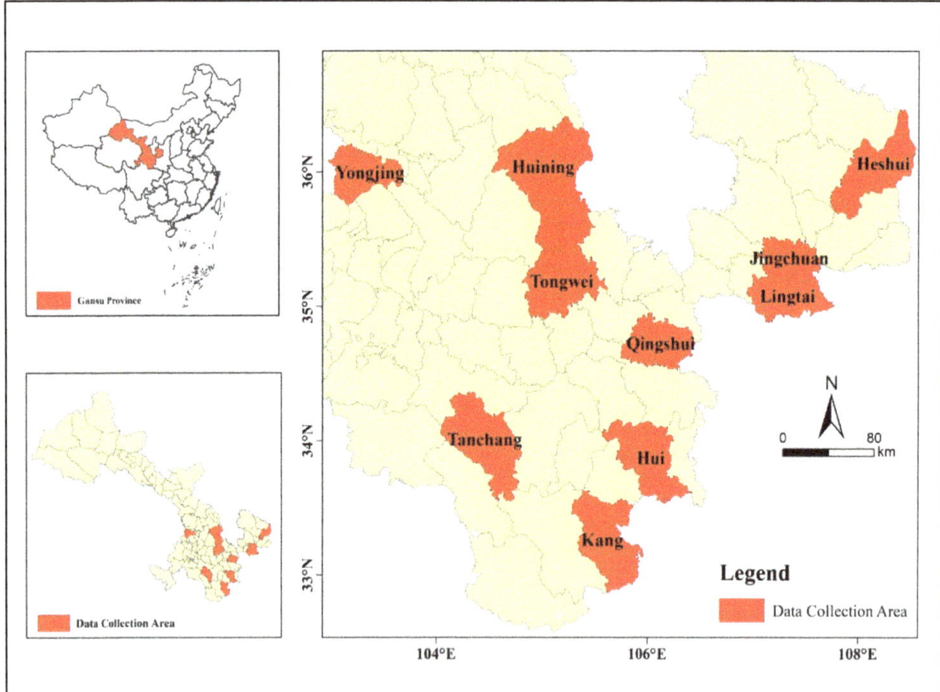

Figure 2. Study Area.

The group leader rechecked the questionnaire to identify any missing components, and a timely callback was made after each day's investigation was completed, which helped the study ensure the quality of the investigation. Before conducting the survey, the study took formal permission from the local forestry administrative authority and the local communist party. Before starting the formal interviews, each respondent was clearly informed and explained that the primary motives of the data collection were just for academic purposes, and the study would not store or share any form of the data for any business purposes. They were well aware that they can opt out anytime from answering any part of the questionnaire. Moreover, in the formal questionnaire, we have also included questions regarding verbal permission. Therefore, strict requirements for taking formal permission from the institutional review board were used, as suggested by Josephson and Smale [94] and Yanow and Schwartz-Shea [95]. Additionally, we have obtained verbal permission from each of the villages' heads, which helps us ensure a higher response rate. The survey was conducted as a part of the research project called "Monitoring of Collective Forest Tenure Reform-Gansu province". This is known as the first initiative to continuously track and monitor large-scale farmers to assess the effectiveness of communal forest property rights reforms. The progress of clarifying the property rights of communal forest lands, deepening the reform, increasing service follow-up, farmers' evaluations of the reform, and the policy needs are part of the monitoring content.

3. Results

3.1. Effect of the Primary Forest Tenure Reform on the Farmers' Investment in Public Welfare Forest Areas

According to the research framework and techniques that are given above, the fixed-effect model is used to quantify the impact of the major collective forest tenure change on farmers' investment. The Hausman test is used to determine the applicability of the fixed-effect model before adopting it. The Hausman test p-value is significant at 1%, showing that the fixed effect model is appropriate. Table 2 shows the results of the primary reform of collective forest tenure on farmers' investment. The primary reform of collective forests can promote the short-term investment in public welfare forest areas. However, the primary reform has an insufficient incentive for long-term investment. The regression coefficient between the primary reform of collective forests and short-term investment in public welfare forest areas is positive, and the p-value is less than 0.05, which means the null hypothesis 1 is rejected. In other words, it demonstrates that the primary reform of collective forests can promote short-term investment in public welfare forest areas. The regression coefficient between the primary reform and long-term investment is insignificant (p-value for four years of confirmation is more than 0.1) or negative (coefficient for six years of confirmation is negative and the p-value is less than 0.01). Therefore, hypothesis 2 is accepted. More specifically, it proved that the primary reform doesn't have a positive incentive for long-term investment, as the confirmation period increases.

Table 2. The impact of primary reform of collective forest on farmers' investment.

	(1)	(2)	(3)	(4)
Confirmation of tenure for two years	0.006(0.002) **	-	-	0.004(0.001) ***
Confirmation of tenure for four years	-	−0.002(0.001)	-	−0.002(0.002)
Confirmation of tenure for six years	-	-	−0.005(0.001) ***	−0.004(0.002) ***
age	−0.001(0.001)	−0.001(0.001)	−0.001(0.001)	−0.001(0.001)
sex	0.002(0.003)	0.002(0.003)	0.002(0.003)	0.003(0.003)
Education	0.001(0.001)	0.005(0.001)	0.001(0.001)	0.001(0.001)
Number of labors	−0001(00.001)	0.002(0.001)	−0.003(0.006)	−0.001(0.001)
Number of migrant labors	0.001(0.001)	0.008(0.001)	−0.005(0.007)	−0.001(0.001)
Forest land area/blocks	0.001(0.001) ***	0.001(0.001) ***	0.001(0.001) ***	0.001(0.001) ***
Household total income	0.234(0.014) ***	0.233(0.014) ***	0.236(0.136) ***	0.240(0.138) ***
Cons	−0.006(0.006)	−0.004(0.006)	−0.004(0.006)	−0.006(0.006)

Notes: **, and *** indicate significance at 5%, and 1% levels, respectively, and the standard errors are reported in parentheses.

The regression coefficient between them does not vary significantly, if at all, as the confirmation period increases. Farmers invest in public welfare forest regions for two reasons: one, to earn matching forestry subsidies by investing in forest resource management and protection, and two, to generate income through the growth of understory planting. Due to the primary change, farmers will benefit from more stable tenure and operating periods.

Farmers have higher hopes for forest land management stability and continued gains from understory planting in the early days of the significant reform. Moreover, they are eager to invest in forestry to increase their earnings [96].

However, with time, farmers discovered that they could still receive most of the payments even if they did not invest in forest management and conservation. Furthermore, the overall amount of forestry subsidies in public benefit forest regions is low (the average amount provided to families is 10 yuan/mu), which is insufficient to entice farmers to invest in long-term forest management and protection. On the other hand, under-forest planting is complex and necessitates advanced technology and capital support. However, technology and market development in the public welfare forest area are not yet mature enough, and there are risks and losses in production and operation, necessitating the use of borrowed funds to complete capital turnover. As a result, proper forest mortgages are becoming increasingly important to farmers.

The primary reform can pique forest households' interest in making a short-term investment, but it fails to address new farmer demands such as with circulation and mortgages, leaving farmers with a small motive to make long-term investments. In addition, the fragmentation of forest land has a significant negative impact on farmers' forestland investment. The smaller area per block rectifies a relatively greater degree of forest fragmentation and possesses less long-term investment of forest households. This is possible because the fragmentation of forest land increases the production loss [97,98], increases forestry production and management costs [99,100], reduces productivity, and weakens the farmer's investment. Besides, household income significantly promoted the forest land investment of forest households. It could be due to China's rural social security system [101]. Forest farmers expect that forest land resources will be one of the sources of livelihood security in the future [102,103], so they will correspondingly increase investment to ensure the sustainable development of forest land when the family income has been raised.

3.2. The Influence of Supplementary Reform on Farmers' Investment in Public Welfare Forest Areas

A parallel trend assumption test should be conducted before the DID test. The results showed that if the individuals in the treatment group did not receive intervention or impact, the changing trend of the results was the same as that in the control group. It illustrates that the premise of the double-difference method is met, and the DID model is appropriate.

Table 3 shows the impact of supplementary reform on farmers' forestry investment. Column (1) shows the results of not introducing the control variable, and column (2) shows the introduced results. The two results show that the long-term investment of forest farmers in counties with successful supplementary reform is higher than that in other counties, and the gap is more significant with time. That is to say, the coefficient between DID and farmers' investment is positive, and the p-value is less than 0.1, so therefore hypothesis 3 has been rejected. The effective promotion of supplementary reform can carry forward farmers' long-term input in management and protection of forest land. The whole forest land disposal right, for example, might save transaction costs and increase farmers' investment excitement. The right to mortgage and circulate forest land is crucial for the disposal tenure. The mortgage has a strong relationship with loan availability [104]. The easier it is for forest farmers to access forest management funds if the mortgage is liberalized correctly as part of the additional reform of forest tenure, the greater their investment and incentive ability. For forestry production's efficient distribution of land elements, forest land circulation is required. The seamless flow of land factors increases marginal productivity, increasing farmers' incentive to produce and invest [23].

Table 3. The impact of supplementary reform of collective forest son farmers' investment.

	(1)	(2)
DID	0.005(0.003) *	0.005(0.003) *
time	0.001(0.003) **	−0.001(0.001)
treat	0.004(0.001) ***	0.002(0.002)
Control variable	not-introduced	introduced
cons	0.001(0.001) ***	−0.046(0.021) **

Notes: *, **, and *** indicate significance at 10%, 5%, and 1% levels, respectively, and the standard errors are reported in parentheses.

3.3. Robustness Test

The study employed Propensity Score Matching (PSM) as a robustness test to verify whether households' investments before and after the forest rights reform have consistent outcomes, as suggested by Song et al. [105]. Referring to Smith and Todd [106] and Caliendo and Kopeining [107], the following model is constructed according to the general steps of PSM to calculate the average treatment effect. This study uses approximate randomization of non-random data to estimate the counterfactual probability of the treatment and control groups. The formula is set to $ATT = E(Y_{1i} + Y_{0i}|D_i = 1)$, where D_i is the treatment variable, Y_{1i} refers to the investment in the treatment group, and Y_{0i} indicates the investment in the control group.

The outcomes shown in Table 4 depict that the primary reform of forest tenure can promote farmers' investment in the short term, but the long-term incentive is insufficient. However, the supplementary reform can effectively improve farmers' long-term input. The results obtained are consistent with the above test results. Therefore, the research conclusion of this paper is relatively stable.

Table 4. Results of the treatment effect of forest tenure reform on farmers' investment.

	Treatment Effect	Treatment Group	Control Group	Gap	Standard Error	T-Value	Sig.
Confirmation of tenure for two years	Unmatched	2.906	2.252	0.655	0.188	3.48	yes
	ATT	2.885	2.162	0.723	0.222	3.25	
Confirmation of tenure for six years	Unmatched	1.126	2.845	−1.720	0.184	−9.31	yes
	ATT	1.126	2.911	−1.785	0.196	−9.12	
Supplementary reform	Unmatched	3.089	2.220	0.869	0.195	4.46	yes
	ATT	3.089	2.234	0.854	0.241	3.55	

4. Discussion

According to the tracking data, the investment of forest households in public welfare forest areas shows two distinct development trends after the reform of forest tenure. On the one hand, farmers' investments in public welfare forest areas have a pattern of increasing within the short term. However, in the long run, it shows declining trends. Farmers' investment is expected to rise in the short term once the primary reform is largely accomplished. In 2013, which is the second year after the completion of the primary reform, overall investment in forest land was 1.43 times what it was before the primary reform. Farmers' investment, on the other hand, shows a substantial drop in volatility as confirmation time increases. As a result, this study believes that the impact of the significant reform's confirmation and certification on farmers' investment has a certain amount of variability. As indicated in the literature review sections, most of the existing literature (such as Yi et al. [23], Holden et al. [29], and [84]) showed that the relationship between forest tenure and investment is one-sided (relevant or not) and that the long-term tracking of forest land investment changes is insufficient. As a result, the study explores if the long-term evolution of forest tenure reform has an uneven impact on forest households' short and long-term investment in public benefit forest regions to build a novel reference policy.

Farmers' investment in diverse regions, on the other hand, exhibits varied development tendencies as a result of the promotion of supplementary forest tenure reform, despite the tracked samples having identical personal and family characteristics. According to the findings of the previous study (for example, Wang et al. [108] and JingWen et al. [109]), the development of forest land circulation and forest right mortgages is causally related to farmers' investment. As a result, this study explores whether the supplemental reform plays a significant role in the distinct development of the investment and whether there is any relationship between supplementary forest tenure reform and farmer investment. Therefore, this study tests the impact of forest tenure's primary and supplementary reform on farmers' investment using the fixed-effect and DID models. The results show that the primary reform of forest tenure significantly affects farmers' short-term investment, which is consistent with the research of Yi et al. [23] and Ren et al. [25].

However, the primary change provides insufficient incentives for long-term investment by farmers, which corresponds to the current state of China's public welfare forest areas. Farmers' confidence and safety perceptions of earning revenue through certification have improved due to the significant reform, which has increased farmers' investment. Seemingly, the integrity of property rights is limited due to the constraints of public welfare forests and the fact that the major reform only provides farmers the right to use and manage them. The benefits to the subject will be lessened if certain of the rights inherent in property rights are lacking or limited. Over time, farmers' imperfect rights will not consistently supply their new product needs, reducing the incentive to invest in property rights. Therefore, it can be assumed that the major reform's encouragement of farmers' long-term investment is limited. The findings also demonstrate that the forest tenure supplementary reform has significantly boosted farmers' long-term investment. Forest farmers are more ready to manage and protect forest land or make other inputs if forest tenure mortgage and circulation develop in a controlled manner. Jacoby and Minten [110] and Melesse and Bulte [111] achieved similar conclusions when exploring the relationship between land property rights and investment, although their research is not focused on forest property rights reform and forestry investment. It is challenging to promote supplementary forest tenure reform, particularly in the loss of forest income in public welfare forest areas, where forest right mortgages and forest land transfers are complex [112]. Financial institutions typically consider that risk control of forest belt loans is still difficult to grasp and that the forest right certificate cannot fully fulfill the function of collateral. Farmers are allowed the right to mortgage forest land, but they are restricted in every step of the process. Farmers have less information about the transfer in forest land tenure reform because the forest land cannot be used for other purposes after the transfer, and standard forest land transfer procedures have not been developed [113]. Hence, farmers have less information about the transfer, and farmers frequently face difficulties such as information asymmetry during the transfer, causing the forest land transfer to fall short of its ideal state.

The whole forest land disposal right might save transaction costs and increase farmers' investment excitement. The right to mortgage and circulate forest land is crucial for the disposal tenure. The mortgage has a strong relationship with loan availability [104]. The easier it is for forest farmers to access forest management funds if the mortgage is liberalized correctly as part of the additional reform of forest tenure, the greater their investment and incentive ability. For forestry production's efficient distribution of land elements, forest land circulation is required. The seamless flow of land factors increases marginal productivity, increasing farmers' incentive to produce and invest [81].

5. Conclusions

The study's fundamental motive is to trace how forest land investment in public welfare forest regions changes over time. Based on the property tenure theory, the research employs the fixed-effect and DID models to examine the effects of forest rights subject reform and supplementary reform on forest land investment. While the existing studies frequently overlook studying forest land in public welfare forest areas, this study portrays the following outcomes (i) The primary reform of forest tenure can promote the forest farmers' forest land investment in the short term, but the long-term incentive is insufficient. (ii) There are inequalities in investment between regions with a stronger development of forest right supplementary reform and those without one, and the disparity increasingly widens over time. This demonstrates that via the continual improvement and promotion of the additional reform of forest tenure, the rights of forest households to transact and dispose of forest land have been increasingly liberalized, successfully stimulating long-term investment by forest farmers.

Based on the above research conclusions, the study puts forward the following policy suggestions: (i) Relying on the primary reform dividend to promote forest producers' investment is insufficient. Government must continue to pay attention to the development of additional reforms and the liberalization of mortgages and forest land circulation. (ii) In the process of supplementary reform, we can improve the participation of forest farmers in social credit and financial connections. (iii) Build a forest land transfer platform and strengthen the supervision and service of the forest rights transfer. (iv) Combine forest right mortgages with circulation reform and have the market play the decisive role in the pricing and disposal of forest right mortgages through circulation. (v) Enhance the government's role by using government funds to establish forest rights collection and storage centers or guarantee institutions provide comprehensive services such as forest rights collection and storage and forest right transfers. (vi) Relying on the primary reform dividend to promote forest producers' investment is insufficient. Government must continue to pay attention to the development of additional reforms. They should develop normative policy documents for forest rights mortgages and initiate the consistent transfer with local circumstances.

Author Contributions: Conceptualization, Y.W. and A.S.; methodology, Y.W. and M.S.H.; software, Y.W., M.L. and Q.M.; validation, A.S., Z.C., A.K.H. and M.A.R.; formal analysis, Y.W. and A.S.; investigation, Y.W. and A.S.; resources, A.S.; data curation, A.K.H., M.L. and M.A.R.; writing—original draft preparation, Y.W. and A.S.; writing—review and editing, Y.W. and A.S.; visualization, M.S.H., M.L. and Z.C.; supervision, Q.M.; project administration, Q.M. and A.S.; funding acquisition, Q.M. All authors have read and agreed to the published version of the manuscript.

Funding: This research was funded by the National forestry administration of China, grant number 2018-R08.

Institutional Review Board Statement: As the study does not involve any personal data and the respondent was well aware that they can opt out anytime during the data collection phase, any written institutional review board statement is not required.

Informed Consent Statement: As the study does not involve any personal data and the respondent was well aware that they can opt out anytime during the data collection phase, any written institutional review board statement is not required.

Data Availability Statement: The data will be provided upon request by the corresponding author.

Acknowledgments: The authors acknowledges the help and support of the anonymous reviewers for their valuable inputs.

Conflicts of Interest: The authors declare no conflict of interest.

References

1. Chen, J.; Innes, J.L. The Implications of New Forest Tenure Reforms and Forestry Property Markets for Sustainable Forest Management and Forest Certification in China. *J. Environ. Manag.* **2013**, *129*, 206–215. [CrossRef] [PubMed]
2. Aggarwal, S.; Larson, A.; McDermott, C.; Katila, P.; Giessen, L. Tenure Reform for Better Forestry: An Unfinished Policy Agenda. *For. Policy Econ.* **2021**, *123*, 102376. [CrossRef]
3. Larson, A.M.; Barry, D.; Dahal, G.R. New Rights for Forest-Based Communities? Understanding Processes of Forest Tenure Reform. *Int. For. Rev.* **2010**, *12*, 78–96. [CrossRef]

4. Xu, T.; Zhang, X.; Agrawal, A.; Liu, J. Decentralizing While Centralizing: An Explanation of China's Collective Forestry Reform since the 1980s. *For. Policy Econ.* **2020**, *119*, 102268. [CrossRef]

5. Valencia, L.M. Uphill Battle: Forest Rights and Restoration on Podu Landscapes in Keonjhar, Odisha. *J. South Asian Dev.* **2021**, *16*, 342–366. [CrossRef]

6. De Paula Pereira, A.S.A.; dos Santos, V.J.; do Carmo Alves, S.; e Silva, A.A.; da Silva, C.G.; Calijuri, M.L. Contribution of Rural Settlements to the Deforestation Dynamics in the Legal Amazon. *Land Use Policy* **2022**, *115*, 106039. [CrossRef]

7. Rai, S.; Dahal, B.; Anup, K.C. Climate Change Perceptions and Adaptations by Indigenous Chepang Community of Dhading, Nepal. *GeoJournal* **2022**, 1–7. [CrossRef]

8. Baul, T.K.; Peuly, T.A.; Nandi, R.; Kar, S.; Karmakar, S. Role of Homestead Forests in Adaptation to Climate Change: A Study on Households' Perceptions and Relevant Factors in Bandarban Hill District, Bangladesh. *Environ. Manag.* **2022**, *69*, 906–918. [CrossRef]

9. Yang, Y.; Li, H.; Cheng, L.; Ning, Y. Effect of Land Property Rights on Forest Resources in Southern China. *Land* **2021**, *10*, 392. [CrossRef]

10. Holden, E.; Linnerud, K.; Banister, D. Sustainable Development: Our Common Future Revisited. *Glob. Environ. Chang.* **2014**, *26*, 130–139. [CrossRef]

11. Larson, A.M. Forest Tenure Reform in the Age of Climate Change: Lessons for REDD+. *Glob. Environ. Chang.* **2011**, *21*, 540–549. [CrossRef]

12. Kashwan, P. What Explains the Demand for Collective Forest Rights amidst Land Use Conflicts? *J. Environ. Manag.* **2016**, *183*, 657–666. [CrossRef] [PubMed]

13. Pacheco, P.; Barry, D.; Cronkleton, P.; Larson, A. The Recognition of Forest Rights in Latin America: Progress and Shortcomings of Forest Tenure Reforms. *Soc. Nat. Resour.* **2012**, *25*, 556–571. [CrossRef]

14. Xie, L.; Berck, P.; Xu, J. The Effect on Forestation of the Collective Forest Tenure Reform in China. *China Econ. Rev.* **2016**, *38*, 116–129. [CrossRef]

15. Xu, J.T. Collective Forest Tenure Reform in China: What Has Been Achieved so Far. In Proceedings of the World Bank Conference on Land Governance, Washington, DC, USA, 26–27 April 2010.

16. Xu, J.; Hyde, W.F. China's Second Round of Forest Reforms: Observations for China and Implications Globally. *For. Policy Econ.* **2019**, *98*, 19–29. [CrossRef]

17. He, M.; Wu, Z.; Li, W.; Zeng, Y. Forest Certification in Collectively Owned Forest Areas and Sustainable Forest Management: A Case of Cooperative-Based Forest Certification in China. *Small-Scale For.* **2015**, *14*, 245–254. [CrossRef]

18. Xu, C.; Li, L.; Cheng, B. The Impact of Institutions on Forestland Transfer Rents: The Case of Zhejiang Province in China. *For. Policy Econ.* **2021**, *123*, 102354. [CrossRef]

19. Dong, J.; Liang, W.; Fu, Y.; Liu, W.; Managi, S. Impact of Devolved Forest Tenure Reform on Formal Credit Access for Households: Evidence from Fujian, China. *Econ. Anal. Policy* **2021**, *71*, 486–498. [CrossRef]

20. Yiwen, Z.; Kant, S. Secure Tenure or Equal Access? Farmers' Preferences for Reallocating the Property Rights of Collective Farmland and Forestland in Southeast China. *Land Use Policy* **2022**, *112*, 105814. [CrossRef]

21. Yu, J.; Wei, Y.; Fang, W.; Liu, Z.; Zhang, Y.; Lan, J. New Round of Collective Forest Rights Reform, Forestland Transfer and Household Production Efficiency. *Land* **2021**, *10*, 988. [CrossRef]

22. Ren, Y.; Kuuluvainen, J.; Toppinen, A.; Yao, S.; Berghäll, S.; Karppinen, H.; Xue, C.; Yang, L. The Effect of China's New Circular Collective Forest Tenure Reform on Household Non-Timber Forest Product Production in Natural Forest Protection Project Regions. *Sustainability* **2018**, *10*, 1091. [CrossRef]

23. Yi, Y.; Köhlin, G.; Xu, J. Property Rights, Tenure Security and Forest Investment Incentives: Evidence from China's Collective Forest Tenure Reform. *Environ. Dev. Econ.* **2014**, *19*, 48–73. [CrossRef]

24. Zhou, Y.; Ma, X.; Ji, D.; Heerink, N.; Shi, X.; Liu, H. Does Property Rights Integrity Improve Tenure Security? Evidence from China's Forest Reform. *Sustainability* **2018**, *10*, 1956. [CrossRef]

25. Ren, Y.; Kuuluvainen, J.; Yang, L.; Yao, S.; Xue, C.; Toppinen, A. Property Rights, Village Political System, and Forestry Investment: Evidence from China's Collective Forest Tenure Reform. *Forests* **2018**, *9*, 541. [CrossRef]

26. Yiwen, Z.; Kant, S.; Long, H. Collective Action Dilemma after China's Forest Tenure Reform: Operationalizing Forest Devolution in a Rapidly Changing Society. *Land* **2020**, *9*, 58. [CrossRef]

27. Zhang, Y.; Song, W.; Nuppenau, E.A. Farmers' Changing Awareness of Environmental Protection in the Forest Tenure Reform in China. *Soc. Nat. Resour.* **2016**, *29*, 299–310. [CrossRef]

28. Xie, Y.; Wen, Y.; Zhang, Y.; Li, X. Impact of Property Rights Reform on Household Forest Management Investment: An Empirical Study of Southern China. *For. Policy Econ.* **2013**, *34*, 73–78. [CrossRef]

29. Holden, S.T.; Yi, Y.; Jiang, X.; Xu, J. Tenure Security and Investment Effects of Forest Tenure Reform in China. In *Land Tenure Reform in Asia and Africa: Assessing Impacts on Poverty and Natural Resource Management*; Holden, S.T., Otsuka, K., Deininger, K., Eds.; Palgrave Macmillan: London, UK, 2013; pp. 256–282, ISBN 978-1-137-34381-9.

30. Segura Warnholtz, G.; Fernández, M.; Smyle, J.; Springer, J. *Securing Forest Tenure Rights for Rural Development: Lessons from Six Countries in Latin America*; World Bank: Washington, DC, USA, 2017; ISBN 978-0-9910407-8-0.

31. Qin, P.; Xu, J. Forest Land Rights, Tenure Types, and Farmers' Investment Incentives in China: An Empirical Study of Fujian Province. *China Agric. Econ. Rev.* **2013**, *5*, 154–170. [CrossRef]

32. Qin, P.; Carlsson, F.; Xu, J. Forest Tenure Reform in China: A Choice Experiment on Farmers' Property Rights Preferences. *Land Econ.* **2011**, *87*, 473–487. [CrossRef]

33. Freitas, F.L.M.; Sparovek, G.; Berndes, G.; Persson, U.M.; Englund, O.; Barretto, A.; Mörtberg, U. Potential Increase of Legal Deforestation in Brazilian Amazon after Forest Act Revision. *Nat. Sustain.* **2018**, *1*, 665–670. [CrossRef]

34. Xie, Y.; Gong, P.; Han, X.; Wen, Y. The Effect of Collective Forestland Tenure Reform in China: Does Land Parcelization Reduce Forest Management Intensity? *J. For. Econ.* **2014**, *20*, 126–140. [CrossRef]

35. World Bank. *China Forest Policy: Deepening the Transition, Broadening the Relationship*; World Bank: Washington, DC, USA, 2010.

36. Tseng, T.-W.J.; Robinson, B.E.; Bellemare, M.F.; BenYishay, A.; Blackman, A.; Boucher, T.; Childress, M.; Holland, M.B.; Kroeger, T.; Linkow, B.; et al. Influence of Land Tenure Interventions on Human Well-Being and Environmental Outcomes. *Nat. Sustain.* **2021**, *4*, 242–251. [CrossRef]

37. Steele, M.Z.; Shackleton, C.M.; Uma Shaanker, R.; Ganeshaiah, K.N.; Radloff, S. The Influence of Livelihood Dependency, Local Ecological Knowledge and Market Proximity on the Ecological Impacts of Harvesting Non-Timber Forest Products. *For. Policy Econ.* **2015**, *50*, 285–291. [CrossRef]

38. Belcher, B.; Ruíz-Pérez, M.; Achdiawan, R. Global Patterns and Trends in the Use and Management of Commercial NTFPs: Implications for Livelihoods and Conservation. *World Dev.* **2005**, *33*, 1435–1452. [CrossRef]

39. Ticktin, T.; Shackleton, C. Harvesting Non-Timber Forest Products Sustainably: Opportunities and Challenges. In *Non-Timber Forest Products in the Global Context*; Shackleton, S., Shackleton, C., Shanley, P., Eds.; Springer: Berlin/Heidelberg, Germany, 2011; pp. 149–169, ISBN 978-3-642-17983-9.

40. Liu, C.; Lu, J.; Yin, R. An Estimation of the Effects of China's Priority Forestry Programs on Farmers' Income. *Environ. Manag.* **2010**, *45*, 526–540. [CrossRef]

41. Lund, J.F. Ecological Sustainability for Non-Timber Forest Products—Dynamics and Case Studies of Harvesting. *Int. For. Rev.* **2015**, *17*, 256–258.

42. Ticktin, T. The Ecological Implications of Harvesting Non-Timber Forest Products. *J. Appl. Ecol.* **2004**, *41*, 11–21. [CrossRef]

43. Duan, W.; Shen, J.; Hogarth, N.J.; Chen, Q. Risk Preferences Significantly Affect Household Investment in Timber Forestry: Empirical Evidence from Fujian, China. *For. Policy Econ.* **2021**, *125*, 102421. [CrossRef]

44. Devisscher, T.; Spies, J.; Griess, V.C. Time for Change: Learning from Community Forests to Enhance the Resilience of Multi-Value Forestry in British Columbia, Canada. *Land Use Policy* **2021**, *103*, 105317. [CrossRef]

45. Yang, L.; Ren, Y. Property Rights, Village Democracy, and Household Forestry Income: Evidence from China's Collective Forest Tenure Reform. *J. For. Res.* **2021**, *26*, 7–16. [CrossRef]

46. Li, M.; Sarkar, A.; Wang, Y.; Khairul Hasan, A.; Meng, Q. Evaluating the Impact of Ecological Property Rights to Trigger Farmers' Investment Behavior—An Example of Confluence Area of Heihe Reservoir, Shaanxi, China. *Land* **2022**, *11*, 320. [CrossRef]

47. Kubitza, C.; Krishna, V.V.; Urban, K.; Alamsyah, Z.; Qaim, M. Land Property Rights, Agricultural Intensification, and Deforestation in Indonesia. *Ecol. Econ.* **2018**, *147*, 312–321. [CrossRef]

48. Gao, X.; Shi, X.; Fang, S. Property Rights and Misallocation: Evidence from Land Certification in China. *World Dev.* **2021**, *147*, 105632. [CrossRef]

49. Banerjee, A.V.; Gertler, P.J.; Ghatak, M. Empowerment and Efficiency: Tenancy Reform in West Bengal. *J. Polit. Econ.* **2002**, *110*, 239–280. [CrossRef]

50. Dao Minh, T.; Yanagisawa, M.; Kono, Y. Forest Transition in Vietnam: A Case Study of Northern Mountain Region. *For. Policy Econ.* **2017**, *76*, 72–80. [CrossRef]

51. Acemoglu, D.; Johnson, S.; Robinson, J.A. The Colonial Origins of Comparative Development: An Empirical Investigation: Reply. *Am. Econ. Rev.* **2012**, *102*, 3077–3110. [CrossRef]

52. Yang, Q.; Zhu, Y.; Liu, L.; Wang, F. Land Tenure Stability and Adoption Intensity of Sustainable Agricultural Practices in Banana Production in China. *J. Clean. Prod.* **2022**, *338*, 130553. [CrossRef]

53. Dalmazzo, A.; Marini, G. Foreign Debt, Sanctions and Investment: A Model with Politically Unstable Less Developed Countries. *Int. J. Finance Econ.* **2000**, *5*, 141–153. [CrossRef]

54. Negi, S.; Pham, T.T.; Karky, B.; Garcia, C. Role of Community and User Attributes in Collective Action: Case Study of Community-Based Forest Management in Nepal. *Forests* **2018**, *9*, 136. [CrossRef]

55. Deininger, K.; Chamorro, J.S. Investment and Equity Effects of Land Regularisation: The Case of Nicaragua. *Agric. Econ.* **2004**, *30*, 101–116. [CrossRef]

56. Ghebru, H.H.; Holden, S.T. Reverse-Share-Tenancy and Agricultural Efficiency: Farm-Level Evidence from Ethiopia. *J. Afr. Econ.* **2015**, *24*, 148–171. [CrossRef]

57. Lanjouw, J.O.; Levy, P.I. Untitled: A Study of Formal and Informal Property Rights in Urban Ecuador. *Econ. J.* **2002**, *112*, 986–1019. [CrossRef]

58. Galiani, S.; Schargrodsky, E. Property Rights for the Poor: Effects of Land Titling. *J. Public Econ.* **2010**, *94*, 700–729. [CrossRef]

59. Markussen, T.; Tarp, F. Political Connections and Land-Related Investment in Rural Vietnam. *J. Dev. Econ.* **2014**, *110*, 291–302. [CrossRef]

60. Lazos-Chavero, E.; Meli, P.; Bonfil, C. Vulnerabilities and Threats to Natural Forest Regrowth: Land Tenure Reform, Land Markets, Pasturelands, Plantations, and Urbanization in Indigenous Communities in Mexico. *Land* **2021**, *10*, 1340. [CrossRef]

61. Haley, D.; Nelson, H. Has the Time Come to Rethink Canada's Crown Forest Tenure Systems? *For. Chron.* **2007**, *83*, 630–641. [CrossRef]
62. Holden, S.; Yohannes, H. Land Redistribution, Tenure Insecurity, and Intensity of Production: A Study of Farm Households in Southern Ethiopia. *Land Econ.* **2002**, *78*, 573–590. [CrossRef]
63. Carter, M.R.; Olinto, P. Getting Institutions "Right" for Whom? Credit Constraints and the Impact of Property Rights on the Quantity and Composition of Investment. *Am. J. Agric. Econ.* **2003**, *85*, 173–186. [CrossRef]
64. Conigliani, C.; Cuffaro, N.; D'Agostino, G. Large-Scale Land Investments and Forests in Africa. *Land Use Policy* **2018**, *75*, 651–660. [CrossRef]
65. Lönnstedt, L.; Sedjo, R.A. Forestland Ownership Changes in the United States and Sweden. *For. Policy Econ.* **2012**, *14*, 19–27. [CrossRef]
66. Borras, S.M. Redistributive Land Reform in 'Public' (Forest) Lands? Lessons from the Philippines and Their Implications for Land Reform Theory and Practice. *Prog. Dev. Stud.* **2006**, *6*, 123–145. [CrossRef]
67. Mullan, K.; Grosjean, P.; Kontoleon, A. Land Tenure Arrangements and Rural–Urban Migration in China. *World Dev.* **2011**, *39*, 123–133. [CrossRef]
68. Giles, J.; Mu, R. Village Political Economy, Land Tenure Insecurity, and the Rural to Urban Migration Decision: Evidence from China. *Am. J. Agric. Econ.* **2018**, *100*, 521–544. [CrossRef]
69. Valsecchi, M. Land Property Rights and International Migration: Evidence from Mexico. *J. Dev. Econ.* **2014**, *110*, 276–290. [CrossRef]
70. Miller, D.C.; Rana, P.; Nakamura, K.; Irwin, S.; Cheng, S.H.; Ahlroth, S.; Perge, E. A Global Review of the Impact of Forest Property Rights Interventions on Poverty. *Glob. Environ. Chang.* **2021**, *66*, 102218. [CrossRef]
71. Binkley, C.S.; Stewart, F.; Power, S. *Pension Fund Investment in Forestry*; World Bank: Washington, DC, USA, 2020.
72. Blancard, S.; Boussemart, J.-P.; Briec, W.; Kerstens, K. Short- and Long-Run Credit Constraints in French Agriculture: A Directional Distance Function Framework Using Expenditure-Constrained Profit Functions. *Am. J. Agric. Econ.* **2006**, *88*, 351–364. [CrossRef]
73. Dries, L.; Swinnen, J.F.M. Foreign Direct Investment, Vertical Integration, and Local Suppliers: Evidence from the Polish Dairy Sector. *World Dev.* **2004**, *32*, 1525–1544. [CrossRef]
74. Bose, P. Land Tenure and Forest Rights of Rural and Indigenous Women in Latin America: Empirical Evidence. *Womens Stud. Int. Forum* **2017**, *65*, 1–8. [CrossRef]
75. Ceddia, M.G.; Gunter, U.; Corriveau-Bourque, A. Land Tenure and Agricultural Expansion in Latin America: The Role of Indigenous Peoples' and Local Communities' Forest Rights. *Glob. Environ. Chang.* **2015**, *35*, 316–322. [CrossRef]
76. Monterroso, I.; Cronkleton, P.; Pinedo, D.; Larson, A.M. *Reclaiming Collective Rights: Land and Forest Tenure Reforms in Peru (1960–2016)*; CIFOR: Bogor Barat, Indonesia, 2017; Volume 224.
77. Deininger, K.; Ali, D.A.; Alemu, T. Impacts of Land Certification on Tenure Security, Investment, and Land Market Participation: Evidence from Ethiopia. *Land Econ.* **2011**, *87*, 312–334. [CrossRef]
78. Alban Singirankabo, U.; Willem Ertsen, M. Relations between Land Tenure Security and Agricultural Productivity: Exploring the Effect of Land Registration. *Land* **2020**, *9*, 138. [CrossRef]
79. Li, L.; Hao, T.; Chi, T. Evaluation on China's Forestry Resources Efficiency Based on Big Data. *J. Clean. Prod.* **2017**, *142*, 513–523. [CrossRef]
80. Benin, S.; Ahmed, M.; Pender, J.; Ehui, S. Development of Land Rental Markets and Agricultural Productivity Growth: The Case of Northern Ethiopia. *J. Afr. Econ.* **2005**, *14*, 21–54. [CrossRef]
81. Baragwanath, K.; Bayi, E. Collective Property Rights Reduce Deforestation in the Brazilian Amazon. *Proc. Natl. Acad. Sci. USA* **2020**, *117*, 20495–20502. [CrossRef] [PubMed]
82. Zhou, Y.; Shi, X.; Ji, D.; Ma, X.; Chand, S. Property Rights Integrity, Tenure Security and Forestland Rental Market Participation: Evidence from Jiangxi Province, China. *Nat. Resour. Forum* **2019**, *43*, 95–110. [CrossRef]
83. Paletto, A.; De Meo, I.; Cantiani, M.G.; Cocciardi, D. Balancing Wood Market Demand and Common Property Rights: A Case Study of a Community in the Italian Alps. *J. For. Res.* **2014**, *19*, 417–426. [CrossRef]
84. Stickler, M.M.; Huntington, H.; Haflett, A.; Petrova, S.; Bouvier, I. Does de Facto Forest Tenure Affect Forest Condition? Community Perceptions from Zambia. *For. Policy Econ.* **2017**, *85*, 32–45. [CrossRef]
85. Lu, S.; Sun, H.; Zhou, Y.; Qin, F.; Guan, X. Examining the Impact of Forestry Policy on Poor and Non-Poor Farmers' Income and Production Input in Collective Forest Areas in China. *J. Clean. Prod.* **2020**, *276*, 123784. [CrossRef]
86. Zhou, B.; Zhang, C.; Song, H.; Wang, Q. How Does Emission Trading Reduce China's Carbon Intensity? An Exploration Using a Decomposition and Difference-in-Differences Approach. *Sci. Total Environ.* **2019**, *676*, 514–523. [CrossRef]
87. Nunn, N.; Qian, N. The Potato's Contribution to Population and Urbanization: Evidence From A Historical Experiment. *Q. J. Econ.* **2011**, *126*, 593–650. [CrossRef]
88. Živojinović, I.; Nedeljković, J.; Stojanovski, V.; Japelj, A.; Nonić, D.; Weiss, G.; Ludvig, A. Non-Timber Forest Products in Transition Economies: Innovation Cases in Selected SEE Countries. *For. Policy Econ.* **2017**, *81*, 18–29. [CrossRef]
89. Mukul, S.A.; Rashid, A.Z.M.M.; Uddin, M.B.; Khan, N.A. Role of Non-Timber Forest Products in Sustaining Forest-Based Livelihoods and Rural Households' Resilience Capacity in and around Protected Area: A Bangladesh Study. *J. Environ. Plan. Manag.* **2016**, *59*, 628–642. [CrossRef]

90. Zhu, H.; Hu, S.; Ren, Y.; Ma, X.; Cao, Y. Determinants of Engagement in Non-Timber Forest Products (NTFPs) Business Activities: A Study on Worker Households in the Forest Areas of Daxinganling and Xiaoxinganling Mountains, Northeastern China. *For. Policy Econ.* **2017**, *80*, 125–132. [CrossRef]

91. Stanek, E.J., III; Motta Singer, J.d.; Lencina, V.B. A Unified Approach to Estimation and Prediction under Simple Random Sampling. *J. Stat. Plan. Inference* **2004**, *121*, 325–338. [CrossRef]

92. Amegnaglo, C.J.; Anaman, K.A.; Mensah-Bonsu, A.; Onumah, E.E.; Amoussouga Gero, F. Contingent Valuation Study of the Benefits of Seasonal Climate Forecasts for Maize Farmers in the Republic of Benin, West Africa. *Clim. Serv.* **2017**, *6*, 1–11. [CrossRef]

93. Sachitra, V.; Chong, S.-C. Relationships between Institutional Capital, Dynamic Capabilities and Competitive Advantage: Empirical Examination of the Agribusiness Sector. *Int. Rev. Manag. Mark.* **2017**, *7*, 389–397.

94. Josephson, A.; Smale, M. What Do You Mean by "Informed Consent"? Ethics in Economic Development Research. *Appl. Econ. Perspect. Policy* **2021**, *43*, 1305–1329. [CrossRef]

95. Yanow, D.; Schwartz-Shea, P. Reforming Institutional Review Board Policy: Issues in Implementation and Field Research. *PS Polit. Sci. Polit.* **2008**, *41*, 483–494. [CrossRef]

96. Luo, Y.; Liu, J.; Zhang, D.; Dong, J. Actor, Customary Regulation and Case Study of Collective Forest Tenure Reform Intervention in China. *Small-Scale For.* **2015**, *14*, 155–169. [CrossRef]

97. Molnar, A.; Gomes, D.; Sousa, R.; Vidal, N.; Hojer, R.F.; Arguelles, L.A.; Kaatz, S.; Martin, A.; Donini, G.; Scherr, S.; et al. Community Forest Enterprise Markets in Mexico and Brazil: New Opportunities and Challenges for Legal Access to the Forest. *J. Sustain. For.* **2008**, *27*, 87–121. [CrossRef]

98. Latruffe, L.; Piet, L. Does Land Fragmentation Affect Farm Performance? A Case Study from Brittany, France. *Agric. Syst.* **2014**, *129*, 68–80. [CrossRef]

99. Wu, Z.; Liu, M.; Davis, J. Land Consolidation and Productivity in Chinese Household Crop Production. *China Econ. Rev.* **2005**, *16*, 28–49. [CrossRef]

100. Manjunatha, A.V.; Anik, A.R.; Speelman, S.; Nuppenau, E.A. Impact of Land Fragmentation, Farm Size, Land Ownership and Crop Diversity on Profit and Efficiency of Irrigated Farms in India. *Land Use Policy* **2013**, *31*, 397–405. [CrossRef]

101. Liu, C.; Mullan, K.; Liu, H.; Zhu, W.; Rong, Q. The Estimation of Long Term Impacts of China's Key Priority Forestry Programs on Rural Household Incomes. *J. For. Econ.* **2014**, *20*, 267–285. [CrossRef]

102. Wei, D.; Chao, H.; Yali, W. Role of Income Diversification in Reducing Forest Reliance: Evidence from 1838 Rural Households in China. *J. For. Econ.* **2016**, *22*, 68–79. [CrossRef]

103. Liu, P.; Yin, R.; Li, H. China's Forest Tenure Reform and Institutional Change at a Crossroads. *For. Policy Econ.* **2016**, *72*, 92–98. [CrossRef]

104. Migliorelli, M.; Dessertine, P. Time for New Financing Instruments? A Market-Oriented Framework to Finance Environmentally Friendly Practices in EU Agriculture. *J. Sustain. Finance Investig.* **2018**, *8*, 1–25. [CrossRef]

105. Song, N.; Aguilar, F.X.; Butler, B.J. Conservation Easements and Management by Family Forest Owners: A Propensity Score Matching Approach with Multi-Imputations of Survey Data. *For. Sci.* **2014**, *60*, 298–307. [CrossRef]

106. Smith, J.A.; Todd, P.E. Does Matching Overcome LaLonde's Critique of Nonexperimental Estimators? *J. Econom.* **2005**, *125*, 305–353. [CrossRef]

107. Caliendo, M.; Kopeinig, S. Some Practical Guidance for the Implementation of Propensity Score Matching. *J. Econ. Surv.* **2008**, *22*, 31–72. [CrossRef]

108. Wang, S.; Grant, R.F.; Verseghy, D.L.; Andrew Black, T. Modelling Carbon-Coupled Energy and Water Dynamics of a Boreal Aspen Forest in a General Circulation Model Land Surface Scheme. *Int. J. Climatol.* **2002**, *22*, 1249–1265. [CrossRef]

109. JingWen, X.; WenJian, H.; HongXiao, Z.; HouQi, H. Literature review on the relationship between credit constraint and farmers' income: A case study of forest right mortgage system. *World For. Res.* **2018**, *31*, 13–18.

110. Jacoby, H.; Minten, B. *Land Titles, Investment, and Agricultural Productivity in Madagascar: A Poverty and Social Impact Analysis*; World Bank: Washington, DC, USA, 2006.

111. Melesse, M.B.; Bulte, E. Does Land Registration and Certification Boost Farm Productivity? Evidence from Ethiopia. *Agric. Econ.* **2015**, *46*, 757–768. [CrossRef]

112. FAO. *Understanding Forest Tenure in Africa: Opportunities and Challenges for Forest Tenure Diversification*; Forestry Policy and Institutions Working Paper; FAO: Rome, Italy, 2008.

113. Parajuli, R.; Lamichhane, D.; Joshi, O. Does Nepal's Community Forestry Program Improve the Rural Household Economy? A Cost–Benefit Analysis of Community Forestry User Groups in Kaski and Syangja Districts of Nepal. *J. For. Res.* **2015**, *20*, 475–483. [CrossRef]

Article

Evaluating the Impact of Ecological Property Rights to Trigger Farmers' Investment Behavior—An Example of Confluence Area of Heihe Reservoir, Shaanxi, China

Min Li [1], Apurbo Sarkar [1], Yuge Wang [1], Ahmed Khairul Hasan [2] and Quanxing Meng [1],*

[1] College of Economics and Management, Northwest A&F University, Xianyang 712100, China; limin1983@nwafu.edu.cn (M.L.); apurbo@nwafu.edu.cn (A.S.); wangyuge@nwafu.edu.cn (Y.W.)
[2] Department of Agronomy, Bangladesh Agricultural University, Mymensingh 2202, Bangladesh; akhasan@bau.edu.bd
* Correspondence: mqsheng@nwafu.edu.cn

Abstract: Property rights of natural resources have been acting as a critical legislative tool for promoting sustainable resource utilization and conservation in various regions of the globe. However, incorporating ecological property rights into the natural resources property rights structure may significantly influence farmers' behavior in forestry investment. It may also trigger forest protection, water conservation, and urban water security. The main aim of the research is to evaluate the impact of ecological property rights and farmers' investment behavior in the economic forest. We have constructed an analytical framework of collective forest rights from two indicators of integrity and stability, by adopting the theory of property rights and ecological capital to fulfill the study's aims. The empirical data has been comprised of the microdata of 708 farmers, collected from the confluence area of the Heihe Reservoir, Shaanxi, China. The study also conducted pilot ecological property rights transactions in the surveyed area. The study utilized the double-hurdle model to test the proposed framework empirically. The results show that forest land use rights, economic products, and eco-product income rights positively affect farmers' forestry investment intensity, and disposal rights (forest land transfer rights) negatively affect farmers' investment intensity. However, in terms of the integrity of property rights, only the right to profit from ecological products affects farmers' forestry investment willingness, and other property rights are insignificant. The study also found that the lower the farmers' forest land expropriation risk is expected, the greater the possibility of investment and the higher the input level. However, we traced that the farmers' forest land adjustment has no significant impact on farmers' willingness to invest. Obtaining the benefits of ecological products has been found as the primary motivation for forestry investment within the surveyed area. The completeness of ownership rights positively impacted farmers' investment intensity. Farmers should realize the ecological value of water conservation forests through the market orientation of the benefit of ecological products. Therefore, the government should encourage farmers and arrange proper training to facilitate a smooth investment. A well-established afforestation program should also be carried out.

Keywords: property rights; ecological property rights; forestry investment; farmers' behavior; reservoir confluence area

Citation: Li, M.; Sarkar, A.; Wang, Y.; Khairul Hasan, A.; Meng, Q. Evaluating the Impact of Ecological Property Rights to Trigger Farmers' Investment Behavior—An Example of Confluence Area of Heihe Reservoir, Shaanxi, China. *Land* **2022**, *11*, 320. https://doi.org/10.3390/land11030320

Academic Editors: Jay Mar D. Quevedo, Norie Tamura, Yuta Uchiyama and Ryo Kohsaka

Received: 26 December 2021
Accepted: 16 February 2022
Published: 22 February 2022

Publisher's Note: MDPI stays neutral with regard to jurisdictional claims in published maps and institutional affiliations.

1. Introduction

The world's land and groundwater reserves have become scarce and have already been overused and exploited [1,2]. The proper management of such crucial resources is the main theme of soil and water conservation. Soil and water are the two prime resources essential for human existence, and these resources are becoming increasingly scarce and massively consumed with the sharp growth of the world's populace [3,4]. As a result, the significance of sustaining soil and water and preserving the integrity of both crucial resources should be considered without sacrificing productivity [5]. Agro-forestry can be a possible solution for humankind, as it helps soil and water conservation [6]. Agro-forestry is a method of land management that includes trees and shrubs in farming, allowing for the growing of trees, crops, and cattle on the same plot of land [7,8]. It provides opportunities to obtain profits from booming commodity markets while also improving the land, water resources, and the environment. It is a platform for developing integrated, diversified, and productive land usage patterns by combining agriculture and forestry technology [9]. Trees contribute to lessening the erosive power of raindrops on crops, allowing more water to reach the crop. Moreover, It reduces soil erosion and significantly raises soil fertility, and it helps preserve water by increasing absorption capacity and hydraulic properties [10].

Interestingly, with the development of the social economy in recent years, the problem of urban water shortage has become increasingly prominent, mainly manifested in insufficient water supply and water quality safety risks [11]. Ecological services, such as fresh-water supply and purified water quality, provided by water conservation forests, are the key to ensuring safe water supply and quality [12]. In this regard, the confluence area of reservoirs can play a vital role in supporting various cities, especially in China. The restoration and protection of water conservation forests are the keys to preventing and dissolving the ecological security risks of the river basin and confluence areas of reservoirs, and ensuring freshwater supply in cities. Moreover, the deterioration of the ecological environment of the river basin is currently the main threat in this regard. However, poverty and ecological fragility worsen the situation [13]. Usually, farmers tend to overuse forest resources for their livelihoods, and the lack of proper management and protection frameworks damage the service functions of the forest ecosystem. Due to weak infrastructure and a lack of essential resources, water conservation in forests has become one of the most prominent tactics for sustaining farmers' agro-production and livelihood in the reservoir's confluence area [14]. Existing studies have shown that fostering a well-structured agro-forestry management system relies on the following three basic criteria: (i) farmers' engagement and investment behavior, (ii) property rights support, and (iii) economic viability [15,16].

Farmers' forestry investment responds to economic signals based on family characteristics, natural conditions, and legal frameworks [17,18]. Research on the forestry investment behavior of farmers has been derived from several aspects. Existing research mainly focuses on resource endowment [19,20], business scale [21,22], transition cost cognition [23,24], and risk preference [25,26]. Some researchers derived the forest investment by analyzing the inherent impact of farmers' behavior factors from local governance [27,28], public policy [29,30], village environment [31,32], market environment [33,34], public governance with community tourism [35,36], and other external constraints of farmers' forestry investment behavior [37,38]. Forestry production is always carried out under established industrial policies and institutional frameworks. Among many policies, collective forest rights are an important means of affecting forestry investment [39]. Clear and stable collective forest rights encourage farmers to invest in forestry by enhancing income expectations and clarifying investment returns [40]. The intensity of the property rights system's incentives to farmers' production and investment behaviors depends on the system's degree of consistency between inputs and returns [41]. In addition to forest products, agro-forestry also supplies ecological products, such as water conservation and water purification [42]. Property rights are central concepts of the Coase Theorem [43]. It argues that, within idealized economic circumstances, when property rights conflict, the participants would bargain or enter negotiations that fully represent the actual expenses and fundamental worth of the property rights in concern, eventually culminating in the most effective solution [44]. Therefore, the rental value of the water conservation of forests should be reflected in the financial and ecological product markets, respectively [45]. At present, farmers engaged in forestry production, in the confluence area of the reservoir, not only obtain income from forest products but also from ecological products, in the form of ecological compensation [46,47].

Seemingly, the protection of water sources and the balance of interests between the upstream and downstream of the river basin have become increasingly prominent [48]. The concentrated manifestation is the shortage of urban water resources and the lack of willingness to protect water conservation forests in the confluence area of the reservoir [49]. The water conservation and forest ecosystem can provide hydrological and ecological services, such as water conservation and water purification, which are the key to ensuring water supply and quality safety [50]. Scholars in this field have done much research to reveal the influence of collective forest rights on farmers' investment behavior (such as Zhang et al., 2011 [50], Kashwan [51], and Wu and Zhang [52]). However, few studies incorporate ecological property rights into the property rights structure and analyze the impact of collective forest rights on the forestry investment behavior of farmers in the confluence area of reservoirs (for example, Nichiforel et al. [53], Wen et al. [54] and Yu and Xu [55]). The lack of ecological property rights has caused market failures in ecological governance in the confluence area of reservoirs, resulting in farmers' lack of willingness to invest in forestry or insufficient investment intensity. Therefore, incorporating ecological property rights into the structure of collective forest rights and exploring how it affects farmers' forestry investment decisions in reservoir confluence areas requires further research. The main aims of the study are to incorporate ecological property rights into the collective forest tenure structure, analyze the impact of collective forest tenure on farmers' forestry investment behavior in the reservoir confluence area, from the perspective of ecological property rights, and explore the influencing factors of farmers' forestry investment.

In the absence of ecological property rights, vertical transfer payment is currently the primary method of forest ecological compensation in the confluence area of reservoirs [56]. Although this approach embodies the principle of fairness, it lacks attention to hydrological and ecological service providers and farmers, and their forest reforestation, management, and protection behaviors have not received the economic incentives they deserve [57]. They lack forestry investment willingness or insufficient investment intensity. Seemingly, the subdivision of property rights is a meaningful way to implement complex property rights [58]. By subdividing forest property rights, the economic property rights of water conservation forests can be separated from ecological property rights [59]. By exercising ecological property rights, farmers might realize the ecological value of water conservation forests through the ecological market and redeem the goodness of "clear water and green mountains" into "sustainable water conservations and ecologically sound forest management" [60,61].

In summary, collective forest rights impact farmers' forestry investment behavior [62,63]. Consequently, with forest tenure reforms on their way in many parts of the world, it is an excellent time to reflect on the experiences so far and rectify the following research questions: (i) Do the reforms have the desired outcomes? (ii) How do ecological property rights foster farmer forestry investment behavior? (iii) Are farmers willing to invest in forestry? (iv) To what extent are farmers willing to invest in the forestry ecosystem? (v) How should a measurement system to measure collective forest rights be constructed? Answering the questions mentioned above will be the main innovations of the study. Moreover, few studies incorporate ecological property rights into the property structure and analyze the impact of collective forest rights on farmers' forestry investment behavior in reservoir confluence areas. The study evaluates the impact of collective forest rights on the forestry investment behavior of farmers, based on ecological property rights from two aspects of integrity and stability, by taking the Heihe Reservoir confluence area as an example. The study provides a comprehensive definition of ecological property rights by establishing an ecological market, promoting farmers' forestry investment by guaranteeing farmers' income, and realizing sustainable development of water conservation forests. We incorporate ecological property rights into the structure of collective forest rights, analyze the impact of collective forest rights on the forestry investment behavior of farmers in the confluence area of reservoirs, based on the perspective of ecological property rights, and explore the influencing factors of farmers' forestry investment.

In addition to forest products, the output products of forestry production also include ecological products, such as water conservation and water purification. Therefore, the rental value of water conservation forests should be reflected in the economic and ecological product's market, respectively. The resources attached to ecological products have unique economic characteristics and are the objects of ecological property rights. According to the ecological and economic value of water conservation forests and their market characteristics, the property rights of water conservation forests can be divided into ecological property rights and economic property rights. Among them, ecological property rights refer to the existence of a certain number and quality of forest trees, when the minimum hydrological, ecological service supply required to ensure the water volume of the reservoir and the water environment health and safety standards is guaranteed. Likewise, economic property rights refer to the right to obtain economic benefits on the premise of ensuring positive externalities. The ecological property rights of water conservation forests require a certain number and quality of trees. The ultimate purpose is to obtain hydrological, ecological services, such as water conservation and water purification, and ensure that hydrological and ecological services can meet the needs of reservoir water volume and water environment health and safety. In this way, the right to benefit from water conservation forests is correspondingly subdivided into the right to benefit from ecological and economic products. The study adopted ecological property rights from the prospectives of the rights of use, benefit, and disposal, which may be deviated by adjusting, and expropriation risk expectations.

2. Theoretical Analysis and Research Hypothesis

Farmers' forestry investment decisions result from balancing costs and benefits [64]. The balancing process is affected by both property rights' integrity and stability [65,66]. The scope, benefits, and the degree of exclusivity of the property rights, and whether the benefits can be sustained are getting much more attention from governments, academics, and farmers [67,68]. Interestingly, China has had a unique experience in ownership transformation, as the authority for forestry management transferred from the community (collective) to individual farmers [19]. The ecological and collective property rights may be derived from the two aspects, integrity [69,70] and stability [71,72]. The prospects of integrity highlight the interrelationship between the vitality of authority within forest ecological property rights and the penetration of moral authority to more ecologically friendly behavior [73,74]. Stability denotes the optimality, continuity, and sustainability of the rights, which can be to the long-term benefit of farmers [75,76].

2.1. Integrity of Collective Forest Rights and Forestry Investment Behavior of Farmers in the Confluence Area of Reservoirs

The integrity of property rights refers to the extent to which the subject of property rights excludes other subjects from interfering with the use and disposal of resources independently and, thus, enjoys exclusive benefits [77,78]. It is generally believed that the more complete the property rights, the stronger the investment incentives. Farmers obtain income through the use and disposal of forest resources, so the use, disposal, and profit include the entire process of resource utilization [79,80]. Kashwan [51] analyzed the demand for community forest rights and found a "close relationship between collective forest rights and the farmer's investment behavior". Yi et al. [81] evaluated 3,180 households in eight provinces, from south to north China, and concluded that there is a positive interaction between China's collective forest protection rights and farmer households' perception towards forestry investment. Lee et al. [82] found that more substantial contracted rights affect investment strongly, after exploring 231 counties in eight states of the Central and Southern Appalachian Region of the United States. Hildebrandt and Knok [83] found that when the perceived benefits of complementary objectives increase with economic impact objective, the property right policies of forestland investment are fostered progressively. Therefore, the current study proposed Hypothesis 1, as follows:

Hypothesis 1: *The integrity of collective forest rights positively affects farmers' forestry investment behavior.*

Existing studies have shown that independent selection of tree species, conversion to other forestry uses, and management of non-wood forest products and other forest land use rights sub-items have a significant role in stimulating forestry investment. Generally, obtaining income is the direct purpose of farmers' forestry investment [84]. The degree to which the marginal return of forestry production can be equal to the marginal output determines the degree of exclusivity of farmers' income rights [81]. The output of water conservation forests is a form of forest products supported by ecological products, such as water conservation and water purification. The long-term neglect of the ecological value of water conservation forests has led to the deviation of the marginal return of forestry products [85]. The essence of this is the deprivation and encroachment of farmers' income rights [86]. According to Ji et al. [87] and Irimie and Essmann [88], using only the right can highly impact farmers' forestry investment. The definition of use and disposal rights is necessary for transiting a smooth investment, and obtaining sufficient income is also considered as a prerequisite of farmer's investment [89]. Interestingly, the intensity of property rights ensures the right to use natural resources (such as water conservations), which could be crucial in facilitating investment decisions [90]. Based on these, we have proposed Hypothesis 2.

Hypothesis 2: *The right of use positively affects the forestry investment behavior of farmers.*

The article distinguishes forest economic products and ecological products and seemingly divides the income rights of water conservation forests into economic product income rights [91,92] and ecological product income rights [93,94]. In this way, incorporating ecological property rights into the property rights structure can more comprehensively analyze the impact of income rights on farmers' forestry investment behavior in the reservoir's confluence area. On the contrary, disposal rights measure the degree of exclusivity of forest land and forest tree disposal behavior, including circulation, mortgage, logging, and inheritance [95]. The central aspect of China's agricultural land contracting is that, usually, used household as a unit of rural households ("Rural Land Contract Law" and its judicial interpretation), and the death of a single family member will not cause the problem of contract inheritance [96]. There are many opportunities for farmers and ranchers to introduce agro-forestry practices on their land, which may open up new income possibilities, while adding conservation benefits [97]. Framers may be willing to invest more if the ecological property right satisfies the prime demands of any farmer, such as income and livelihood opportunities. Therefore, the study proposed Hypothesis 3, as follows:

Hypothesis 3: *The right to income positively affects the forestry investment behavior of farmers.*

The confluence area of the reservoir is a quasi-protection area for water source protection, where damage from cropping and vegetation is prohibited, and farmers generally do not have logging rights [98,99]. Therefore, the best option within these areas is an investment in forest and ecosystem restorations [100]. In this regard, the Forestry Bureau of China and the Banking Regulatory Commission of China formulated a new mortgage loan policy. They stated that "Banking financial institutions should not accept water conservation forests and other non-disposable forest rights as mortgage properties". Thus, the ecological property right enjoyed by farmers in the confluence area of the reservoir could act as the only right to transfer forest land [101]. As a result, the right of circulation has a positive impact on farmers' forestry investment, in that the circulation of forest land provides farmers with a way to recover investment and obtain income [102,103]. Understanding patterns of change across disposal rights is essential for farmers that foster healthy and resilient forests for the future. Based on the above discussion, the study proposes Hypothesis 4, as follows:

Hypothesis 4: *The right of disposal positively affects the forestry investment behavior of farmers.*

2.2. The Stability of Collective Forest Rights and the Forestry Investment Behavior of Farmers in the Confluence Area of Reservoirs

The payback period of forestry production investment is prolonged and often influenced by several externalities [104]. Therefore, the long-term stability of collective forest rights is the key to whether farmers can recover their investment and make profits within the term of property rights [105], which has a significant impact on farmers' forestry investment [106]. Stable collective forest rights encourage farmers to invest in forestry through three methods; ensuring that investment income is not encroached with facilitating access to credit funds and promoting the transfer of property rights to recover investment [107]. On the other hand, unstable collective forest rights can reduce farmers' investment recovery expectations [108]. Unpredictable forest land adjustment or collection will take away farmers' long-term investment in forest land, like a random tax, and weaken farmers' investment capabilities [109]. According to Kumar and Kerr [110], well-structured collective laws and regulations should have influenced the investment behavior of Indian forest dwellers' grassroots formations. It is apparent that if the collective forestry rights can be maintained consistently and stably, it may foster a favorable condition for farmers' investment [111,112]. Thus the study proposes Hypothesis 5, as follows:

Hypothesis 5: *Unstable collective forest rights negatively affect farmers' agro-forestry investment behavior.*

Collective forest investment's cash flows come from payment for ecosystem services, land appreciation, land preservation tax credits, the sale of land rights, and other fees, such as hunting or fishing [113]. Therefore, forestry investment willingness considers different risk sources that may impact farmers' forestry investment behavior [114]. The development of agro-forestry to increase its effectiveness requires massive capital and capital is always associated with several markets, policy-related and external risk factors [115]. Increasing risks and uncertainties related to stochastic agro-ecological and institutional factors, and the deterioration of land due to unsustainable farming, are among the significant constraints to agricultural development in developing countries [116]. Perceived risk and risk management strategies could be crucial for the investment facilitation of farmers [117]. Existing studies showed that positive perceived risk expectations and risk management could foster positive responses from the prospects of agro-forestry [115,118,119]. Do et al. [120] identified that adjusting risk perceptions, associated with farmers' time preference, crop yields, and crop prices, appeared to have the most significant influence on whether to invest in agro-forestry. By evaluating family farmers in Brazil, Martinelli et al. [121] found that unpredictable environmental and macroeconomic factors mainly determine the return on investment in agro-forestry. Jerneck and Olsson [122] revealed that small-scale Kenyan farmers' behavior is derived mainly by the degree of expected uncertainty and risk associated with the return on investments. However, it is apparent that if farmers foster any risks associated with a long growth period, they often choose not to invest [123,124]. Therefore, hypotheses 6 and 7 have been proposed, as follows:

Hypothesis 6: *Adjusting risk expectations negatively affects farmers' forestry investment behavior.*

Hypothesis 7: *The expropriation risk expectation negatively affects the forestry investment behavior of farmers.*

3. Materials and Methods

Based on the theory of property rights and ecological capital, an analytical framework of collective forest rights is constructed from two aspects of completeness and stability. Completeness includes three dimensions of use rights, disposal rights, and income rights, and stability includes two dimensions of adjustment expectations and expropriation expectations. According to the characteristics of forest economic and ecological products, the income rights of water conservation forests are divided into economic and ecological product income rights. Based on the perspective of the separation of economic and ecological property rights, the collective forest rights investigate the forestry investment behavior of farmers. Using the micro-data of 708 farmers, in the confluence area of Heihe Reservoir, the double-hurdle model is used to empirically test the differential impact of collective forest rights on farmers' forestry investment willingness and investment intensity.

3.1. Data Source

The sample dataset used in the article has been extracted from a field survey conducted by the research team of well-trained postgraduate-level students in the confluence area of Heihe Jinpen Reservoir from June to July 2019 on the subject of "property rights cognition, perceived value, and forestry investment behavior of farmers". The Jinpen Dam is a rock-fill embankment dam situated in Zhouzhi County of Shaanxi Province, China, where a tributary channel of the Weihe River flows into the Yellow River. It is situated north of the Qinling Mountains, 90 km away from Xi'an City. The Heihe River, which originates from the Qinling mountain, is the main water supply for the Jinpen Reservoir. The Heihe River is a first-level tributary of the Wei River, and the Jinpen Reservoir is the primary water source of Xi'an city [125]. The confluence area of the reservoir covers an area of 1481 km^2, and it mainly flows through the three towns of Chenhe, Banfangzi, and Houzhenzi in Zhouzhi County of Shaanxi Province. According to the geographical distribution and population ratio, stratified and simple random sampling methods selected 13, 8, and 4 administrative villages in Chenhe, Banfangzi, and Houzhenzi Towns. Figure 1 portrays the study area map. After, we randomly selected 27–30 farmers from each village to conduct a household survey with face-to-face interview tactics accompanied by a structured questionnaire. Interviewers asked the farmers about the questionnaire's content and recorded the responses in written form. It includes demographic information (control variables) and the content regarding the dependent and independent variables. A total of 743 responses have been gathered, with 708 valid responses, and the efficiency was 95.29%. Prior to the formal interviews, the interviewers briefly described the aims and content of the questionnaire to the interviewee, which improved the response rate. Moreover, verbal permission was taken before starting the survey. The interviewee was informed that the information collected via the interviews would be used solely for research purposes, and they can opt-out at any time for any responses.

3.2. Pre-Processing of Variables

The study uses the average value of other households' knowledge of property rights in the same village as an instrumental variable to eliminate possible endogenous estimation biases, as suggested by Liu and Jia [126] and Ma et al. [127]. There may be an endogenous problem between farmers' collective forest rights perception and forestry investment behaviour in the formula (1). Because, in the same administrative village, the perception of a farmer's collective forest rights may be affected by other farmers' property rights [128]. At the same time, the perception of property rights of other farmers in the same village is not directly related to the forestry investment behaviour of the sample [129]. In addition, the subdivided property rights indicators affect different aspects of farmers' forestry production decisions. In order to reduce the impact of multicollinearity and avoid the randomness of subjective assignment, the entropy method is used to calculate the index weight as suggested by Luo et al. [130].

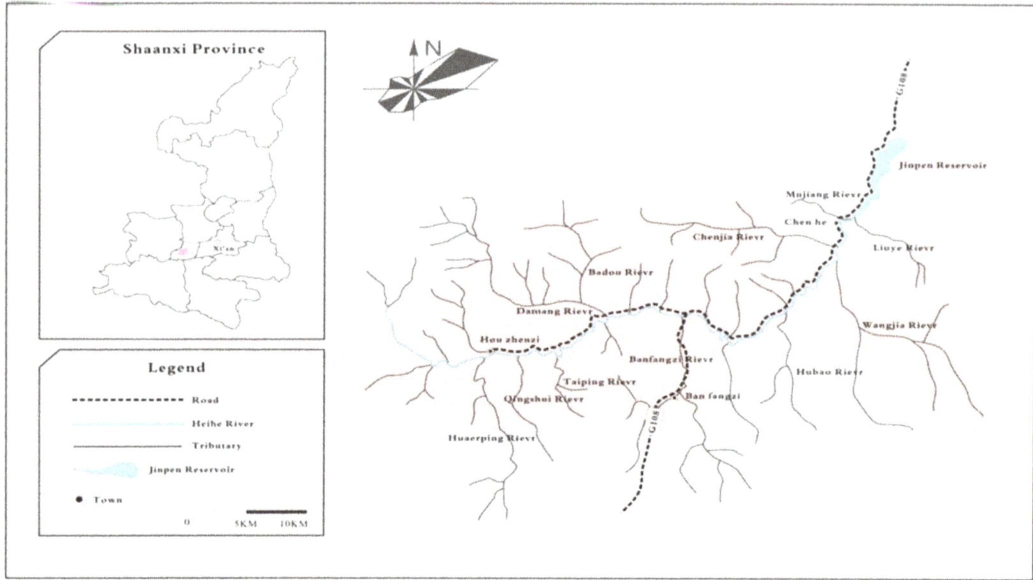

Figure 1. Study area map.

In the study, we have chosen tree species and operating non-wood forest products, which are weighted together to measure the level of use right, and the income of economic products and ecological products are weighted together to measure the level of income right. Finally, the three rights indicators of rights of use, benefits, and disposal are introduced into the model. The independent variables in the study are divided into the following two categories: core independent variable and control variables. Collective forest property rights acted as core independent variables, including property rights of integrity and stability. The study uses age, education level, health status, status as a Communist party member or not (village cadre) to reflect the characteristics of the sampled individuals (control variables). In contrast, we used the family population and the size of fixed assets to reflect the characteristics of the sample households; the average single forest area, forest land distance, forest land quality, and forest trees reflect the characteristics of agro-forestry.

3.3. Variable Selection and Descriptive Statistics

3.3.1. Dependent Variable: Forestry Investment Bbehavior of Farmers

Labor and capital are the main factors of production for farmers' forestry production and operation, and there is a specific time interval for significant forestry capital investment, such as seedlings and fertilizers. The article uses the five-year cumulative sum of funds for farmers' households per unit of forest land from 2015 to 2019 to measure farmers' investment behavior. The factors affecting the willingness of forestry investment and investment intensity of farmers in the reservoir confluence area may not be the same. Therefore, the study divides the forestry investment behavior of farmers in the reservoir confluence area into the following two stages: participation decision-making and quantitative decision-making, as suggested by Assé and Lassoie [131] and Zeng et al. [132]. Participation in decision-making to examine whether farmers are willing to invest in forestry is a binary dummy variable; quantitative decision-making examines how much farmers invest in forestry and is a continuous variable.

Participation in decision-making is used to examine whether farmers are willing to invest in forestry as a dual dummy variable. If the forestry investment during the investigation period is 0, the farmers have no willingness to invest in forestry, and the assumption value is 0. On the contrary, if the farmers have made forestry investments, the value is 1. Quantitative decision-making examines how much farmers invest in forestry as a continuous variable. In the study, the amount of forestry investment incurred by farmers during the investigation period is used to express the forestry investment intensity of farmers.

3.3.2. Core Independent Variable: Collective Forest Rights

For a long time, forestry departments and village administration affected farmers' forestry production directly or indirectly [133]. When farmers exercise property rights such as rights of use, disposal, and income, the actual degree of exclusivity may effectively determine property rights [134]. Therefore, the article refers to the logic of "content of property rights-government (village collective) intervention-degree of exclusivity" proposed by Li et al. [135] and uses farmers' perception of the degree of exclusivity of property rights (government departments and administrative villages) to measure collective forests. The study adopts the definition of property rights from the analysis of Ma et al. [136], van Gelder [137], and Nguyen et al. [138] and constructs a measurement system to assess collective forest rights from two indicators of completeness and stability. Right to use, Usufruct, Right of disposal have been used as indicators of collective forest rights (core independent variable). Right to use means a non-exclusive license for the farmer to access or use the property right services [139]. Fructus (fruit, in a figurative sense) is the right to derive profit from a thing possessed, for instance, by selling crops, leasing immovables or annexed movables, taxing for entry, and so on [140]. The right of disposal of goods, including retention of ownership and retention of the right to sell the goods, might have a crucial impact on farmers' investment behavior [141]. The specific indicators associated with all the variables are stated in Table 1.

3.3.3. Control Variables

The study selects control variables from the following three aspects: individual sample characteristics, sample family characteristics, and woodland tree characteristics. This article uses age, education level, health status, and status as a Communist party member or not (village cadre) to reflect the characteristics of the sampled individuals. In contrast, we used the family population and the size of fixed assets to reflect the characteristics of the sample households; the average single forest area, forest land distance, forest land quality, and forest trees reflect the characteristics of agro-forestry. The meaning and descriptive statistics of the variables are shown in Table 1.

3.4. Model Construction

Water conservation and lack of ecological property rights within forestry have caused market failures in ecological governance in reservoir confluence areas. Incorporating ecological property rights into the property structure to study the impact of collective forest rights on farmers' investment in forestry is significant in rectifying water conservation forest protection and urban water safety. At the same time, subdivisions of property rights are crucial for implementing complex property rights systems [142,143]. The property rights approach suggests that if exclusive property rights are adequately defined, the public good prospects of environmental quality can be transformed into a private good, and optimal environmental allocation will be reached [144]. According to the theory of property rights, a subdivision of property rights is a meaningful way to implement complex property rights [145]. By subdividing forest property rights, the economic property rights of water conservation of forests can be separated from ecological property rights [146], thereby defining ecological property rights. The definition is thereby adopted in the study. When the ecological property rights are clearly defined and farmers are given the right to exchange property rights, farmers can realize the ecological value of water conservation forests through the ecological market and obtain the benefits of ecological products.

Table 1. Variable definition and descriptive statistics.

Variable Type	Variable Name	Meaning And Assignment	Mean	Standard Deviation
Dependent Variable	Willingness to Invest	No Willingness to Invest = 0; Willingness To Invest = 1	0.93	0.26
	Investment Intensity	2015–2019 Cumulative Investment per Unit Area (Yuan)	935.82	615.76
Core Independent Variable	Right to Use			
	Conversion to Other Forestry Uses	No Right = 0; Uncertainty = 1; Right, Subject to Partial Consent of the Village or Government = 2; Right, and Free to Exercise = 3	1.76	0.63
	Choose Tree Species		1.94	0.72
	Operating Non-Wood Forest Products		2.20	1.17
	Usufruct			
	Economic Product Income	No Right = 0; Not Sure = 1; Right but Not Exclusive, Part of the Income Is Invaded by the Village or the Government = 2; Right, and Exclusive Income = 3	2.50	0.79
	Ecological Product Benefits		1.39	0.56
	Right of Disposal			
	Circulation Right	No Right = 0; Uncertainty = 1; Right, Subject to Partial Consent of the Village or Government = 2; Right, and Free to Exercise = 3	2.12	0.57
	Adjustment Risk	Possibility of Adjustment within the Woodland Village: Impossible = 0, Uncertain = 1, Possible = 2	1.40	0.61
	Levy Risk	Possibility of Expropriation of Forest Land: Impossible = 0, Uncertain = 1, Possible = 2	1.38	0.75
Control Variable	Age	Age of Respondents in 2019 (Years)	49.50	11.35
	Education Level	Illiterate (No School) = 1; Elementary School = 2; Junior High School = 3; High School = 4; College = 5; Bachelor's Degree and Above = 6	2.31	0.97
	Health Status	Very Poor = 1; Relatively Poor = 2; General = 3; Relatively Healthy = 4; Very Healthy = 5	3.47	1.13
	Whether or Not a Party Member (Village Cadre)	No = 0; Yes = 1	0.18	0.39
	Family Population	Total Family Population (Person)	4.28	1.23
	Fixed Assets	The Total Value of Family Fixed Assets (Ten Thousand Yuan)	24.50	15.69
	Forest Area	Farmer Households Contracted Forest Land, the Average Area of Single Piece of Forest Land (Mu)	10.67	6.29
	Woodland Distance	The Time Required from Home to Woodland Rounded up to 10 min	96.94	110.56
	Woodland Quality	Very Poor = 1; Poor = 2; General = 3; Better = 4; Very Good = 5	2.50	0.95
	Tree Type	No Forest Land = 1; Pure Timber Forest = 2; Mixed Timber Forest and Economic Forest = 3; Pure Economic Forest = 4	3.61	0.74

Interestingly, obtaining income is the direct purpose of farmers' forestry investment. This article divides forest income rights from economic product and ecological product income rights and analyzes the effect of collective forest rights on farmers' forestry investment behavior. Impact analysis has highlighted the critical role of farmers' forestry investment decision-making. Based on this, we propose to define ecological property rights, establish an ecological property rights trading market, and realize the ecological value of water conservation forests through property rights exchange as an effective way to protect farmers' income rights and encourage farmers to invest in forestry.

In the study, the forestry investment of farmers in the confluence area acts as a dependent variable, and the reservoir is similar to a continuous variable (as the dependent variable, farmers' forestry investment in the confluence area is similar to a continuous variable). However, for this part of the data without forestry investment willingness, the dependent variable is compressed at 0. The dependent variable's probability distribution includes a discrete point of 0 and is based on a continuous distribution. At the same time, the factors that affect farmers' forestry investment willingness and intensity may not be the same as suggested by Duan et al. [25]. Therefore, the production mechanism is set as a dependent variable derived by 0, and the continuous variable may be different. In addition, there may be a correlation between forestry investment willingness and investment intensity, and deciding whether or not an investment has a tail-end effect on investment intensity will lead to selection bias. Thus, the double-hurdle model is more suitable than the tobit model [147]. Therefore, the study uses the Heckman model to estimate the sample selection, and the estimation results show that the inverse Mills ratio is insignificant. The null hypothesis that investment willingness and intensity are independent of each other cannot be rejected. Therefore, the article uses the probit and truncated double-hurdle model to estimate forestry investment willingness and investment intensity independently in two stages. The study sets the basic model as follows:

$$I_i = \alpha + \beta_1 PI_i + \beta_2 PS_i + \gamma X_i + \varepsilon_i \tag{1}$$

Among them, I_i is the forestry input of the ith farmer household in the confluence area of the reservoir (the natural logarithm of the farmer's actual forestry investment), and PI_i and PS_i represent the farmers' complete knowledge and understanding of the collective forest rights they hold, respectively. Seemingly, PI_i and PS_i represent the farmers' integrity cognition and stability cognition of the collective forest rights they hold, respectively and X_i is the control variable, and ε_i is the random disturbance term. There may be an endogenous selection bias problem between farmers' collective forest rights perception and forestry investment behavior in the formula. This study uses the average value of other households' knowledge of property rights in the same village as an instrumental variable to eliminate possible endogenous estimation biases, as suggested by Liu and Jia [126] and Ma et al. [127]. This is because in the same administrative village, the perception of collective forest rights in a sample may be affected by the perception of other farmers' property rights [128]. At the same time, the perception of property rights of other farmers in the same village is not directly related to the forestry investment behavior of the sample [129].

4. Results

Model Estimation Results

Table 2 presents the regression results of the impact of collective forest rights on farmers' forestry investment, and it shows that in terms of completeness, only income rights significantly affect farmers' forestry investment willingness, and other property rights are not significant. Based on farmers' willingness to invest, rights to use and income rights positively impact farmers' forestry investment intensity, while the impact of disposal rights is not significant. In terms of stability, expropriation risk significantly negatively impacts farmers' forestry investment willingness and intensity. Seemingly, adjustment risk also significantly negatively affects farmers' forestry investment intensity but has no significant impact on investment willingness. It could happen as the adjustment of forest land rights is subject to adjustment of land within the village, due to population changes, and the timing of adjustments generally avoids the harvest season. Therefore, adjustment of expectations will not affect farmers' investment participation in decision-making, but when farmers expect that forest land rights may be adjusted, long-term investment will not be recovered, which will reduce the intensity of forestry investment.

Table 2. Regression results of the impact of collective forest rights on farmers' forestry investment.

Project	Investment Willingness (Probit Model)		Investment Intensity (Truncated Model)	
	Coefficient	Z Value	Coefficient	Z-Value
The integrity of property rights				
Right to use	0.320 (0.778)	0.41	1.514 (0.324)	4.68 ***
Usufruct	1.738 (0.867)	2.00 **	1.095 (0.314)	3.49 ***
Right of disposal	−0.767 (0.630)	−1.22	−0.336 (0.206)	−1.63
Stability of property rights				
Redistribution risk	0.6515 (0.828)	0.79	−1.101 (0.347)	−3.17 ***
Expropriation risk	−1.575 (0.693)	−2.27 **	−0.683 (0.291)	−2.35 **
Control variable				
Age	0.0047 (0.008)	0.56	0.016 (0.004)	4.31 ***
education level	−0.355 (0.103)	−3.45 ***	−0.026 (0.045)	−0.59
Health status	0.098 (0.083)	1.18	0.040 (0.037)	1.09
Whether or not a party member (village cadre)	−0.260 (0.229)	−1.14	0.100 (0.104)	0.96
Family population	−0.011 (0.077)	−0.14	0.086 (0.031)	2.77 ***
Family fixed assets	0.377 (0.171)	2.22 **	0.137 (0.093)	1.47
Forest area	0.172 (0.032)	5.26 ***	−0.015 (0.006)	−2.57 **
Woodland distance	−0.002 (0.0008)	−2.63 ***	−0.0007 (0.0003)	−2.09 **
Woodland quality	0.386 (0.103)	3.75 ***	0.225 (0.041)	5.49 ***
Tree type	0.051 (0.130)	0.40	0.054 (0.052)	1.03
LR	116.76 ***			
Wald			134.90 ***	
Sample size	708		657	
Mean VIF	1.53		1.53	

Note: **, *** mean significant at the statistical level of 5%, and 1%, respectively.

The level of education negatively affects the investment willingness of farmers. A higher level of education can foster non-forest employment choices of farmers and lower the willingness to invest in forestry. The age of the household head has a positive impact on the investment intensity. It is generally believed that based on the willingness to invest, as the age increases, the farmer has accumulated more forestry management experience, and at the same time, the opportunities for non-forest jobs are also reduced, and they are more inclined to increase forestry investment. The number of family members has a positive impact on the forestry investment intensity of farmers. As the number of family members increases, more labor will be available for forestry production. Household fixed assets positively impact farmers' willingness to invest in forestry. Similarly, when the forestry production cycle is long, it could bring many uncertainties and investment risks. Therefore, the greater the total fixed assets of farmers, the stronger the ability to resist risks, and the more likely they are to invest in forestry.

Seemingly, the average land plot area positively affects the willingness to invest and negatively affects the investment intensity. With the increase in the land plot area, the increase in the benefits of the scale effect will encourage farmers to invest in forestry. However, if the income level of farmers in the confluence area of the reservoir is low, the funds that can be used for forestry investment are limited and will not increase with the increase in the plot area. Therefore, the investment per unit area will decrease with the increase in the plot area. The distance from forest to home negatively affects forestry investment willingness and intensity. The increase in the distance from home to the forest will lead to an increase in forestry input costs, which will inhibit farmers' willingness and intensity of forestry investment. However, forest quality has a positive impact on investment willingness and intensity. Better forest quality influences the possibility of profitability and higher income, and it will positively influence the willingness of farmers to invest in forestry and eventually increase the investment intensity. The plantation forests in the confluence area of the Heihe Reservoir are mainly economic forests, so the type of tree has no significant impact on farmers' willingness to invest in forestry and investment intensity.

It can be seen from Table 2 that the income right significantly affects the forestry investment willingness of farmers at the 5% significance level and significantly affects the forestry investment intensity of farmers at the 1% significance level. Forestry production by farmers in the confluence area of reservoirs can benefit from both forest products and ecological products. In order to clarify the impact of income rights on farmers' forestry investment decisions, this article further explores the impact of farmers' forestry investment from the perspective of the separation of economic and ecological property rights. Table 3 denotes the regression results of decision-making factors. It can be seen from Table 3 that, in terms of the integrity of property rights, only the right to earn from ecological products affects the willingness of farmers to invest in forestry, and other property rights are not significant. However, based on the willingness of farmers to invest, the right to use forest land, economical products, and ecological product income rights positively affects farmers' forestry investment intensity, and disposal rights (forest land transfer rights) negatively affect farmers' forestry investment intensity.

Table 3. Regression results of the impact of collective forest rights on farmers' forestry investment, based on separation of economic property rights and ecological property rights.

Project	Investment Willingness (Probit Model)		Investment Intensity (Truncated Model)	
	Coefficient	Z Value	Coefficient	Z-Value
The integrity of property rights				
Right to use	0.280 (0.787)	0.36	1.556 (0.323)	4.81 ***
Economic product income	0.5450 (0.643)	0.85	0.904 (0.258)	3.50 ***
Ecological product benefits	1.201 (0.632)	1.90 *	0.484 (0.219)	2.22 **
Right of disposal	−0.775 (0.631)	−1.23	−0.414 (0.208)	−1.99 **
Stability of property rights				
Redistribution risk	0.606 (0.838)	0.72	−0.980 (0.351)	−2.79 ***
Expropriation risk	−1.470 (0.836)	−1.76 *	−1.048 (0.337)	−3.11 ***
Control variable				
Age	0.005 (0.008)	0.54	0.016 (0.004)	4.23 ***
Education level	−0.355 (0.103)	−3.45 ***	−0.027 (0.044)	−0.60
Health status	0.098 (0.083)	1.17	0.038 (0.037)	1.01
Whether or not a party member (village cadre)	−0.263 (0.230)	−1.15	0.113 (0.104)	1.09
Family population	−0.011 (0.077)	−0.15	0.084 (0.031)	2.72 ***
Family fixed assets	0.385 (0.172)	2.24 **	0.148 (0.093)	1.60
Forest area	0.172(0.033)	5.26 ***	−0.015 (0.006)	−2.48 **
Woodland distance	−0.002 (0.001)	−2.64 ***	−0.001 (0.000)	−2.19 **
Woodland quality	0.386 (0.103)	3.74 ***	0.217 (0.041)	5.27 ***
Tree type	0.054 (0.132)	0.41	0.031 (0.053)	0.58
LR	117.08 ***			
Wald			140.58 ***	
Sample size	708		657	
Mean VIF	1.59		1.59	

Note: *, **, *** mean significant at the statistical level of 10%, 5%, and 1%, respectively.

5. Discussion

With the development of society and economy, the problems of water resource protection and the balance of benefits between upstream and downstream of the river basin have become increasingly prominent. The forest basin in the reservoir's confluence area is considered a crucial source of clean water. China's socio-economic growth depends on efficient watershed stewardship. While having immense investments in watershed governance and infrastructures, relatively stronger and integrated water governance at the municipal and federal tiers should be required to formulate practical and innovative water resources protection trends. Vital strategies for supporting the sharply expanding economy include offering more water for environmental usage, intensifying market instruments to foster water use efficiency, and accepting transformative behavioral measures to fight against water contamination. In this regard, farmers' involvement via active participation and collectiveness, in the forms of ecological property rights, can act as sophisticated approaches. However, the lack of forest ecological property rights in the confluence area of reservoirs has caused the externalities of hydro-ecological services to be unable to be internalized, leading to market failures in ecological governance and farmers lacking forestry investment willingness or insufficient investment intensity.

The forestry production of farmers in the confluence area of the Heihe Reservoir originated from the return of farmland to forests in 1998. Before that, traditional agriculture was the primary livelihood for farmers in this area, and there were few forestry producers [148]. In 1998, farmers in this area returned farmland to forests, to obtain ecological compensation, and started forestry production [149]. The confluence area of the reservoir is a quasi-protection area for water source protection, and the right to use forest land is more restricted than in general areas. The forestry production behavior of farmers is mainly to implement the policies of the local forestry department, and there is not much room for independent decision-making. At the same time, due to the geographical environment of the mountainous area, it is not favorable to use machinery. All the core farming work, such as preparing soil, sowing, weed and pest control, and harvesting, are done manually, and the income of forest products is limited. Therefore, obtaining ecological compensation is the primary motivation of farmers' forestry investment in this area, which is consistent with the ecological value of the forest trees in the reservoir confluence area.

The current trends and assessment of ecological property rights in contemporary policy-oriented literature, by legislative bodies and other researchers, are inadequate. There is an emergent need for an innovative assessment of ecological property rights within the aspects of farmers' agroforestry investment behavior. The impact of ecological property rights is being emphasized greatly in developmental and ecological programs because of its importance in responsible natural resource stewardship, effective governance, and impoverished community empowerment. Thus, the study evaluates the potential role of ecological property rights within the core concepts of property rights. Ecological property rights may also influence land-use strategies, including identifying various motivating factors or drivers and managing arrangements in agroforestry systems, as well as facilitating greater ecological systems. Seemingly, developmental organizations progressively recognize the importance of ecological property rights as a key role in deciding how land and natural resources are utilized and maintained, and how the benefits of those resources are dispersed. The study also formulates a pilot transactions framework to rectify the on-hand effects and provide an overview of the critical ecological property rights concepts involved in designing and implementing natural resource management programs. As the confluence area of the Heihe Reservoir is restricted for usual farming, farmers' investment in forestry within the area can facilitate proper usage of the land, livelihood facilities, and economic solvency of farmers. Thus, the current study design rectifies the innovativeness and significance of this crucial topic.

The key factor that affects the willingness of farmers to invest in forestry in the confluence area of the reservoir is whether it is "profitable". Based on the farmers' decision to invest, the integrity of the right to use, and other owners, will affect the amount of investment. Therefore, it is necessary to define forest ecological property rights and protect farmers' right to income. There was no significant effect on willingness to invest, and thus, Hypotheses 2–4 are partially verified. This research conclusion contradicts the theoretical hypothesis that farmers obtain benefits through the use and disposal of forest resources, and the more complete the property rights, the stronger the investment incentives. However, it is consistent with the fact that farmers in the confluence area of the reservoir invest in economic forestry. The forest land use and disposal rights in this area are strictly restricted, and the benefits of forest products are meager [149]. Obtaining ecological compensation is the main purpose of farmers participating in ecological projects, such as returning farmland to forests [150]. Based on obtaining reasonable ecological compensation, other rights, such as rights to use and disposal rights, will impact the investment intensity of farmers. The results show that farmers are likely to give up forestry investment directly if reasonable ecological compensation is not guaranteed. Therefore, the following assumptions could be made:

Assumption 1: *The integrity of collective forest tenure positively affects the forestry investment behavior of farmers (Accepted).*

Assumption 2: *The right of use positively affects the forestry investment behavior of farmers (partially accepted, only affects investment intensity and has no significant impact on investment willingness).*

Assumption 3: *The right to income positively affects the forestry investment behavior of farmers (partially accepted, the right to benefit from ecological products has a significant impact on investment willingness and intensity, while the right to benefit from economic products only affects investment intensity and has no significant impact on investment willingness).*

Assumption 4: *The right of disposal positively affects the forestry investment behavior of farmers (partially accepted, only affects investment intensity and has no significant impact on investment willingness).*

In the confluence area of reservoirs, obtaining the benefits of ecological products is the primary motivation for farmers' forestry investment and has a significant positive impact on the intensity of farmers' forestry investment. It is different from the research conclusions of Ji et al. [87], Yi et al. [81], Holden and Otsuka [151]. The forestry investment intensity of farmers does not significantly impact investment willingness. This research conclusion contradicts the theoretical hypothesis that farmers obtain income through the use and disposal of forest resources. The exclusive property rights found fostering, the stronger the investment incentives. However, these findings are consistent with the fact that the right to use, and disposal of, forests in the confluence area of the reservoir is stringently restricted, the income of forest products is meager, and the ecological compensation based on the extent of farmers' participation in environmental projects, such as returning farmland to forests is insufficient. The outcome is consistent with the results reported by Suleiman et al. [152] and Nerfa et al. [153].

However, due to the geographical environment of the mountainous area, it is impossible to use heavy machinery, and the income by-product is limited, not even enough to cover the cost in many cases. However, farmers in this area generally receive ecological product benefits in ecological compensation [154]. In the confluence area of the Heihe Reservoir, only 34.04% of the rural households in the sample participated in the survey, received income from forest products in 2018, and the households receiving ecological compensation income accounted for 98.73% of the total sample. In the absence of ecological property rights, vertical transfer payment is currently the primary method for forest ecological compensation in the confluence area of reservoirs. Although this approach embodies the principle of fairness, it lacks attention to the farmers as ecological service providers and does not reflect the supply and demand relationship of ecological products. As a result, farmers' forest reforestation and management behaviors do not receive the economic incentives they deserve. The findings show that the adjustment risk has a significant negative impact on farmers' forestry investment intensity but impacts investment willingness. The effect of adjusting risk expectations on farmers' forestry investment willingness is insignificant, inconsistent with the existing research that generally found that property rights security significantly impacts investment willingness and intensity [155,156]. It may be because the adjustment of forest land is the adjustment of land within the village, due to population changes in administrative villages, and the adjustment implementation time node generally avoids the harvest season. Therefore, adjusting expectations will not affect farmers' investment participation in decision-making, but when farmers expect that forest land may be adjusted, long-term investment will not be recovered, which will reduce the input intensity of forestry investment. Thus, the following assumptions could be made:

Assumption 5: *Unstable collective forest rights negatively affect farmers' agro-forestry investment behavior (Accepted).*

Assumption 6: *Adjusting risk expectations negatively affects farmers' forestry investment behavior (partially accepted, negatively affects investment intensity and has no significant impact on investment willingness).*

Assumption 7: *The expropriation risk expectation negatively affects the forestry investment behavior of farmers (accepted, has a significant negative impact on investment willingness and investment intensity).*

6. Conclusions

The study uses the survey data of farmers in the confluence area of the Heihe Jinpen Reservoir, based on the perspective of ecological property rights, to study the impact of collective forest rights on the forestry investment behavior of farmers in the reservoir area. Because water source protection restricts farmers' production and livelihood in the confluence area, farmers require ecological compensation. Although the current vertical ecological compensation reflects the principle of fairness, it ignores the efficiency of resource allocation and does not reflect the supply–demand relationship of ecological products. Therefore, based on the divisibility of property rights, the income rights of water conservation forests are divided into economic product income rights and ecological product income rights. Moreover, a well-structured pilot test of ecological property rights transactions is carried out in the confluence areas of reservoirs, where conditions permit, and farmers can realize the benefits of water conservation forests through the ecological market. Ecological value encourages farmers to invest in forestry, carry out afforestation and reforestation, and realize water conservation forests' sustainable development.

The study portrays the following outcomes: (i) Incentive received for forestry, profitable forestry investment, and obtaining ecological product income and rights act as the primary motivation for farmers' forestry investment within the reservoir confluence area. (ii) The rights to use and disposal were the central assumptions for farmers' willingness to invest. The completeness of property rights of other owners impacted the investment amount intensity. Specifically, in terms of the integrity of property rights, the right to profit and income rights from ecological products affect farmers' willingness to invest in forestry, and other property rights are insignificant, whereas the income right has a positive impact and the disposal right (forest land circulation right) negatively affects the forestry investment intensity of farmers. (iii) Regarding property rights stability, the lower the farmers' expectation of forest land acquisition risk, the greater the possibility of investment and the higher the input level. Since the forest land adjustment usually avoids the harvesting period, the farmers' forest land adjustment expectation will only negatively affect the input level and, therefore, affect investment willingness negatively. (iv) In contrast, forest land use rights, financial products, and ecological products are crucial for farmers' willingness to invest. The income right positively affects the forestry investment intensity of farmers, and the disposal right (forest land circulation right) negatively affects the forestry investment intensity of farmers. (v) Regarding the stability of property rights, the lower the farmers' forest land expropriation risk is expected, the greater the possibility of investment and the higher the input level. The right of use, right of income, and rights to disposal are the necessary conditions for farmers' forestry investment.

Based on the above research conclusions, the study puts forward the following policy suggestions: (i) Sustainable development of water conservation forests should be highlighted, to encourage farmers to invest in forestry. (ii) However, farmers' subjective perception of collective forest tenure affects their forestry investment behavior. Therefore, more attention should be paid to improving farmers' subjective cognition, where agricultural extension offices and demonstration zones should extend their support. (iii) The government should carry out collective forest rights publicity, by arranging frequent visits by village cadres, village meetings, and technical training, which could effectively improve farmers' awareness of property rights and promote forestry investment. (iv) In addition, subjective perception of farmers' collective forest rights affects their forestry investment behavior. Therefore, while improving collective forest rights in reservoir confluence areas at the legal level, attention should be paid to farmers' subjective perceptions of collective forest rights. (v) Government should realize the actual demand for the sustainability of natural resources and ensure well-balanced conservations. In contrast, they should simplify obtaining property rights within the context of ecological property rights.

However, the following issues still need further consideration: (i) The study included a limited area, which may hinder the application of the model and validity of the outcomes for other forest regions. Thus, future research should use multiple areas to test the ecological property rights transactions for better reliability and valid assumptions (ii) Issues like ecological property rights policies, regulations, and ecological ethics support should be explored further. (iii) The establishment and effective operation of the ecological market guarantee framework should be explored critically, with different forest zones. (iv) Future studies should include the issue of the behavioral capacity of the farmers' ecological property rights transactions within the confluence area of the reservoir, to get more robust results. (v) The potential studies should present the key variable of interest within separate results subsections to provide more comprehensive outlines.

Author Contributions: Conceptualization, M.L. and A.S.; methodology, M.L. and A.S.; software, M.L.; validation, M.L., A.S. and A.K.H.; formal analysis, M.L. and A.S.; investigation, M.L. and Y.W.; resources, Y.W.; data curation, M.L. and Y.W.; writing—original draft preparation, M.L. and A.S.; writing—review and editing, M.L., A.S. and Q.M.; visualization, A.K.H. and Y.W.; supervision, Q.M.; project administration, A.S., A.K.H. and Q.M.; funding acquisition, Q.M. All authors have read and agreed to the published version of the manuscript.

Funding: The study was funded by "Cross-regional compensation system of key ecological functional areas in Shaanxi Province", Funding No. 2018KRM011.

Institutional Review Board Statement: Ethical review and approval were waived for this study as the study did not collect any personal data from the respondents, and respondents were informed that they could opt out any time if they wanted.

Informed Consent Statement: Informed consent was obtained from all subjects involved in the study.

Data Availability Statement: The data will be provided upon request by the corresponding author.

Acknowledgments: The authors acknowledge that respectable reviewers employed their utmost expertise, wisdom, and careful evaluation to make the study well presentable and ensure the quality of the writing. We also extend our gratitude to respondents for their valuable inputs and the interviewer's teams for their rigorous support in the data collection process.

Conflicts of Interest: The authors declare no conflict of interest.

References

1. Heri-Kazi, A.B.; Bielders, C.L. Erosion and Soil and Water Conservation in South-Kivu (Eastern DR Congo): The Farmers' View. *Land Degrad. Dev.* **2021**, *32*, 699–713. [CrossRef]
2. Raj, A.; Jhariya, M.K.; Yadav, D.K.; Banerjee, A. Agroforestry Systems in the Hills and Their Ecosystem Services. In *Environmental and Sustainable Development through Forestry and Other Resources*; Apple Academic Press: Waretown, NJ, USA, 2020; ISBN 978-0-429-27602-6.
3. De Carvalho, A.F.; Fernandes-Filho, E.I.; Daher, M.; de Carvalho Gomes, L.; Cardoso, I.M.; Fernandes, R.B.A.; Schaefer, C.E.G.R. Microclimate and Soil and Water Loss in Shaded and Unshaded Agroforestry Coffee Systems. *Agrofor. Syst.* **2021**, *95*, 119–134. [CrossRef]
4. Nath, A.J.; Lal, R.; Das, A.K. Ethnopedology and Soil Properties in Bamboo (*Bambusa* Sp.) Based Agroforestry System in North East India. *Catena* **2015**, *135*, 92–99. [CrossRef]

5. Deltour, P.; França, S.C.; Pereira, O.L.; Cardoso, I.; De Neve, S.; Debode, J.; Höfte, M. Disease Suppressiveness to Fusarium Wilt of Banana in an Agroforestry System: Influence of Soil Characteristics and Plant Community. *Agric. Ecosyst. Environ.* **2017**, *239*, 173–181. [CrossRef]

6. Wu, Q.; Liang, H.; Xiong, K.; Li, R. Eco-Benefits Coupling of Agroforestry and Soil and Water Conservation under KRD Environment: Frontier Theories and Outlook. *Agrofor. Syst.* **2019**, *93*, 1927–1938. [CrossRef]

7. Quinkenstein, A.; Wöllecke, J.; Böhm, C.; Grünewald, H.; Freese, D.; Schneider, B.U.; Hüttl, R.F. Ecological Benefits of the Alley Cropping Agroforestry System in Sensitive Regions of Europe. *Environ. Sci. Policy* **2009**, *12*, 1112–1121. [CrossRef]

8. Sharma, R.; Chauhan, S.K.; Tripathi, A.M. Carbon Sequestration Potential in Agroforestry System in India: An Analysis for Carbon Project. *Agrofor. Syst.* **2016**, *90*, 631–644. [CrossRef]

9. Stöcker, C.M.; Bamberg, A.L.; Stumpf, L.; Monteiro, A.B.; Cardoso, J.H.; de Lima, A.C.R. Short-Term Soil Physical Quality Improvements Promoted by an Agroforestry System. *Agrofor. Syst.* **2020**, *94*, 2053–2064. [CrossRef]

10. Wang, S.; Olatunji, O.A.; Guo, C.; Zhang, L.; Sun, X.; Tariq, A.; Wu, X.; Pan, K.; Li, Z.; Sun, F.; et al. Response of the Soil Macrofauna Abundance and Community Structure to Drought Stress under Agroforestry System in Southeastern Qinghai-Tibet Plateau. *Arch. Agron. Soil Sci.* **2020**, *66*, 792–804. [CrossRef]

11. Karman, C.C. The Role of Time in Environmental Risk Assessment. *Spill Sci. Technol. Bull.* **2000**, *6*, 159–164. [CrossRef]

12. De Mello, K.; Taniwaki, R.H.; de Paula, F.R.; Valente, R.A.; Randhir, T.O.; Macedo, D.R.; Leal, C.G.; Rodrigues, C.B.; Hughes, R.M. Multiscale Land Use Impacts on Water Quality: Assessment, Planning, and Future Perspectives in Brazil. *J. Environ. Manag.* **2020**, *270*, 110879. [CrossRef] [PubMed]

13. Yan, T.; Qian, W.Y. Environmental Migration and Sustainable Development in the Upper Reaches of the Yangtze River. *Popul. Environ.* **2004**, *25*, 613–636. [CrossRef]

14. Wang, X.; Liu, X.; Tang, W.; Pang, H. Evaluation on stand quality of water conservation forest in Longkou Forest Farm on Danjiangkou reservoir area. *J. Nanjing For. Univ. Nat. Sci. Ed.* **2013**, *37*, 63–68.

15. Duguma, L.A. Financial Analysis of Agroforestry Land Uses and Its Implications for Smallholder Farmers Livelihood Improvement in Ethiopia. *Agrofor. Syst.* **2013**, *87*, 217–231. [CrossRef]

16. Quandt, A.; Neufeldt, H.; McCabe, J.T. Building Livelihood Resilience: What Role Does Agroforestry Play? *Clim. Dev.* **2019**, *11*, 485–500. [CrossRef]

17. Lu, S.; Sun, H.; Zhou, Y.; Qin, F.; Guan, X. Examining the Impact of Forestry Policy on Poor and Non-Poor Farmers' Income and Production Input in Collective Forest Areas in China. *J. Clean. Prod.* **2020**, *276*, 123784. [CrossRef]

18. Owubah, C.E.; Le Master, D.C.; Bowker, J.M.; Lee, J.G. Forest Tenure Systems and Sustainable Forest Management: The Case of Ghana. *For. Ecol. Manag.* **2001**, *149*, 253–264. [CrossRef]

19. Qin, P.; Xu, J. Forest Land Rights, Tenure Types, and Farmers' Investment Incentives in China: An Empirical Study of Fujian Province. *China Agric. Econ. Rev.* **2013**, *5*, 154–170. [CrossRef]

20. Tan, Z.; Chen, K.; Liu, P. Possibilities and Challenges of China's Forestry Biomass Resource Utilization. *Renew. Sustain. Energy Rev.* **2015**, *41*, 368–378. [CrossRef]

21. Assa, B.S.K. Foreign Direct Investment, Bad Governance and Forest Resources Degradation: Evidence in Sub-Saharan Africa. *Econ. Polit.* **2018**, *35*, 107–125. [CrossRef]

22. Li, J.; Bluemling, B.; Dries, L. Property Rights Effects on Farmers' Management Investment in Forestry Projects: The Case of Camellia in Jiangxi, China. *Small-Scale For.* **2016**, *15*, 271–289. [CrossRef]

23. Tikkanen, J.; Isokääntä, T.; Pykäläinen, J.; Leskinen, P. Applying Cognitive Mapping Approach to Explore the Objective–Structure of Forest Owners in a Northern Finnish Case Area. *For. Policy Econ.* **2006**, *9*, 139–152. [CrossRef]

24. Xie, Y.; Gong, P.; Han, X.; Wen, Y. The Effect of Collective Forestland Tenure Reform in China: Does Land Parcelization Reduce Forest Management Intensity? *J. For. Econ.* **2014**, *20*, 126–140. [CrossRef]

25. Duan, W.; Shen, J.; Hogarth, N.J.; Chen, Q. Risk Preferences Significantly Affect Household Investment in Timber Forestry: Empirical Evidence from Fujian, China. *For. Policy Econ.* **2021**, *125*, 102421. [CrossRef]

26. Andersson, M.; Gong, P. Risk Preferences, Risk Perceptions and Timber Harvest Decisions—An Empirical Study of Nonindustrial Private Forest Owners in Northern Sweden. *For. Policy Econ.* **2010**, *12*, 330–339. [CrossRef]

27. Van der Jagt, A.P.N.; Lawrence, A. Local Government and Urban Forest Governance: Insights from Scotland. *Scand. J. For. Res.* **2019**, *34*, 53–66. [CrossRef]

28. Maraseni, T.N.; Bhattarai, N.; Karky, B.S.; Cadman, T.; Timalsina, N.; Bhandari, T.S.; Apan, A.; Ma, H.O.; Rawat, R.S.; Verma, N.; et al. An Assessment of Governance Quality for Community-Based Forest Management Systems in Asia: Prioritisation of Governance Indicators at Various Scales. *Land Use Policy* **2019**, *81*, 750–761. [CrossRef]

29. Cashore, B.; Vertinsky, I. Policy Networks and Firm Behaviours: Governance Systems and Firm Reponses to External Demands for Sustainable Forest Management. *Policy Sci.* **2000**, *33*, 1–30. [CrossRef]

30. Gough, A.D.; Innes, J.L.; Allen, S.D. Development of Common Indicators of Sustainable Forest Management. *Ecol. Indic.* **2008**, *8*, 425–430. [CrossRef]

31. Nayak, P.K.; Berkes, F. Politics of Co-Optation: Community Forest Management Versus Joint Forest Management in Orissa, India. *Environ. Manag.* **2008**, *41*, 707–718. [CrossRef]

32. Wulandari, C.; Inoue, M. The Importance of Social Learning for the Development of Community Based Forest Management in Indonesia: The Case of Community Forestry in Lampung Province. *Small-Scale For.* **2018**, *17*, 361–376. [CrossRef]

33. Sotirov, M.; Sallnäs, O.; Eriksson, L.O. Forest Owner Behavioral Models, Policy Changes, and Forest Management. An Agent-Based Framework for Studying the Provision of Forest Ecosystem Goods and Services at the Landscape Level. *For. Policy Econ.* **2019**, *103*, 79–89. [CrossRef]

34. Blanc, S.; Accastello, C.; Bianchi, E.; Lingua, F.; Vacchiano, G.; Mosso, A.; Brun, F. An Integrated Approach to Assess Carbon Credit from Improved Forest Management. *J. Sustain. For.* **2019**, *38*, 31–45. [CrossRef]

35. Sgroi, F. Forest Resources and Sustainable Tourism, a Combination for the Resilience of the Landscape and Development of Mountain Areas. *Sci. Total Environ.* **2020**, *736*, 139539. [CrossRef] [PubMed]

36. Zong, C.; Cheng, K.; Lee, C.-H.; Hsu, N.-L. Capturing Tourists' Preferences for the Management of Community-Based Ecotourism in a Forest Park. *Sustainability* **2017**, *9*, 1673. [CrossRef]

37. Nguyen, T.V.; Lv, J.H.; Ngo, V.Q. Factors Determining Upland Farmers' Participation in Non-Timber Forest Product Value Chains for Sustainable Poverty Reduction in Vietnam. *For. Policy Econ.* **2021**, *126*, 102424. [CrossRef]

38. Gelo, D. Forest Commons, Vertical Integration and Smallholder's Saving and Investment Responses: Evidence from a Quasi-Experiment. *World Dev.* **2020**, *132*, 104962. [CrossRef]

39. Krul, K.; Ho, P.; Yang, X. Incentivizing Household Forest Management in China's Forest Reform: Limitations to Rights-Based Approaches in Southwest China. *For. Policy Econ.* **2020**, *111*, 102075. [CrossRef]

40. Ren, Y.; Kuuluvainen, J.; Yang, L.; Yao, S.; Xue, C.; Toppinen, A. Property Rights, Village Political System, and Forestry Investment: Evidence from China's Collective Forest Tenure Reform. *Forests* **2018**, *9*, 541. [CrossRef]

41. Miyake, Y.; Kimoto, S.; Uchiyama, Y.; Kohsaka, R. Income Change and Inter-Farmer Relations through Conservation Agriculture in Ishikawa Prefecture, Japan: Empirical Analysis of Economic and Behavioral Factors. *Land* **2022**, *11*, 245. [CrossRef]

42. Kohsaka, R.; Miyake, Y. The Politics of Quality and Geographic Indications for Non-Timber Forest Products: Applying Convention Theory beyond Food Contexts. *J. Rural Stud.* **2021**, *88*, 28–39. [CrossRef]

43. Robson, A.; Skaperdas, S. Costly Enforcement of Property Rights and the Coase Theorem. *Econ. Theory* **2008**, *36*, 109–128. [CrossRef]

44. Hervés-Beloso, C.; Moreno-García, E. Revisiting the Coase Theorem. *Econ. Theory* **2021**, *3*, 1–18. [CrossRef]

45. O'Donnell, F.C.; Flatley, W.T.; Springer, A.E.; Fulé, P.Z. Forest Restoration as a Strategy to Mitigate Climate Impacts on Wildfire, Vegetation, and Water in Semiarid Forests. *Ecol. Appl.* **2018**, *28*, 1459–1472. [CrossRef] [PubMed]

46. Pei, S.; Zhang, C.; Liu, C.; Liu, X.; Xie, G. Forest Ecological Compensation Standard Based on Spatial Flowing of Water Services in the Upper Reaches of Miyun Reservoir, China. *Ecosyst. Serv.* **2019**, *39*, 100983. [CrossRef]

47. Yang, Y.; Zhang, X.; Chang, L.; Cheng, Y.; Cao, S. A Method of Evaluating Ecological Compensation Under Different Property Rights and Stages: A Case Study of the Xiaoqing River Basin, China. *Sustainability* **2018**, *10*, 615. [CrossRef]

48. Zieminska-Stolarska, A.; Zbicinski, I.; Imbierowicz, M.; Skrzypski, J. Waterpraxis as a Tool Supporting Protection of Water in the Sulejow Reservoir. *Desalination Water Treat.* **2013**, *51*, 4194–4206. [CrossRef]

49. Zhao, Y.; Zheng, B.; Wang, L.; Qin, Y.; Li, H.; Cao, W. Characterization of Mixing Processes in the Confluence Zone between the Three Gorges Reservoir Mainstream and the Daning River Using Stable Isotope Analysis. *Environ. Sci. Technol.* **2016**, *50*, 9907–9914. [CrossRef]

50. Zhang, B.; Xie, G.; Yan, Y.; Yang, Y. Regional Differences of Water Conservation in Beijing's Forest Ecosystem. *J. For. Res.* **2011**, *22*, 295. [CrossRef]

51. Kashwan, P. What Explains the Demand for Collective Forest Rights amidst Land Use Conflicts? *J. Environ. Manag.* **2016**, *183*, 657–666. [CrossRef]

52. Wu, S.; Zhang, S. Game Analysis of Chinese Stakeholders in Collective Forest Rights System Reform. *Chin. J. Popul. Resour. Environ.* **2014**, *12*, 330–337. [CrossRef]

53. Nichiforel, L.; Keary, K.; Deuffic, P.; Weiss, G.; Thorsen, B.J.; Winkel, G.; Avdibegović, M.; Dobšinská, Z.; Feliciano, D.; Gatto, P.; et al. How Private Are Europe's Private Forests? A Comparative Property Rights Analysis. *Land Use Policy* **2018**, *76*, 535–552. [CrossRef]

54. Wen, W.; Xu, G.; Wang, X. Spatial Transferring of Ecosystem Services and Property Rights Allocation of Ecological Compensation. *Front. Earth Sci.* **2011**, *5*, 280. [CrossRef]

55. Yu, B.; Xu, L. Review of Ecological Compensation in Hydropower Development. *Renew. Sustain. Energy Rev.* **2016**, *55*, 729–738. [CrossRef]

56. Xie, G.; Cao, S.; Lu, C.; Zhang, C.; Xiao, Y. Current Status and Future Trends for Eco-Compensation in China. *J. Resour. Ecol.* **2015**, *6*, 355–362. [CrossRef]

57. Mudombi-Rusinamhodzi, G.; Thiel, A. Property Rights and the Conservation of Forests in Communal Areas in Zimbabwe. *For. Policy Econ.* **2020**, *121*, 102315. [CrossRef]

58. Di Corato, L.; Gazheli, A.; Lagerkvist, C.-J. Investing in Energy Forestry under Uncertainty. *For. Policy Econ.* **2013**, *34*, 56–64. [CrossRef]

59. Adger, W.N.; Luttrell, C. Property Rights and the Utilisation of Wetlands. *Ecol. Econ.* **2000**, *35*, 75–89. [CrossRef]

60. Portillo-Quintero, C.; Sanchez-Azofeifa, A.; Calvo-Alvarado, J.; Quesada, M.; do Espirito Santo, M.M. The Role of Tropical Dry Forests for Biodiversity, Carbon and Water Conservation in the Neotropics: Lessons Learned and Opportunities for Its Sustainable Management. *Reg. Environ. Chang.* **2015**, *15*, 1039–1049. [CrossRef]

61. Ferraz, S.F.B.; de Paula Lima, W.; Rodrigues, C.B. Managing Forest Plantation Landscapes for Water Conservation. *For. Ecol. Manag.* **2013**, *301*, 58–66. [CrossRef]
62. Yu, J.; Wei, Y.; Fang, W.; Liu, Z.; Zhang, Y.; Lan, J. New Round of Collective Forest Rights Reform, Forestland Transfer and Household Production Efficiency. *Land* **2021**, *10*, 988. [CrossRef]
63. Wang, C.; Wen, Y.; Wu, J. The Socio-Economic Effect of the Reform of the Collective Forest Rights System in Southern China: A Case of Tonggu County, Jiangxi Province. *Small-Scale For.* **2014**, *13*, 425–444. [CrossRef]
64. Ha, T.T.P.; van Dijk, H.; Visser, L. Impacts of Changes in Mangrove Forest Management Practices on Forest Accessibility and Livelihood: A Case Study in Mangrove-Shrimp Farming System in Ca Mau Province, Mekong Delta, Vietnam. *Land Use Policy* **2014**, *36*, 89–101. [CrossRef]
65. Yan, J.; Yang, Y.; Xia, F. Subjective Land Ownership and the Endowment Effect in Land Markets: A Case Study of the Farmland "Three Rights Separation" Reform in China. *Land Use Policy* **2021**, *101*, 105137. [CrossRef]
66. Lu, Z.-H.; Wu, G.; Ma, X.; Bai, G.-X. Current Situation of Chinese Forestry Tactics and Strategy of Sustainable Development. *J. For. Res.* **2002**, *13*, 319–322. [CrossRef]
67. Miller, D.C.; Rana, P.; Nakamura, K.; Irwin, S.; Cheng, S.H.; Ahlroth, S.; Perge, E. A Global Review of the Impact of Forest Property Rights Interventions on Poverty. *Glob. Environ. Chang.* **2021**, *66*, 102218. [CrossRef]
68. Fang, J.; Lu, S.; Yu, X.; Rao, L.; Niu, J.; Xie, Y.; Zhag, Z. Forest ecosystem service and its evaluation in China. *J. Appl. Ecol.* **2005**, *16*, 1531–1536.
69. Tierney, G.L.; Faber-Langendoen, D.; Mitchell, B.R.; Shriver, W.G.; Gibbs, J.P. Monitoring and Evaluating the Ecological Integrity of Forest Ecosystems. *Front. Ecol. Environ.* **2009**, *7*, 308–316. [CrossRef]
70. Ordóñez, C.; Duinker, P.N. Ecological Integrity in Urban Forests. *Urban Ecosyst.* **2012**, *15*, 863–877. [CrossRef]
71. Chakraborty, R.N. Stability and Outcomes of Common Property Institutions in Forestry: Evidence from the Terai Region of Nepal. *Ecol. Econ.* **2001**, *36*, 341–353. [CrossRef]
72. Dorren, L.K.A.; Berger, F.; Imeson, A.C.; Maier, B.; Rey, F. Integrity, Stability and Management of Protection Forests in the European Alps. *For. Ecol. Manag.* **2004**, *195*, 165–176. [CrossRef]
73. Yang, L.; Ren, Y. Property Rights, Village Democracy, and Household Forestry Income: Evidence from China's Collective Forest Tenure Reform. *J. For. Res.* **2021**, *26*, 7–16. [CrossRef]
74. Yang, L.; Ren, Y. Has China's New Round of Collective Forestland Tenure Reform Caused an Increase in Rural Labor Transfer? *Land* **2020**, *9*, 284. [CrossRef]
75. Schallau, C.H. Community Stability: Issues, Institutions, and Instruments. In *Community and Forestry*; Routledge: London, UK, 1990; ISBN 978-0-429-04325-3.
76. Yang, M.; Zhu, Y.; Zheng, X.; Wu, W. The Influence of the Stability of Forestland Property Right on the Farmers' Afforestation Investment. *For. Resour. Manag.* **2017**, *2*, 1. [CrossRef]
77. Kruger, C.; Boxall, P.C.; Luckert, M.K. Preferences of Community Public Advisory Group Members for Characteristics of Canadian Forest Tenures in Pursuit of Sustainable Forest Management Objectives. *For. Policy Econ.* **2013**, *26*, 121–130. [CrossRef]
78. Zhou, Y.; Ma, X.; Ji, D.; Heerink, N.; Shi, X.; Liu, H. Does Property Rights Integrity Improve Tenure Security? Evidence from China's Forest Reform. *Sustainability* **2018**, *10*, 1956. [CrossRef]
79. Han, T.; Lu, H.; Ren, H.; Wang, J.; Song, G.; Hui, D.; Guo, Q.; Zhu, S. Are Reproductive Traits of Dominant Species Associated with Specific Resource Allocation Strategies during Forest Succession in Southern China? *Ecol. Indic.* **2019**, *102*, 538–546. [CrossRef]
80. Egu, E.C.; Nwankwo, E.C.; Offiong, E.E. Assessment of Forest Investment, Financial Flows and Revenue Collection in the Abia State Forest Sector, Nigeria. *J. Appl. Sci. Environ. Manag.* **2021**, *25*, 763–771. [CrossRef]
81. Yi, Y.; Köhlin, G.; Xu, J. Property Rights, Tenure Security and Forest Investment Incentives: Evidence from China's Collective Forest Tenure Reform. *Environ. Dev. Econ.* **2014**, *19*, 48–73. [CrossRef]
82. Lee, Y.G.; Zhu, G.; Sharma, B.P.; English, B.C.; Cho, S.-H. Role of Complementary and Competitive Relationships among Multiple Objectives in Conservation Investment Decisions. *For. Policy Econ.* **2021**, *131*, 102569. [CrossRef]
83. Hildebrandt, P.; Knoke, T. Investment Decisions under Uncertainty—A Methodological Review on Forest Science Studies. *For. Policy Econ.* **2011**, *13*, 1–15. [CrossRef]
84. Wang, L.; Zhang, F.; Wang, Z.; Tan, Q. The Impact of Rural Infrastructural Investment on Farmers' Income Growth in China. *China Agric. Econ. Rev.* **2021**, online ahead of print. [CrossRef]
85. Xingchang, Z.; Mingan, S.; Shiqing, L.; Keshan, P. A Review of Soil and Water Conservation in China. *J. Geogr. Sci.* **2004**, *14*, 259–274. [CrossRef]
86. Li, Y.; Wang, B.; Rao, L.; Wang, Y. Research on Water Conservation Function of Typical Forests in Jinyun Mountain. *Agric. Sci. Amp Technol.—Hunan* **2012**, *13*, 181–188.
87. Ji, D.; Ma, X.; Shi, X. The Impact of Forest Property Rights on Forestland Investments: From the Perspective of Property Rights Integrity and Security—A Case from Suichuan and Fengcheng of Jiangxi Province. *Issues Agric. Econ.* **2015**, *3*, 54–61.
88. Irimie, D.L.; Essmann, H.F. Forest Property Rights in the Frame of Public Policies and Societal Change. *For. Policy Econ.* **2009**, *11*, 95–101. [CrossRef]
89. Dorji, L.; Webb, E.L.; Shivakoti, G.P. Forest Property Rights under Nationalized Forest Management in Bhutan. *Environ. Conserv.* **2006**, *33*, 141–147. [CrossRef]

90. Cartagena, P.; Peralta, C. Effects of Public Agricultural and Forestry Policies on the Livelihoods of Campesino Families in the Bolivian Amazon. In *Socio-Environmental Regimes and Local Visions: Transdisciplinary Experiences in Latin America*; Arce Ibarra, M., Parra Vázquez, M.R., Bello Baltazar, E., de Araujo, L.G., Eds.; Springer International Publishing: Cham, Switzerland, 2020; pp. 381–408, ISBN 978-3-030-49767-5.

91. Mamo, G.; Sjaastad, E.; Vedeld, P. Economic Dependence on Forest Resources: A Case from Dendi District, Ethiopia. *For. Policy Econ.* **2007**, *9*, 916–927. [CrossRef]

92. Campos, P.; Álvarez, A.; Mesa, B.; Oviedo, J.L.; Caparrós, A. Linking Standard Economic Account for Forestry and Ecosystem Accounting: Total Forest Incomes and Environmental Assets in Publicly-Owned Conifer Farms in Andalusia-Spain. *For. Policy Econ.* **2021**, *128*, 102482. [CrossRef]

93. Shaanker, R.U.; Ganeshaiah, K.N.; Krishnan, S.; Ramya, R.; Meera, C.; Aravind, N.A.; Kumar, A.; Rao, D.; Vanaraj, G.; Ramachandra, J.; et al. Livelihood Gains and Ecological Costs of Non-Timber Forest Product Dependence: Assessing the Roles of Dependence, Ecological Knowledge and Market Structure in Three Contrasting Human and Ecological Settings in South India. *Environ. Conserv.* **2004**, *31*, 242–253. [CrossRef]

94. Mandle, L.; Ticktin, T.; Nath, S.; Setty, S.; Varghese, A. A Framework for Considering Ecological Interactions for Common Non-Timber Forest Product Species: A Case Study of Mountain Date Palm (Phoenix Loureiroi Kunth) Leaf Harvest in South India. *Ecol. Process.* **2013**, *2*, 21. [CrossRef]

95. Xie, L.; Berck, P.; Xu, J. The Effect on Forestation of the Collective Forest Tenure Reform in China. *China Econ. Rev.* **2016**, *38*, 116–129. [CrossRef]

96. Feng, L.; Bao, H.X.H.; Jiang, Y. Land Reallocation Reform in Rural China: A Behavioral Economics Perspective. *Land Use Policy* **2014**, *41*, 246–259. [CrossRef]

97. Kassie, G.W. Agroforestry and Farm Income Diversification: Synergy or Trade-off? The Case of Ethiopia. *Environ. Syst. Res.* **2017**, *6*, 8. [CrossRef]

98. Li, Q.; Guo, F.; Guan, Y. A GIS-Based Evaluation of Environmental Sensitivity for an Urban Expressway in Shenzhen, China. *Engineering* **2018**, *4*, 230–234. [CrossRef]

99. Song, H.; Shi, Y.; Wang, L. Method of Regional Environmental Risk Assessment for Reservoir Type Drinking-Water Source. In Proceedings of the 2012 International Symposium on Geomatics for Integrated Water Resource Management, Lanzhou, China, 19–21 October 2012; pp. 1–6.

100. Chai, B.; Li, Y.; Huang, T.; Zhao, X. Pollution Characteristics of Thermally-Stratified Reservoir: A Case Study of the Heihe Reservoir in Xi'an City, China. *J. Chem. Pharm. Res.* **2014**, *6*, 1231–1240.

101. Bo, G. Study on the Ecological Property Right System of Forest Resources in China. *Ecol. Econ.* **2014**, *9*, 91–109.

102. Cao, S.; Wang, X.; Song, Y.; Chen, L.; Feng, Q. Impacts of the Natural Forest Conservation Program on the Livelihoods of Residents of Northwestern China: Perceptions of Residents Affected by the Program. *Ecol. Econ.* **2010**, *69*, 1454–1462. [CrossRef]

103. Hogarth, N.J.; Belcher, B.; Campbell, B.; Stacey, N. The Role of Forest-Related Income in Household Economies and Rural Livelihoods in the Border-Region of Southern China. *World Dev.* **2013**, *43*, 111–123. [CrossRef]

104. Xiong, L.; Wang, F.; Cheng, B.; Yu, C. Identifying Factors Influencing the Forestry Production Efficiency in Northwest China. *Resour. Conserv. Recycl.* **2018**, *130*, 12–19. [CrossRef]

105. Ben-xin, X.U. The Legislative Perfection about the Pricing System of the Collective Forest Rights Transaction in China. *J. Kunming Univ. Sci. Technol. Soc. Sci. Ed.* **2012**, *4*, 46–51.

106. Wang, L.; Dong, W.; Shen, W. Performance analysis on the reformation of collective forest rights. *J. Nanjing For. Univ. Nat. Sci. Ed.* **2010**, *34*, 133–136.

107. He, J. Rights to Benefit from Forest? A Case Study of the Timber Harvest Quota System in Southwest China. *Soc. Nat. Resour.* **2016**, *29*, 448–461. [CrossRef]

108. Zhang, D. Policy Reform and Investment in Forestry. In *China's Forests*; Routledge: New York, NY, USA, 2003; ISBN 978-1-936331-23-9.

109. Ren, Y.; Kuuluvainen, J.; Toppinen, A.; Yao, S.; Berghäll, S.; Karppinen, H.; Xue, C.; Yang, L. The Effect of China's New Circular Collective Forest Tenure Reform on Household Non-Timber Forest Product Production in Natural Forest Protection Project Regions. *Sustainability* **2018**, *10*, 1091. [CrossRef]

110. Kumar, K.; Kerr, J.M. Democratic Assertions: The Making of India's Recognition of Forest Rights Act. *Dev. Chang.* **2012**, *43*, 751–771. [CrossRef]

111. Beisner, B.; Haydon, D.; Cuddington, K. Alternative Stable States in Ecology. *Front. Ecol. Environ.* **2003**, *1*, 376–382. [CrossRef]

112. Home, R.; Balmer, O.; Jahrl, I.; Stolze, M.; Pfiffner, L. Motivations for Implementation of Ecological Compensation Areas on Swiss Lowland Farms. *J. Rural Stud.* **2014**, *34*, 26–36. [CrossRef]

113. Jiang, S.; Lewis, B.J.; Dai, L.; Jia, W.; An, Y. The Reform of Collective Forest Rights in China and Its Implementation in the Fushun City Region. *Ann. For. Res.* **2014**, *57*, 319–332. [CrossRef]

114. Lu, S.; Chen, N.; Zhong, X.; Huang, J.; Guan, X. Factors Affecting Forestland Production Efficiency in Collective Forest Areas: A Case Study of 703 Forestland Plots and 290 Rural Households in Liaoning, China. *J. Clean. Prod.* **2018**, *204*, 573–585. [CrossRef]

115. Mercer, D.E. Adoption of Agroforestry Innovations in the Tropics: A Review. *Agrofor. Syst.* **2004**, *61*, 311–328. [CrossRef]

116. Jha, S.; Kaechele, H.; Sieber, S. Factors Influencing the Adoption of Agroforestry by Smallholder Farmer Households in Tanzania: Case Studies from Morogoro and Dodoma. *Land Use Policy* **2021**, *103*, 105308. [CrossRef]

117. Caveness, F.A.; Kurtz, W.B. Agroforestry Adoption and Risk Perception by Farmers in Sénégal. *Agrofor. Syst.* **1993**, *21*, 11–25. [CrossRef]
118. Neupane, R.P.; Sharma, K.R.; Thapa, G.B. Adoption of Agroforestry in the Hills of Nepal: A Logistic Regression Analysis. *Agric. Syst.* **2002**, *72*, 177–196. [CrossRef]
119. Krčmářová, J.; Kala, L.; Brendzová, A.; Chabada, T. Building Agroforestry Policy Bottom-Up: Knowledge of Czech Farmers on Trees in Farmland. *Land* **2021**, *10*, 278. [CrossRef]
120. Do, H.; Luedeling, E.; Whitney, C. Decision Analysis of Agroforestry Options Reveals Adoption Risks for Resource-Poor Farmers. *Agron. Sustain. Dev.* **2020**, *40*, 20. [CrossRef]
121. do Carmo Martinelli, G.; Schlindwein, M.M.; Padovan, M.P.; Gimenes, R.M.T. Decreasing Uncertainties and Reversing Paradigms on the Economic Performance of Agroforestry Systems in Brazil. *Land Use Policy* **2019**, *80*, 274–286. [CrossRef]
122. Jerneck, A.; Olsson, L. Food First! Theorising Assets and Actors in Agroforestry: Risk Evaders, Opportunity Seekers and 'the Food Imperative' in Sub-Saharan Africa. *Int. J. Agric. Sustain.* **2014**, *12*, 1–22. [CrossRef]
123. Bettles, J.; Battisti, D.S.; Cook-Patton, S.C.; Kroeger, T.; Spector, J.T.; Wolff, N.H.; Masuda, Y.J. Agroforestry and Non-State Actors: A Review. *For. Policy Econ.* **2021**, *130*, 102538. [CrossRef]
124. Damianidis, C.; Santiago-Freijanes, J.J.; den Herder, M.; Burgess, P.; Mosquera-Losada, M.R.; Graves, A.; Papadopoulos, A.; Pisanelli, A.; Camilli, F.; Rois-Díaz, M.; et al. Agroforestry as a Sustainable Land Use Option to Reduce Wildfires Risk in European Mediterranean Areas. *Agrofor. Syst.* **2021**, *95*, 919–929. [CrossRef]
125. Zhou, Z.; Huang, T.; Ma, W.; Li, Y.; Zeng, K. Impacts of Water Quality Variation and Rainfall Runoff on Jinpen Reservoir, in Northwest China. *Water Sci. Eng.* **2015**, *8*, 301–308. [CrossRef]
126. Liu, R.; Jia, Y. Resilience and Circularity: Revisiting the Role of Urban Village in Rural-Urban Migration in Beijing, China. *Land* **2021**, *10*, 1284. [CrossRef]
127. Ma, X.; Heerink, N.; van Ierland, E.; Shi, X. Land tenure insecurity and rural-urban migration in rural China. *Pap. Reg. Sci.* **2016**, *95*, 383–406. [CrossRef]
128. Xu, J.; He, W.; Liu, Z.J.; Zhang, H. How Do Property Rights Affect Credit Restrictions? Evidence from China's Forest Right Mortgages. *Small-Scale For.* **2021**, *20*, 235–262. [CrossRef]
129. Mullan, K.; Grosjean, P.; Kontoleon, A. Land Tenure Arrangements and Rural–Urban Migration in China. *World Dev.* **2011**, *39*, 123–133. [CrossRef]
130. Luo, X.; Lone, T.; Jiang, S.; Li, R.; Berends, P. A Study of Farmers' Flood Perceptions Based on the Entropy Method: An Application from Jianghan Plain, China. *Disasters* **2016**, *40*, 573–588. [CrossRef] [PubMed]
131. Assé, R.; Lassoie, J.P. Household Decision-Making in Agroforestry Parklands of Sudano-Sahelian Mali. *Agrofor. Syst.* **2011**, *82*, 247–261. [CrossRef]
132. Zeng, X.; Li, T.; Chen, C.; Si, Z.; Huang, G.; Guo, P.; Zhuang, X. A Hybrid Land-Water-Environment Model for Identification of Ecological Effect and Risk under Uncertain Meteorological Precipitation in an Agroforestry Ecosystem. *Sci. Total Environ.* **2018**, *633*, 1613–1628. [CrossRef]
133. Zubair, M.; Garforth, C. Farm Level Tree Planting in Pakistan: The Role of Farmers' Perceptions and Attitudes. *Agrofor. Syst.* **2006**, *66*, 217–229. [CrossRef]
134. Verbist, B.; Dinata Putra, A.E.; Budidarsono, S. Factors Driving Land Use Change: Effects on Watershed Functions in a Coffee Agroforestry System in Lampung, Sumatra. *Agric. Syst.* **2005**, *85*, 254–270. [CrossRef]
135. Li, N.; He, W.; Qiu, T.; Chen, L. Farmland Property Right Structure, Production Factor Efficiency and Agricultural Performance. Manag. *World* **2017**, *3*, 44–62.
136. Ma, X.; Heerink, N.; Feng, S.; Shi, X. Farmland Tenure in China: Comparing Legal, Actual and Perceived Security. *Land Use Policy* **2015**, *42*, 293–306. [CrossRef]
137. Van Gelder, J.-L. What Tenure Security? The Case for a Tripartite View. *Land Use Policy* **2010**, *27*, 449–456. [CrossRef]
138. Nguyen, T.T.; Duong, T.H.; Dinh, M.T.T.; Pham, T.H.H.; Truong, T.M.A. The Impact of Trust on Intellectual Property Right Protection: A Cross-National Study. *J. Econ. Dev.* **2021**. online ahead of print. [CrossRef]
139. Mutiani, C.; Febriamansyah, R.; Mahdi. Chapter 3—Agroforestry Management Practices in Relation to Tenure Security in Koto Tangah Subdistrict, West Sumatra, Indonesia. In *Natural Resource Governance in Asia*; Ullah, R., Sharma, S., Inoue, M., Asghar, S., Shivakoti, G., Eds.; Elsevier: Amsterdam, The Netherlands, 2021; pp. 27–37, ISBN 978-0-323-85729-1.
140. Rasul, G.; Thapa, G.B. Financial and Economic Suitability of Agroforestry as an Alternative to Shifting Cultivation: The Case of the Chittagong Hill Tracts, Bangladesh. *Agric. Syst.* **2006**, *91*, 29–50. [CrossRef]
141. Dieudonné, B.S.; Espoir, B.B.; Philippe, L. Dynamics of Customary Land Rights and Its Impact on the Agronomic Choices for Small Farmers in the South Kivu Province, Eastern DR Congo. *Acad. J. Interdiscip. Stud.* **2021**, *10*, 199–212.
142. Sá-Sousa, P. The Portuguese Montado: Conciliating Ecological Values with Human Demands within a Dynamic Agroforestry System. *Ann. For. Sci.* **2014**, *71*, 1–3. [CrossRef]
143. Lockie, S. Market Instruments, Ecosystem Services, and Property Rights: Assumptions and Conditions for Sustained Social and Ecological Benefits. *Land Use Policy* **2013**, *31*, 90–98. [CrossRef]
144. Hanna, S.; Munasinghe, M. *Property Rights and the Environment: Social and Ecological Issues*; World Bank Publications: Washington, DC, USA, 1995; Volume 94, ISBN 0-8213-3415-8.

145. Demsetz, H. Toward a Theory of Property Rights. In *Classic Papers in Natural Resource Economics*; Gopalakrishnan, C., Ed.; Palgrave Macmillan UK: London, UK, 2000; pp. 163–177, ISBN 978-0-230-52321-0.
146. Siebert, H. Property-Rights Approach to the Environmental Problem. In *Economics of the Environment: Theory and Policy*; Siebert, H., Ed.; Springer: Berlin/Heidelberg, Germany, 2008; pp. 97–104, ISBN 978-3-540-73707-0.
147. Chen, L.; Guan, X.; Zhuo, J.; Han, H.; Gasper, M.; Doan, B.; Yang, J.; Ko, T.-H. Application of Double Hurdle Model on Effects of Demographics for Tea Consumption in China. *J. Food Qual.* **2020**, *2020*, e9862390. [CrossRef]
148. Lu, J.; Li, Z. Seasonal Effects of Thermal Stratification on the Water Quality of Deep Reservoirs: A Case Study of Heihe Reservoir, Xi'an City. *J. Lake Sci.* **2014**, *26*, 698–706.
149. Zhang, J. Barriers to Water Markets in the Heihe River Basin in Northwest China. *Agric. Water Manag.* **2007**, *87*, 32–40. [CrossRef]
150. Wang, J.; Liu, M.; Yang, L.; Min, Q. Factors Affecting the Willingness of Farmers to Accept Eco-Compensation in the Qianxi Chestnut Agroforestry System, Hebei. *J. Resour. Ecol.* **2018**, *9*, 407–415. [CrossRef]
151. Holden, S.T.; Otsuka, K. The Roles of Land Tenure Reforms and Land Markets in the Context of Population Growth and Land Use Intensification in Africa. *Food Policy* **2014**, *48*, 88–97. [CrossRef]
152. Suleiman, M.S.; Wasonga, V.O.; Mbau, J.S.; Suleiman, A.; Elhadi, Y.A. Non-Timber Forest Products and Their Contribution to Households Income around Falgore Game Reserve in Kano, Nigeria. *Ecol. Process.* **2017**, *6*, 23. [CrossRef]
153. Nerfa, L.; Rhemtulla, J.M.; Zerriffi, H. Forest Dependence Is More than Forest Income: Development of a New Index of Forest Product Collection and Livelihood Resources. *World Dev.* **2020**, *125*, 104689. [CrossRef]
154. Glavan, M.; Cvejić, R.; Zupanc, V.; Knapič, M.; Pintar, M. Agricultural Production and Flood Control Dry Detention Reservoirs: Example from Lower Savinja Valley, Slovenia. *Environ. Sci. Policy* **2020**, *114*, 394–402. [CrossRef]
155. Lu, H.; Zhang, P.; Hu, H.; Xie, H.; Yu, Z.; Chen, S. Effect of the Grain-Growing Purpose and Farm Size on the Ability of Stable Land Property Rights to Encourage Farmers to Apply Organic Fertilizers. *J. Environ. Manag.* **2019**, *251*, 109621. [CrossRef] [PubMed]
156. Yegbemey, R.N.; Yabi, J.A.; Tovignan, S.D.; Gantoli, G.; Haroll Kokoye, S.E. Farmers' Decisions to Adapt to Climate Change under Various Property Rights: A Case Study of Maize Farming in Northern Benin (West Africa). *Land Use Policy* **2013**, *34*, 168–175. [CrossRef]

Article

External Benefits of Irrigation in Mountain Areas: Stakeholder Perceptions and Water Policy Implications

Silvia Novelli [1,*], **Francesca Moino** [2] **and Patrizia Borsotto** [2]

[1] Department of Agricultural, Forest and Food Sciences (DiSAFA), University of Torino, 10095 Grugliasco, Italy
[2] CREA Research Centre for Agricultural Policies and Bioeconomy, 00198 Rome, Italy
* Correspondence: silvia.novelli@unito.it; Tel.: +39-011-6708723

Abstract: Irrigation contributes to land and ecosystem degradation, especially in intensive farming areas. However, in marginal areas, long-established irrigation systems also supply agroecosystem services. This study aimed to identify and prioritize the external benefits provided by irrigation in extensive grazing farms in an Italian alpine region (Aosta Valley, NW Italy). Three local stakeholder groups (land irrigation consortia members, non-farmer users of the irrigation water service, and non-user citizens) engaged in focus group discussions. The transcriptions were analyzed with an integrated subjective and computer-assisted approach. The main result of the study showed that a convergence of stakeholder opinions led to prioritization of the same four benefits, i.e., hydro-geological and land maintenance, traditional agricultural landscape conservation, biodiversity conservation, and leisure recreational activities provision. Incorporating this information into decision-making processes is relevant in marginal mountain areas, especially in light of the implementation of the water pricing policy laid down in the EU Water Framework Directive. To this end, the economic value of the external benefits should be considered along with the recovery costs for water services. Such information is essential to balance the environmental costs of irrigation and to compare the resource cost of alternative water uses.

Keywords: agroecosystem services; alpine areas; extensive livestock farming; stakeholder participation; focus group; water resources; water framework directive

Citation: Novelli, S.; Moino, F.; Borsotto, P. External Benefits of Irrigation in Mountain Areas: Stakeholder Perceptions and Water Policy Implications. *Land* **2022**, *11*, 1395. https://doi.org/10.3390/land11091395

Academic Editors: Norie Tamura, Yuta Uchiyama, Ryo Kohsaka and Jay Mar D. Quevedo

Received: 27 July 2022
Accepted: 22 August 2022
Published: 25 August 2022

Publisher's Note: MDPI stays neutral with regard to jurisdictional claims in published maps and institutional affiliations.

1. Introduction

Land irrigation water is a key agricultural production factor throughout Europe; its relevance is significantly higher in Southern Europe than in Northern Europe [1]. Different water requirements depend on different climate conditions and crop types, e.g., horticulture needs more water than cereals and grown crops need more water than those just planted [2]. In Europe, irrigated lands represent 6% of the total utilized agricultural area (UAA) while lands able to be irrigated account for 10% of UAA. In Southern Europe (France, Greece, Italy, Malta, Portugal, Spain) the irrigated area represents 12% of UAA, whereas in Northern Europe it comprises only 2% [3].

Irrigation is known to contribute to land and ecosystem degradation by altering the quantity and quality of water in aquatic and terrestrial systems, especially in intensive farming areas [4,5]. The expansion of irrigation over the last century has altered the hydrologic cycle and impacted global climate [6,7]. In fact, recent decreases in water availability have directly affected agricultural productivity and indirectly affected ecosystem services provisioning [8].

Alternatively, the perspective that long-established irrigation systems provide flows of agroecosystem services for biodiversity, wildlife habitats, landscape aesthetics, and more [1] in some areas has prompted investigation in to how proper water management affects other subjects and social welfare. Those who have analyzed proper irrigation water management have all spoken of its multi-faceted complexity: (i) Boelee [9] encouraged stakeholders to manage water based on the needs of food security, farmer livelihood, and ecosystem conservation; (ii) Falkenmark et al. [4] recognized the multiplicity of water uses; (iii) Alcon et al. [10] discussed major contributions of irrigation water as not only for land productivity, but also for food security, rural livelihood, and agroecosystem services provision; (iv) Natali and Branca [11] reviewed the positive externalities from irrigated agriculture. This point-of-view is particularly poignant for agricultural activities in mountain areas, where the highest potential to generate such flows of positive externalities is due to their environmental qualities and low-intensity farming systems [12]. Specifically, most farms in the mountain areas of Southern Europe are small in size and the availability of irrigation water is critical to their economic viability. Irrigation of alpine meadows and pastures is part of the history of agriculture. In fact, there is evidence that in pre-industrial times it had a significant impact on rural community land management that affected visual and ecological aspects [13].

The aim of this study was to identify the external benefits provided by irrigation in a mountain region taken as a case study. External benefits are related to the provision of positive externalities. They occur when production processes increase the welfare of a third party [14]. Usually, external effects are not intended, but rather are incidental outcomes of production decisions [15]. They are fully- or partially-unpriced, hence providers are not or not adequately compensated, and the market fails [16].

The analysis was carried out in the Aosta Valley Region, an almost entirely-mountainous region located in northwestern Italy classified by the European Union as 'rural area with development problems' (Directive 75/268/EEC, art. 3 (3) and Regulation (EC) 1257/99, art. 18). The regional territory is constrained by natural and environmental factors; therefore, it is considered as being a disadvantaged area for agriculture. More than 96% of the regional UAA is devoted to extensive grazing and about 28% of its meadows and pastures are irrigated [17]. Traditionally, much of the irrigation in Aosta Valley was managed with gravity-fed systems, where water is transported from surface sources via small ditches called *rus* that flood or furrow agricultural lands with irrigation water. Irrigation water services are managed by 176 regional land improvement consortia, with a total of 2833 members [18]. In the region, such consortia are in charge also of natural resource conservation and water regulation, hence we refer to them as "water consortia" going forward.

To identify the external benefits of irrigation, a participatory approach seemed best to collect the perceptions of local stakeholders and residents who are the primary beneficiaries of the social and environmental benefits of agricultural water use [19,20]. Specifically, our research considered three questions: (i) meadow and pasture external benefits of irrigation, as perceived by local stakeholders; (ii) stakeholder prioritization of identified benefits; (iii) opinions and perceptions on potential conflicts of regional water use. We employed a qualitative approach. Data were collected through focus group discussions and analyzed using subjective and content analysis methods.

Content analysis is an objective technique to provide systematic and quantitative description of texts or other communication contents [21]. Recently, it was employed to investigate water resource management issues in different environments, using semi-structured in-depth interviews with local stakeholders, experts, and key informants to collect data [22–25]. We experimented with this approach using focus groups to enable viewpoint exchange, disagreement, and discussion between participants. Unlike other studies that generically collect information from heterogeneous stakeholder groups [26], we first categorized the stakeholders. As some studies have shown, this allows for greater accuracy in data collection [27,28].

Outcomes of the analysis were used to discuss some policy implications stemming from the recognition of the non-monetary benefits of irrigation in marginal areas. Specifically, how can key external benefits and beneficiaries affected by decisions be incorporated into the main EU water policy. In this regard, the European Commission stresses both the need for protection of water resources and the socio-economic relevance of irrigated agriculture [1]. The European Water Policy is regulated under the 6th Environment

Action Programme (EAP) (1600/2002/EC) and the Water Framework Directive (WFD, 2000/60/EC), aimed at ensuring a sustainable use of water resources. Sustainable management of natural resources (including water) was also one of the three policy objectives of the Common Agricultural Policy (CAP) 2014–2020, alongside viable food production and balanced territorial development. In 2018, the Commission published a proposal for the post-2020 CAP, by including the promotion of sustainable development and the efficient management of natural resources—such as water, soil, and air—among its nine specific objectives.

The study was carried out as part of the Interreg Italy–Switzerland Co-operation Programme 'Reservaqua'. The general objective is to develop an integrated management strategy for mountain regions to ensure sustainable use and quality protection of alpine water resources.

2. Materials and Methods

A participatory approach was adopted to collect data from regional residents who have a stake in the use of irrigation water and/or who benefit from irrigation water use. The study was conducted in two steps to combine different qualitative techniques such as brainstorming and focus group discussions (Figure 1).

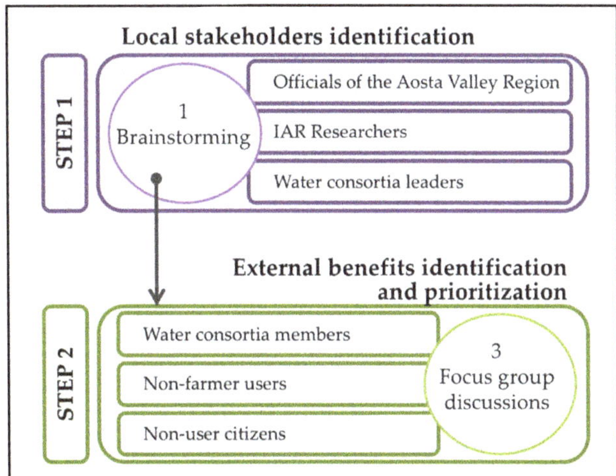

Figure 1. Diagram of methodological approach.

In step one, a brainstorming session was set to identify irrigation water management practices adopted in the region, the stakeholders involved in the use of irrigation water and/or those accruing the external benefits of meadow and pasture irrigation. Brainstorming is a methodology to foster creative thinking through a process of generating group ideas and problem-solving activities [29,30]. The meeting was held in December 2019 and involved nine participants: two officials of the Aosta Valley Region in charge of regional rural development policies, four researchers of the Regional Agricultural Institute (IAR—Institut Agricole Régional, a regional professional training, research and experimentation institute in the agricultural field), and three water consortia leaders. The discussion allowed differentiation of three local stakeholder types: water consortia members (farmers), non-farmer users of the irrigation water service, and non-user citizens. Non-farmer users are citizens and hobbyists who use irrigation water to irrigate their private lawns and gardens, without commercial purposes.

In step two, the three types of stakeholders took part in a focus group, where participants were asked to discuss the research topic from their perspective with the group, and to share personal opinions and experiences [31,32]. The main objective of the discussion was to identify the social and environmental benefits provided by the irrigation of meadows and pastures as perceived by the participants, and to prioritize them. Besides, the alternative uses of water were discussed to highlight any competing or conflicting uses of the resource.

Participants were selected by purposeful sampling, a non-random sampling technique that utilizes identified purposes and criteria to select particular samples. The basic principle is to select participants based on their relevance to the research questions under evaluation [33]. To this end, participants were recruited by local partners of the Reservaqua project, i.e., by the regional administration and the Regional Agricultural Institute. The selection criteria were developed to include various ages and places of residence. Non-user citizens are those with less knowledge of water management technical issues. However, they were selected from among long-time residents, who are very familiar with the area, local natural resources, and local environmental problems. Tourists and second-home owners were excluded. As for the number of participants, the literature suggests six to eight persons per focus group and not to exceed 12 participants [34,35]. Smaller groups of three or four participants are also acceptable when the group shares specialized knowledge or experience [36]. The number of participants involved in the survey was between six and nine.

Water consortia members, non-farmer users, and non-user citizens met in July 2020, July 2021, and September 2020 respectively. Table 1 summarizes the number of participants and the specific questions discussed in each group.

Table 1. Number of focus group participants and questions discussed (marked by an X).

	Water Consortia Members	Non-Farmer Users	Non-User Citizens
Number of participants	8	7	6
Q1—Does irrigation water use positively impacts the territory and/or on the local community? If yes, how?	X	X	X
Q2—Can you rank those benefits on a scale from most important to least important?		X	X
Q3—Besides irrigation use, what are the other water uses in the region? Are there conflicts between alternative water uses and the relevant stakeholders?	X	X	

Each session lasted about two hours and was audio recorded. Transcriptions of the audio recordings were analyzed using a subjective scissor-and-sort technique integrated with a computer-assisted approach [37,38]. With the subjective analysis we identified and classified the major topics and issues discussed over the three focus groups with question Q1, and analyzed question Q2 and Q3. Results from the subjective analysis were validated through a computer-assisted analysis, carried out using QDA Miner software, a qualitative and mixed method analysis tool [39]. We used the software to assign transcription passages to codes that reflect concepts and issues of interest and to analyze code similarities and frequencies with the coding frequency statistical tool [40]. In particular, this approach was used to compare responses across different types of stakeholders and to establish a priority among the external benefits discussed, using the transcription of research question Q1.

3. Results

3.1. External Benefits of Irrigation

The subjective analysis of the focus group transcriptions allowed identification of the relevant sections to question Q1 and grouping of the major topics discussed. The external benefits identified by the participants spanned both general environmental and social aspects as well as very fine features of water management. Four types of benefits were mentioned by all stakeholder groups: hydro-geological and land maintenance, conservation of the traditional agricultural landscape, conservation of biodiversity, and provision of leisure activities and recreational opportunities (tourism).

Steep-slopes and natural constraints make difficult the management of agricultural land in the Aosta Valley Region. Irrigation networks and water infrastructures require a continuous effort for their routine maintenance. All stakeholders recognized that constant monitoring of the territory and routine maintenance by farmers contribute to prevent hydrogeological instability, reducing risks and damage caused by natural disasters, such as floods, landslides, and debris flows. Stakeholders also stated that agronomic management activities associated with irrigation in mountain pastures preserve the functionality of the soils.

The region is traditionally characterized by grazed pastures and grassland for forage production that, in addition to the production role, increase the aesthetic value of the territory. In the opinion of all participants, irrigation contributes to the maintenance of these typical landscape traits. In particular, green meadows and grazing cows are landscape features much appreciated by tourists, especially in the summer season when visitors reach these areas for hiking and outings. Stakeholders recognized the role of irrigation in biodiversity conservation. In their perception, open-air irrigation canals ensure the conservation of wetlands and enrich the biodiversity of typical permanent meadows.

Specific groups discussed other benefits. Both water consortia members and non-user citizens mentioned the role of irrigation for the maintenance of economic and social vitality of disadvantaged areas. In mountain areas, a viable agricultural sector prevents depopulation and its negative consequences, such as the loss of typical local products and land maintenance. Water consortia members also emphasized land stewardship activities provided by irrigation water users. Through a traditional and well-established practice called *corvée*, farmers associated with the consortia volunteer their labor to maintain the irrigation water networks, ensuring at the same time a territorial monitoring service.

Consortia members pointed out that provision of irrigation services guarantee professional jobs that manage the entire water cycle from the source to the users, and has the potential to create new professional roles to increase water service efficiency and coordinate *corvée* activities. In their opinion, irrigation services also preserve water quality through systematic resource monitoring and adoption of filtering systems to maintain overall water chemical and ecological quality. Moreover, water-supply infrastructures (such as open-air irrigation canals, tanks, and duct systems) are important for fire-fighting, especially in mountain areas difficult to reach by emergency vehicles.

Non-user citizens identified some specific benefits not mentioned by other stakeholders. Specifically, they named hydroelectric power production as a secondary consortia activity capable of bringing local economic/employment benefits and artificial recharge of groundwater reserves.

3.2. Prioritization of the External Benefits

Non-farmer users and non-user citizens were directly asked to rank the above-named benefits from most to least important (question Q2). Even if the categories of benefits were shared among respondents, each stakeholder group developed their own criteria to classify them. Non-farmer users prioritized their top five benefits and non-user citizens identified and ranked three groups, consisting of two to three items equal in importance to them (Figure 2).

Non-farmer users ranked benefits according to the magnitude of their spatial effects and the number of beneficiaries involved. In the top position, they ranked biodiversity conservation. In their opinion, its positive effects go beyond the region to non-users who also benefit from it. In other words, they recognized the existence value of biodiversity as prevalent. Hydro-geological and land maintenance and traditional landscape conservation follow, as they mainly affect regional territory land stability and local resident use values. Benefits affecting specific sectors, such as tourism or agriculture, were placed at the bottom of the ranking.

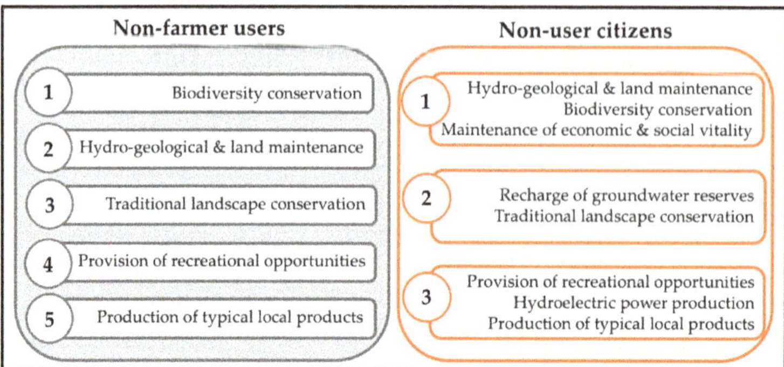

Figure 2. Benefits ranked based on question Q2.

Non-user citizens gave more importance to the effects of irrigation on local community well-being. The first group of benefits includes those, both environmental and economic, that are essential to ensuring a good quality of life for regional residents. The second group refers to environmental benefits having minor impacts on local community short-term livelihood. As with non-farmer users, benefits affecting specific economic sectors followed. There are several similarities between the two rankings, in particular land maintenance and biodiversity conservation rank high in both.

Computer-assisted analysis was employed to support the subjective analysis of Q1 with measurable indicators and to validate the stated priorities from Q2. In particular, we used this approach to prioritize the external benefits and compare responses across the three stakeholder types. We grouped the external benefits identified by subjective analysis with question Q1 into 12 categories: land maintenance, landscape, biodiversity, recreation-tourism, typical production, economic vitality, new jobs, water quality, *corvée*, fire-fighting, hydroelectric and groundwater reserves. Using QDA Miner, each transcribed Q1 text portion was assigned to one of the 12 thematic categories (codes). The software used two key counts, number of words used (Figure 3a) and number of discussion participants (Figure 3b)—as proxy indicators of the extent to which each benefit category was fully discussed among respondents and how widely it was shared among respondents, respectively (the output data provided by the software are reported in Appendix A).

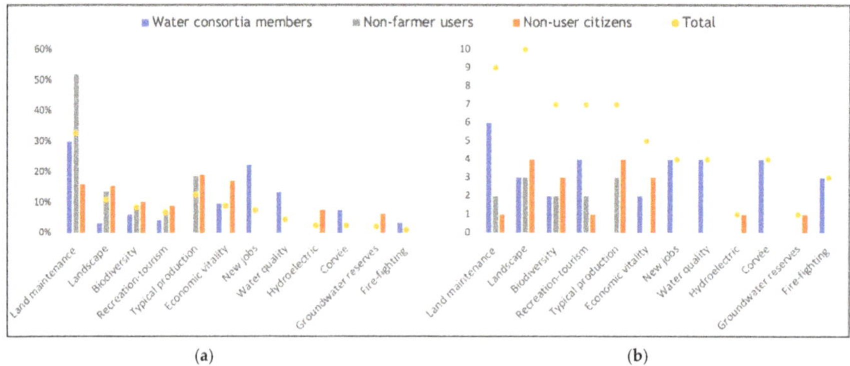

Figure 3. Discussion topics compared across different stakeholder types: (**a**) Spoken words on each topic (%); (**b**) participants involved in the discussion (n).

The measurable indicators confirmed the shared preferences found in the subjective analysis; the same four external benefits were highlighted in all three groups. A relatively sizeable word count (33%) centered on the importance of irrigation for hydro-geological and land maintenance. It was boosted by non-farmer users who devoted more than half of their count of words to this issue. As for the breadth of participant discussions, no topic rose to the level of being discussed by 50% of the 21 participants. The most widely discussed topics were traditional landscape conservation and land maintenance (10 and 9 participants, respectively).

The analysis also confirmed different perceptions across different types of stakeholders. Consortia members discussed a wider variety of benefits than other stakeholders (followed by non-user citizens), including technical issues related to irrigation water management (new jobs, water quality, *corvée*, and fire-fighting). Non-user citizens were concerned for socioeconomic and environmental issues not mentioned by other groups, such as energy and groundwater reserves.

Based on these results, the benefits were grouped and ranked using three criteria: (i) number of focus groups in which the benefits had been discussed; (ii) quartile of positioning based on spoken word percentages; (iii) number of participants involved in the discussion of that benefit (Table 2).

Table 2. Group of benefits ranked based on the selected criteria for question Q1.

Ranking	Group of Benefits	Criteria
1	Landscape Land maintenance	Benefits: (i) mentioned in all 3 focus groups; (ii) included in the fourth quartile for spoken words; (iii) discussed by more than 8 people
2	Biodiversity Typical production Economic vitality	Benefits: (i) mentioned at least in 2 focus groups; (ii) included in the fourth or third quartile for spoken words; (iii) discussed by more than 5 people
3	Recreation-tourism New jobs Water quality	Benefits: (i) mentioned at least in 1 focus group, (ii) included in the fourth, third or second quartile for spoken words; (iii) discussed for more than 2 people
4	Hydroelectric *Corvée* Groundwater reserves Fire-fighting	All remaining benefits

The computer-based prioritization from the discussion of Q1 differs slightly from the rank order of benefits stated explicitly with Q2. The ranking from content analysis of question Q1 is objectively based on the attention and effort dedicated by the participants for each topic. The discussion of the benefit order originated from Q2 brought forward by differing opinions among participants within each of the focus groups. As a result of a heated debate, the final ranking reflected the mediation between respondent points of view, mainly driven by the most determined ones, rather than a collective idea. Nevertheless, land maintenance and landscape and biodiversity conservation were placed at the top of all rankings.

3.3. Alternative Water Use

Consortia leaders and non-farmer users discussed other uses of water (Q3). Although few alternative uses were identified, participants considered those uses as competitors to water irrigation. In the case of drinking water, participants recognized potable water as a legal priority over other uses and considered it a fully-justified essential function. Hydroelectric energy production was named as another alternative use of water and participants varied in their view of it. Among non-farmer users, some participants considered hydropower generation as a benefit for the region, while others identified it as a competing water use. Specifically, competition arises when both irrigation and energy generation are simultaneously required that then results in a reduced water flow rate for agricultural purposes. Despite these divergent opinions, the entire group of non-farmer users agreed that water for hydroelectric purposes is necessary to transition to a sustainable energy source, which makes it increasingly important for the future of the region. Consortia members were unanimously concerned by what they perceived as private high-production hydropower plant inefficient water use that reduces irrigation water availability.

4. Discussion

This section discusses the qualitative analysis of the priority benefits identified. For each, we highlight stakeholder perceptions and opinions endorsed by scientific literature and those not supported by research findings. Second, we present some policy implications stemming from the recognition of such benefits.

4.1. Stakeholder Perceptions

Natural resource management is complex because natural systems are often characterized by competing users and uses, issues that cut across social, economic, administrative, and political units, unclear property rights, and market prices that do not reflect their full value [41]. The brainstorming session in which participants discussed the regional irrigation system, illuminated the existence of three distinct stakeholder groups and roles. Stakeholders may also be categorized according to the degree to which each is involved with or affected by water resource use [42,43]. Farmers and non-farmer users are primary stakeholders. Along with external benefits, they each receive direct private benefits—farmers from higher farm productivity and profitability and non-farmer users from product self-consumption and the utility associated with well-kept gardens. Non-user citizens represent secondary stakeholders, who consume local agricultural products and receive external social and environmental benefits.

Farmers access irrigation services as members of the water consortia. Membership is mandatory for landowners located within the territory of the consortia, and members are eligible for election to consortia management bodies (Regional Law no. 3/2001). For this reason, we involved water consortia members (most of whom are farmers) in the focus group discussions. Compared to other stakeholders, they command a legitimacy as dominant powerbrokers to manage the resource [44]. They affect water management decisions and actions as they are essential to irrigation service planning and provisioning.

Although the influences and technical skills differ among the three stakeholder types, all focus participants recognized and prioritized the same key irrigation benefits. Those benefits relate to activities to maintain extensive livestock farming in the region that involve grazing and herd management [45], agronomic forage management [46,47], and irrigation network management [48]. Some benefits are closely linked. Historic landscape shaped by traditional livestock farming activities represents one of the most-appreciated attractions and inspirations for summer recreation and tourism. The traditional practice of *corvées* performed by the members of water consortia is one activity that contributes to land monitoring and natural hazard protection. Land economic vitality and the conservation of typical local products demand livestock farms remain profitable. The dairy product from these farms is Fontina, a protected designation of origin (PDO) cheese. Some authors confirm the link between most of the external benefits discussed by the stakeholders and traditional agricultural practices (see Vidaller and Dutoit, and Palomo-Campesino et al. [49,50] for a review), mountain farming [51–53], and irrigation activities [5,10,11,13].

On the contrary, science findings at times diverge from stakeholder perceptions of the effects of irrigation on plant diversity. All stakeholders associated the greenness of irrigated meadows and pastures with higher biodiversity. However, scientific evidence does not validate this common belief. In fact, species selection due to irrigation and fertilizing irrigation has actually reduced grassland ecosystem biodiversity in mountain areas [54–56]. Actually, research conducted in the region shows a link between extensive grazing with local livestock breeds, soil conservation, and plant diversity [45,57]; however, these positive effects on biodiversity do not depend on irrigation. Another science-based fact did not arise in the focus group discussion. Only one participant mentioned the effects of irrigation on the conservation of faunal biodiversity, whereas a large body of literature highlights the habitat functions provided to insects, birds, amphibians, and reptiles by structures associated with irrigation (e.g., traditional ditch systems) (see Leibundgut and Kohn [58] for a review). The divergent views between scientists and stakeholders on biodiversity confirms that social learning is fundamental to sustainably manage natural resources provided that there is an information exchange to integrate stakeholder perceptions and opinions with technical information and expertise [59–61].

As for the quality of these findings, they could be refined by expanding the number of focus group discussions. Engaging different types of stakeholders helped to expand the range of different interest and views, although the small number of focus groups may have limited the analysis of conflicting positions discussed in Q2 and Q3. As for the priority external benefits that emerged in Q1, the convergence of opinions in all groups suggests that saturation (i.e., the point at which gathering more information reveals no new important issues) may have been achieved [62].

Identification of the multiple benefits of irrigation is the first step for evidence-based management of irrigation water. Nevertheless, sound decision-making requires additional information. From an economic standpoint, such benefits are positive externalities or public goods without market price. For cost-benefit analysis and trade-off evaluation between alternative uses of the resource, their economic value should be estimated in monetary terms [4,63]. Further analysis would be needed, using the outcomes of this study to inform which principal external benefits need to be included in the estimates [64].

4.2. Policy Implications

In 2000, European member states were provided a pan-EU water policy with two goals—to achieve a good ecological and chemical status for all ground and surface water bodies, and to promote long-term protection and sustainable use of available water resources [65]. The WFD recognized the complexity of water resource management by calling for a public participatory process and overcoming of the usual command-and-control approach to design efficient water policies [66]. The directive requires interested groups and the public-at-large to be actively involved in sustainable water management decisions [61]. To the best of our knowledge, few studies have tested this approach in Europe [61,67,68], and only Ricart and Clarimont [59] analyzed the perceptions and preferences for irrigation use and management in a mountain area. Evidence from the Aosta Valley Region provides insight into the stakeholder groups to be involved in making decisions that affect the irrigation service and related external benefit provisioning in agriculturally-disadvantaged areas. A regional peculiarity is the presence of non-farmer users of irrigation water. Currently, these private actors have a demand for irrigation water, yet do not compete with farmers. However, the need for negotiation may arise if the effects of climate change make water scarcer. Therefore, any approach to decision-making should integrate the roles of all users and the interactions between them.

The WFD introduced water pricing as an economic instrument to achieve the efficient use of water resources. The directive calls for a pricing policy based on the 'polluter-pays' principle to recover the total costs of water services, including environmental and resource costs (WFD, article 9). Environmental costs are the costs water users impose on ecosystems and other users through damaged or negatively-impacted aquatic environments [69]. Results showed that irrigating Aosta Valley meadows and pastures produces a wide range of social and environmental external benefits, some of which are prioritized by all stakeholders. This evidence contributes to the discussion of opportunities and constraints of WFD implementation [66,70]. Marginal and less-favorable lands in danger of depopulation (e.g., mountain areas) are in crucial need of conservation of agricultural activities to provide private as well as external benefits to their communities. Therefore, pricing policy design should consider the monetary value of such external benefits along with the external costs for the use of irrigation water. Other authors have emphasized that external benefits give rise to a socially derived demand for water and to a need to internalize environmental amenities via water allocation policies [71].

Resource costs refer to foregone opportunities of alternative water uses from exploitation or depletion of the resource beyond its natural rate of recovery or recharge [69]. In economic terms, forgone benefit of the alternative use is the opportunity cost for the current use of the resource. In Aosta Valley, hydropower generation is the main alternative water use versus irrigation that requires decision-making among competing users. Resource costs arise if the use of water for hydropower generates a higher economic value than current or future use of water for irrigation. This justifies the transfer from agricultural to hydroelectric use if the allocation process increases the net social benefits for the community. Again, the economic value of external benefits is necessary information for decision-making, as comparisons between alternative uses are only correct if they include estimates of the monetary value of both private and external benefits from irrigation.

Economic and governance approaches have been studied to achieve efficient and socially acceptable water allocation solutions. Tilmant et al. [72] modeled an economic mechanism to compensate farmers who have forgone some (or all) of their private benefits to increase the availability of water for hydropower plants. Crook [73] analyzed the consultation process that led communes and irrigation consortia to ratify long-term conventions for ceding water to hydropower companies in a mountainous Swiss canton. The governance arrangement protects water provisioning, optimizes the use of the resource, maintains the economic benefits supplied by the hydropower companies to the territory, and simultaneously improves water security where irrigation is economically feasible. For stakeholders in the Aosta Valley Region, the issue is controversial; they prefer a small-scale solution. Water consortia members cited the poor economic benefits to local communities and territories provided by current large electric power company management teams. They suggested that smaller plants directly managed by water consortia are able to allocate water resources deftly among different uses following seasonal needs and shortages. Furthermore, local consortia believe that they are capable of managing water resources more efficiently than private hydropower managers by reducing water waste and maximizing economic benefits for the region, which secures additional income for land improvement investment and job creation.

5. Conclusions

Following the European Commission recommendation for public participation and social learning in the design of land use and natural resource management polices [65], this study opened a space for the principal stakeholders in an Italian alpine region to discuss the external benefits of irrigation water services.

In particular, the study allowed local stakeholders to be categorized, described their priority benefits, and identified water uses competing with irrigation. Farmers, non-farmer users of irrigation water, and non-user citizens unanimously prioritized four categories of key benefits. Three of them (hydro-geological and land maintenance, conservation of the traditional landscape, and provision recreational opportunities) closely relate to irrigation of meadows and pastures in traditional livestock farming. Biodiversity conservation turned out to be a controversial item, due to agronomic considerations about the actual effects of irrigation on plant biodiversity in grassland ecosystems. As for the alternative water uses, hydropower generation was indicated as the major competitor to irrigation.

The benefits recognized as priorities are regulating, supporting, and cultural agroecosystem services without market price. Their differentiation provides policymakers with new information on the effects of irrigation water management and pricing decisions and on the trade-offs among policy actions affecting their provision. Incorporating this information into decision-making processes is essential in marginal mountain areas, where the balance between maintenance of agricultural practices and actions to optimize the use of environmental resources should be carefully considered.

Previous research to estimate irrigation water costs under different scenarios in the Aosta Valley raises concerns for implementation of cost recovery principles for water services laid down in the WFD [74]. Farmers in the Region may not be able or willing to pay increased operating expenses. Indeed, introduction of a water pricing policy, as defined by the directive, may hasten declines in current traditional farming practices and produce negative effects on the provision of the related external benefits. Efficient resource use should be incentivized in all European regions, but any economic analysis to define water recovery costs should include area-specific spatial analysis of the provision of external benefits of irrigation. Different environments and types of agricultural systems, as well as lands with differing degrees of economic marginalization have various requirements.

Operationally, internalizing the monetary value of external benefits to make decisions in cost-benefit analysis requires further research. The outcomes of this study can inform economic evaluation. External benefits identified as priorities should be valued in commensurable monetary units and aggregated over the affected stakeholders. Such estimations are essential to balance the environmental costs of irrigation and to estimate the net social benefits of alternative water uses in mountain areas.

Author Contributions: Conceptualization, S.N. and P.B.; methodology, S.N. and P.B.; formal analysis, F.M.; investigation, S.N. and P.B.; writing—original draft preparation, S.N., F.M., and P.B.; writing—review and editing, S.N. and P.B.; visualization, S.N.; project administration, P.B. All authors have read and agreed to the published version of the manuscript.

Funding: RESERVAQUA project was co-financed by the European Union, European Regional Development Fund, the Italian State, the Swiss Confederation and the Cantons, under the Interreg V-A Italy-Switzerland Cooperation Programme 2014–2020.

Institutional Review Board Statement: Ethical review and approval were waived for this study because data and information collected through focus group discussions were not sensitive. Moreover, personal data (name, date of birth, gender etc.) were not collected and personal opinions on the research topics were treated in an aggregated form.

Informed Consent Statement: Informed consent was obtained from all subjects involved in the study.

Data Availability Statement: Not applicable.

Acknowledgments: The authors would like to thank all those who have engaged with this work in some way—as providers of expertise and opinion through the focus groups or various gatherings and meetings through the project.

Conflicts of Interest: The authors declare no conflict of interest. The funders had no role in the design of the study; in the collection, analyses, or interpretation of data; in the writing of the manuscript, or in the decision to publish the results.

Appendix A

Table A1. Spoken words on each topic (%).

	Water Consortia Members	Non-Farmer Users	Non-User Citizens	Total
Land maintenance	30.0	52.0	16.0	32.7
Landscape	3.3	13.6	15.4	10.8
Biodiversity	6.2	8.8	10.2	8.4
Recreation-tourism	4.2	7.0	8.9	6.7
Typical production		18.6	18.9	12.5
Economic vitality	9.6		17.0	8.9
New jobs	22.4			7.5
Water quality	13.4			4.5
Hydroelectric			6.3	2.1
Corvée	7.5			2.5
Groundwater reserves			7.6	2.5
Fire-fighting	3.4			1.1

Table A2. Participants involved in the discussion (n).

	Water Consortia Members	Non-Farmer Users	Non-User Citizens	Total
Land maintenance	6	1	2	9
Landscape	3	4	3	10
Biodiversity	2	3	2	7
Recreation-tourism	4	1	2	7
Typical production		4	3	7
Economic vitality	2	3		5
New jobs	4			4
Water quality	4			4
Hydroelectric		1		1
Corvée	4			4
Groundwater reserves		1		1
Fire-fighting	3			3

References

1. Baldock, D.; Caraveli, H.; Dwyer, J.; Einschütz, S.; Peteresen, J.E.; Sumpsi-Vinas, J.; Varela-Ortega, C. *The Environmental Impacts of Irrigation in the European Union*; Institute for European Environmental Policy: London, UK, 2000; Volume 5210, p. 147.
2. Brouwer, C.; Heibloem, M. *Irrigation Water Management: Irrigation Water Needs*; FAO: Food and Agriculture Organization of the United Nations: Rome, Italy, 1986; Volume 3.
3. Eurostat Data Base. Available online: https://ec.europa.eu/eurostat/databrowser/view/tag00025/default/table?lang=en (accessed on 24 June 2022).
4. Falkenmark, M.; Finlayson, C.M.; Gordon, L.J.; Bennett, E.M.; Chiuta, T.M.; Coates, D.; Ghosh, N.; Gopalakrishnan, M.; de Groot, R.S.; Jacks, G.; et al. *Water for Food Water for Life*; Molden, D., Ed.; Routledge: Oxfordshire, UK, 2013; ISBN 9781136548536.
5. Gordon, L.J.; Finlayson, C.M.; Falkenmark, M. Managing Water in Agriculture for Food Production and Other Ecosystem Services. *Agric. Water Manag.* **2010**, *97*, 512–519. [CrossRef]
6. Lobell, D.; Bala, G.; Mirin, A.; Phillips, T.; Maxwell, R.; Rotman, D. Regional Differences in the Influence of Irrigation on Climate. *J. Clim.* **2009**, *22*, 2248–2255. [CrossRef]
7. Puma, M.J.; Cook, B.I. Effects of Irrigation on Global Climate during the 20th Century. *J. Geophys. Res.* **2010**, *115*, D16120. [CrossRef]
8. Mayers, J.; Batchelor, C.; Bond, I.; Hope, R. *Water Ecosystem Services and Poverty under Climate Change Key Issues and Research Priorities Water Ecosystem Services and Poverty under Climate Change Key Issues and Research Priorities*; Natural Resource Issues 17; International Institute for Environment and Development: London, UK, 2009; ISBN 9781843696872.
9. Boelee, E. *Managing Water and Agroecosystems for Food Security*; CABI: Wallingford, UK, 2013; ISBN 9781780640884.
10. Alcon, F.; Zabala, J.A.; Martínez-García, V.; Albaladejo, J.A.; López-Becerra, E.I.; De-Miguel, M.D.; Martínez-Paz, J.M. The Social Wellbeing of Irrigation Water. A Demand-Side Integrated Valuation in a Mediterranean Agroecosystem. *Agric. Water Manag.* **2022**, *262*, 107400. [CrossRef]
11. Natali, F.; Branca, G. On Positive Externalities from Irrigated Agriculture and Their Policy Implications: An Overview. *Econ. Agro-Aliment.* **2020**, *22*, 1–25. [CrossRef]
12. Robinson, R. Discussion Paper for the Third Meeting of the Adelboden Group. In *Positive Mountain Externalities: Valorisation through Policies and Markets*; Euromontana: Fortrose, UK, 2007; Volume 51.
13. Leibundgut, C.; Kohn, I. European Traditional Irrigation in Transition Part I: Irrigation in Times Past-A Historic Land Use Practice Across Europe. *Irrig. Drain.* **2014**, *63*, 273–293. [CrossRef]
14. Baumol, W.J.; Oates, W.E. *The Theory of Environmental Policy*; Cambridge University: Cambridge, UK, 1988.
15. Mishan, E.J. The Postwar Literature on Externalities: An Interpretative Essay. *J. Econ. Lit.* **1971**, *9*, 1–28.
16. Turner, K.; Pearce, D.W.; Bateman, I. *Environmental Economics. An Elementary Introduction*; John Hopkins University Press: Baltimore, MD, USA, 1994.
17. Istat Datawarehouse Agricultural Census 2010. Available online: http://dati-censimentoagricoltura.istat.it/Index.aspx?lang=en (accessed on 24 June 2022).
18. Regione Valle d'Aosta Consorzi di Miglioramento Fondiario. Available online: https://www.regione.vda.it/agricoltura/CMF/default_i.aspx (accessed on 24 June 2022).
19. Berthet, E.T.; Bretagnolle, V.; Gaba, S. Place-Based Social-Ecological Research Is Crucial for Designing Collective Management of Ecosystem Services. *Ecosyst. Serv.* **2022**, *55*, 101426. [CrossRef]
20. Zoderer, B.M.; Tasser, E.; Carver, S.; Tappeiner, U. Stakeholder Perspectives on Ecosystem Service Supply and Ecosystem Service Demand Bundles. *Ecosyst. Serv.* **2019**, *37*, 100938. [CrossRef]
21. Bengtsson, M. How to Plan and Perform a Qualitative Study Using Content Analysis. *Nurs. Open* **2016**, *2*, 8–14. [CrossRef]
22. Adams, K.J.; Metzger, M.J.; Macleod, C.J.A.; Helliwell, R.C.; Pohle, I. Understanding Knowledge Needs for Scotland to Become a Resilient Hydro Nation: Water Stakeholder Perspectives. *Environ. Sci. Policy* **2022**, *136*, 157–166. [CrossRef]

23. Okumah, M.; Yeboah, A.S.; Amponsah, O. Stakeholders' Willingness and Motivations to Support Sustainable Water Resources Management: Insights from a Ghanaian Study. *Conserv. Sci. Pract.* **2020**, *2*, e170. [CrossRef]
24. Es'haghi, S.R.; Karimi, H.; Rezaei, A.; Ataei, P. Content Analysis of the Problems and Challenges of Agricultural Water Use: A Case Study of Lake Urmia Basin at Miandoab, Iran. *SAGE Open* **2022**, *12*, 21582440221091247. [CrossRef]
25. Ricart, S.; Kirk, N.; Ribas, A. Ecosystem Services and Multifunctional Agriculture: Unravelling Informal Stakeholders' Perceptions and Water Governance in Three European Irrigation Systems. *Environ. Policy Gov.* **2019**, *29*, 23–34. [CrossRef]
26. Martín-López, B.; Iniesta-Arandia, I.; García-Llorente, M.; Palomo, I.; Casado-Arzuaga, I.; Amo, D.G.D.; Gómez-Baggethun, E.; Oteros-Rozas, E.; Palacios-Agundez, I.; Willaarts, B.; et al. Uncovering Ecosystem Service Bundles through Social Preferences. *PLoS ONE* **2012**, *7*, e38970. [CrossRef] [PubMed]
27. Martín-López, B.; Montes, C.; Ramírez, L.; Benayas, J. What Drives Policy Decision-Making Related to Species Conservation? *Biol. Conserv.* **2009**, *142*, 1370–1380. [CrossRef]
28. Garrido, P.; Elbakidze, M.; Angelstam, P. Stakeholders' Perceptions on Ecosystem Services in Östergötland's (Sweden) Threatened Oak Wood-Pasture Landscapes. *Landsc. Urban Plan.* **2017**, *158*, 96–104. [CrossRef]
29. Osborn, A.F. *Applied Imagination: Principles and Procedures of Creative Thinking*; Charles Scribner's Sons: New York, NY, USA, 1953.
30. Isaksen, S.G. *A Review of Brainstorming Research: Six Critical Issues for Inquiry*; Creative Research Unit, Creative Problem Solving Group-Buffalo: New York, NY, USA, 1998; pp. 1–28.
31. Bloor, M.; Frankland, J.; Thomas, M.; Robson, K. *Focus Groups in Social Research*; SAGE Publications: Thousand Oaks, CA, USA, 2001; ISBN 9780761957423.
32. Gibbs, A. *Focus Groups*; Social Research Update: Guilford, UK, 1997.
33. Yin, R.K. *Qualitative Research from Start to Finish*; Guilford Press: New York, NY, USA, 2011.
34. Krueger, R.A.; Casey, M.A. *Focus Groups: A Practical Guide for Applied Research*; SAGE Publications: Thousand Oaks, CA, USA, 2014.
35. Nind, M. A New Era in Focus Group Research: Challenges, Innovation and Practice. *Int. J. Res. Method Educ.* **2019**, *42*, 220–222. [CrossRef]
36. Onwuegbuzie, A.J.; Dickinson, W.B.; Leech, N.L.; Zoran, A.G. A Qualitative Framework for Collecting and Analyzing Data in Focus Group Research. *Int. J. Qual. Methods* **2019**, *8*, 1–21. [CrossRef]
37. Oprandi, N. *Focus Group: Breve Compendio Teorico-Pratico*; Emme&Erre Libri: Padova, Italy, 2001.
38. Stewart, D.; Shamdasani, P.; Rook, D. *Focus Groups*; SAGE Publications: Thousand Oaks, CA, USA, 2007; ISBN 9780761925835.
39. Lewis, R.B.; Maas, S.M. QDA Miner 2.0: Mixed-Model Qualitative Data Analysis Software. *Field Methods* **2007**, *19*, 87–108. [CrossRef]
40. Yamaguchi, C.K.; Stefenon, S.F.; Ramos, N.K.; dos Santos, V.S.; Forbici, F.; Klaar, A.C.R.; Ferreira, F.C.S.; Cassol, A.; Marietto, M.L.; Yamaguchi, S.K.F.; et al. Young People's Perceptions about the Difficulties of Entrepreneurship and Developing Rural Properties in Family Agriculture. *Sustainability* **2020**, *12*, 8783. [CrossRef]
41. Grimble, R.; Wellard, K. Stakeholder Methodologies in Natural Resource Management: A Review of Principles, Contexts, Experiences and Opportunities. *Agric. Syst.* **1997**, *55*, 173–193. [CrossRef]
42. ODA. *Guidance Note on Stakeholder Analysis for Aid Projects and Programmes*; HMSO: London, UK, 1995; pp. 1–10.
43. Clarkson, M.B.E. A Stakeholder Framework for Analyzing and Evaluating Corporate Social Performance. *Acad. Manag. Rev.* **1995**, *20*, 92. [CrossRef]
44. Mitchell, R.K.; Agle, B.R.; Wood, D.J. Toward a Theory of Stakeholder Identification and Salience: Defining the Principle of Who and What Really Counts. *Acad. Manag. Rev.* **1997**, *22*, 853–886. [CrossRef]
45. Borsotto, P. Quali Beni Pubblici Dai Prati e Dai Pascoli Montani? In *Beni Pubblici dai Prati e dai Pascoli della Valle D'Aosta*; INEA: Rome, Italy, 2013; ISBN 978-88-8145-256-9.
46. Bassignana, M.; Clementel, F.; Kasal, A.; Peratoner, G. The Forage Quality of Meadows under Different Management Practices in the Italian Alps. In Proceedings of the 16th Symposium of the European Grassland Federation, Gumpenstein, Austria, 29–31 August 2011.
47. Bassignana, M.; Curtaz, A.; Journot, F.; Poggio, L.; Bovio, M. The Relationship between Farming Systems and Grassland Diversity in Dairy Farms in Valle d'Aosta, Italy. *Grassl. Sci. Eur.* **2009**, *16*, 205–207.
48. Briner, S.; Huber, R.; Bebi, P.; Elkin, C.; Schmatz, D.R.; Grêt-Regamey, A. Trade-Offs between Ecosystem Services in a Mountain Region. *Ecol. Soc.* **2013**, *18*, 35. [CrossRef]
49. Vidaller, C.; Dutoit, T. Ecosystem Services in Conventional Farming Systems. A Review. *Agron. Sustain. Dev.* **2022**, *42*, 22. [CrossRef]
50. Palomo-Campesino, S.; González, J.A.; García-Llorente, M. Exploring the Connections between Agroecological Practices and Ecosystem Services: A Systematic Literature Review. *Sustainability* **2018**, *10*, 4339. [CrossRef]
51. Schirpke, U.; Tasser, E.; Leitinger, G.; Tappeiner, U. Using the Ecosystem Services Concept to Assess Transformation of Agricultural Landscapes in the European Alps. *Land* **2022**, *11*, 49. [CrossRef]
52. Bernués, A.; Rodríguez-Ortega, T.; Ripoll-Bosch, R.; Alfnes, F. Socio-Cultural and Economic Valuation of Ecosystem Services Provided by Mediterranean Mountain Agroecosystems. *PLoS ONE* **2014**, *9*, e102479. [CrossRef]
53. Assandri, G.; Bogliani, G.; Pedrini, P.; Brambilla, M. Beautiful Agricultural Landscapes Promote Cultural Ecosystem Services and Biodiversity Conservation. *Agric. Ecosyst. Environ.* **2018**, *256*, 200–210. [CrossRef]

54. Lonati, M.; Probo, M.; Gorlier, A.; Pittarello, M.; Scariot, V.; Lombardi, G.; Ravetto Enri, S. Plant Diversity and Grassland Naturalness of Differently Managed Urban Areas of Torino (NW Italy). *Acta Hortic.* **2018**, *1215*, 247–253. [CrossRef]

55. Pittarello, M.; Lonati, M.; Ravetto Enri, S.; Lombardi, G. Environmental Factors and Management Intensity Affect in Different Ways Plant Diversity and Pastoral Value of Alpine Pastures. *Ecol. Indic.* **2020**, *115*, 106429. [CrossRef]

56. Pittarello, M.; Lonati, M.; Gorlier, A.; Perotti, E.; Probo, M.; Lombardi, G. Plant Diversity and Pastoral Value in Alpine Pastures Are Maximized at Different Nutrient Indicator Values. *Ecol. Indic.* **2018**, *85*, 518–524. [CrossRef]

57. OECD. *Providing Agri-Environmental Public Goods through Collective Action*; OECD Publishing: Paris, France, 2013; Volume 9789264197. ISBN 9789264197213.

58. Leibundgut, C.; Kohn, I. European Traditional Irrigation in Transition Part II: Traditional Irrigation in Our Time-Decline, Rediscovery and Restoration Perspectives. *Irrig. Drain.* **2014**, *63*, 294–314. [CrossRef]

59. Ricart, S.; Clarimont, S. Modelling the Links between Irrigation, Ecosystem Services and Rural Development in Pursuit of Social Legitimacy: Results from a Territorial Analysis of the Neste System (Hautes-Pyrénées, France). *J. Rural Stud.* **2016**, *43*, 1–12. [CrossRef]

60. Broderick, K. Communities in Catchments: Implications for Natural Resource Management. *Geogr. Res.* **2005**, *43*, 286–296. [CrossRef]

61. Tippett, J.; Searle, B.; Pahl-Wostl, C.; Rees, Y. Social Learning in Public Participation in River Basin Management—Early Findings from HarmoniCOP European Case Studies. *Environ. Sci. Policy* **2005**, *8*, 287–299. [CrossRef]

62. Hennink, M.; Kaiser, B.N. Sample Sizes for Saturation in Qualitative Research: A Systematic Review of Empirical Tests. *Soc. Sci. Med.* **2022**, *292*, 114523. [CrossRef]

63. National Ecosystem Services Partnership. *Federal Resource Management and Ecosystem Services Guidebook*, 2nd ed.; National Ecosystem Services Partnership: Durham, UK, 2016.

64. Davies, A.M.; Laing, R. Designing Choice Experiments Using Focus Groups: Results from an Aberdeen Case Study. *Forum Qual. Sozialforsch.* **2002**, *3*, 3.

65. Kurrer, C. European Parliament Environment Policy: General Principles and Basic Framework. Fact Sheets on the European Union—2021. Available online: https://www.europarl.europa.eu/factsheets/en/sheet/71/environment-policy-general-principles-and-basic-framework (accessed on 16 June 2022).

66. Voulvoulis, N.; Arpon, K.D.; Giakoumis, T. The EU Water Framework Directive: From Great Expectations to Problems with Implementation. *Sci. Total Environ.* **2017**, *575*, 358–366. [CrossRef] [PubMed]

67. Hophmayer-Tokich, S.; Krozer, Y. Public Participation in Rural Area Water Management: Experiences from the North Sea Countries in Europe. *Water Int.* **2008**, *33*, 243–257. [CrossRef]

68. Lienert, J.; Schnetzer, F.; Ingold, K. Stakeholder Analysis Combined with Social Network Analysis Provides Fine-Grained Insights into Water Infrastructure Planning Processes. *J. Environ. Manage.* **2013**, *125*, 134–148. [CrossRef] [PubMed]

69. WATECO group WATer ECOnomics Working Group (WATECO Group). *European Commission, Economics and Environment, The Implementation Challenge of the Water Framework Directive Policy*; Office for Official Publications of the European Communities: Luxembourg, 2003; ISBN 9289441445.

70. Klauer, B.; Sigel, K.; Schiller, J. Disproportionate Costs in the EU Water Framework Directive-How to Justify Less Stringent Environmental Objectives. *Environ. Sci. Policy* **2016**, *59*, 10–17. [CrossRef]

71. Thiene, M.; Tsur, Y. Agricultural Landscape Value and Irrigation Water Policy. *J. Agric. Econ.* **2013**, *64*, 641–653. [CrossRef]

72. Tilmant, A.; Goor, Q.; Pinte, D. Agricultural-to-Hydropower Water Transfers: Sharing Water and Benefits in Hydropower-Irrigation Systems. *Hydrol. Earth Syst. Sci.* **2009**, *13*, 1091–1101. [CrossRef]

73. Crook, D.S. The Historical Impacts of Hydroelectric Power Development on Traditional Mountain Irrigation in the Valais, Switzerland. *Mt. Res. Dev.* **2001**, *21*, 46–53. [CrossRef]

74. Borsotto, P.; Moino, F.; Novelli, S. Modeling Change in the Ratio of Water Irrigation Costs to Farm Incomes under Various Scenarios with Integrated Fadn and Administrative Data. *Econ. Agro-Aliment.* **2021**, *23*, 1–19. [CrossRef]

 *land*

Article

Nexus between Coping Strategies and Households' Agricultural Drought Resilience to Food Insecurity in South Africa

Yonas T. Bahta

Department of Agricultural Economics, The University of the Free State, P.O. Box 339, Bloemfontein 9300, South Africa; bahtay@ufs.ac.za; Tel.: +27-514-019-050

Abstract: Farmers in Africa, including those in South Africa, rely on rain-fed agriculture, which exposes them to the risks of agricultural drought. Agricultural drought has become a major threat to agricultural production, including the extreme mortality of livestock in recent years, thus negatively impacting household food security. Hence, this paper is aimed at (i) assessing the coping strategies employed by smallholding livestock-farming households during food insecurity shocks, and (ii) assessing the relationship between coping strategies and agricultural drought resilience to food insecurity in the Northern Cape Province of South Africa. Interviews, more specifically survey interviews, were conducted with 217 smallholder livestock farmers. The data was analyzed using the agricultural drought resilience index (ADRI), the household food insecurity access scale (HFIAS), and structural equation modeling. Smallholder livestock farming households utilized various coping strategies, ranging from selling livestock (21%) to leasing out their farms (1%). The coping strategies of farming households included using alternative land (20%), storing food (20%), requesting feed for their animals (16%), searching for alternative employment (6%), migrating (6%), raising drought-tolerant breeds (5%), receiving relief grants (3%) and using savings and investments (2%). A statistically significant relationship between coping strategies and agricultural drought resilience to food insecurity means that these strategies have important policy implications. Implementing strategies that encourage households to protect their livelihood and utilize their assets (selling livestock) to increase their resilience is crucial for reducing food insecurity and achieving the Sustainable Development Goals (SDGs) to end hunger and poverty.

Keywords: migration; drought tolerant breeds; adaptation; relief grants; policy intervention; smallholder livestock farmers

Citation: Bahta, Y.T. Nexus between Coping Strategies and Households' Agricultural Drought Resilience to Food Insecurity in South Africa. *Land* **2022**, *11*, 893. https://doi.org/10.3390/land11060893

Academic Editor: Elisa Marraccini

Received: 4 May 2022
Accepted: 9 June 2022
Published: 11 June 2022

Publisher's Note: MDPI stays neutral with regard to jurisdictional claims in published maps and institutional affiliations.

1. Introduction

Feeding the world's future population is one of the United Nations' Sustainable Development Goals and is also a major global challenge [1]. Food access is one of the fundamental human rights that ensures a person's freedom from hunger [2]. Globally, large portions of the population continue to struggle with food insecurity. According to the Food and Agricultural Organization of the United Nations (FAO) et al. [3], and their report on the state of food and nutrition globally, the prospect itself is not as bright as expected. The target of zero hunger, one of the Sustainable Development Goals, is not on track to be met by 2030, despite some progress. The FAO et al. [3] predict that the effect of the COVID-19 pandemic on world health and socioeconomics will worsen the most vulnerable population groups' food security and nutritional situation. Currently, billions of people cannot afford healthy and nutritious food because of high costs.

Despite being food-secure on a national level, South Africa remains food-insecure on a household level because not all households have access to sufficient food. Nearly 20% of South African households do not have sufficient food and the proportions differ according to province, population group, and household size [4]. Conflict and insecurity, climate change, poverty, and an aging population are the main causes of hunger and food insecurity in South Africa [5]. Since COVID-19 broke out, hunger ratios in South Africa have increased. More than 23% of the households in South Africa went hungry during the summer of 2020, and 70% of the households relied on government assistance. In addition, unemployment reached a record high of 32.8%, up by 2% since the pandemic began [5].

The mechanism by which food insecurity shocks are responded to differs depending on the objectives of the agents involved, as well as their levels of targeting. When faced with shocks such as a drought that threatens food security, households use a variety of coping strategies. Sassi [6] and Farzana et al. [7] pointed out that the unintended consequences of such strategies can undermine households' ability to cope with future food insecurity shocks.

Existing national and international studies, such as those by Van Dijk et al. [1], Lehmann-Uschner and Kraehnert [8], Masipa [9], Bahta [10], Meyeki and Bahta [11], and others, assessed household asset dynamics in the aftermath of severe environmental shocks, reviewed the impacts of climate change on food security and projections in sub-Saharan Africa, assessed coping strategies, and identified factors affecting livestock farmers' food security and resilience to drought. Debessa et al. [12] examined how households dealt with shocks resulting from food insecurity, as well as the relationships between coping strategies and food insecurity resilience in one of Ethiopia's food-stressed districts. The authors found a statistically significant relationship between the strategy used and food insecurity resilience.

Van Dijk et al. [1] conducted a systematic literature review and meta-analysis to assess the range of changes in global food security, projected until 2050. The authors discovered that total global food demand will increase by 56% by 2050, while the number of people at risk of hunger will increase by 8%. Looking at the asset dynamics of households when faced with environmental shock, Lehmann-Uschner and Kraehnert [8] found that the poorest households experience the most difficulty in adapting to shocks, adopting coping strategies that are costly to both short- and long-term well-being.

Masipa [9] quantified the effect of climate change on food security in sub-Saharan countries, from crop production to food distribution and consumption. Furthermore, Masipa [9] discovered that climate change, particularly global warming, affected food security through food availability, accessibility, utilization, and affordability. To mitigate these risks, there is a need for an integrated policy approach to protect arable land against global warming.

Bahta [10] investigated the strategies used by smallholder livestock farmers to cope with agricultural drought in South Africa and found that most livestock farmers sold their livestock. Furthermore, the author found that socioeconomic and institutional factors influenced smallholder livestock farmers' coping strategies. Meyeki and Bahta [11] discovered that most smallholder livestock farmers were not resistant to agricultural drought and that assets, social safety nets, and adaptive capacity indicators positively affected household resilience to food insecurity.

To the best of the author's knowledge, very few studies specifically focused on empirical evidence on the relationship between coping strategies and households' agricultural drought resilience to food insecurity. Ansha et al. [13] used a recursive framework to examine how coping strategies relate to household food security. Ado et al. [14] investigated households' coping strategies in the face of food insecurity, using a survey and the Probit model. Amoah and Simatele [15] used semi-structured questionnaires, interviews, and the sustainable livelihood framework (SLF) to investigate the coping strategies used by the rural poor to build resilience against food insecurity. Therefore, this study examined the coping strategies used by smallholder livestock farming households during food insecurity shocks and the relationship between the types of coping strategies and agricultural drought resilience in the face of food insecurity in the Northern Cape Province, South Africa. This paper employed the agricultural drought resilience index (ADRI), the household food insecurity access scale (HFIAS), and structural equation modeling. Previous studies [1,9] focused on the influence of climate change on global food security and the asset dynamics of households [8], as well as the strategies and resilience of smallholder livestock farmers [10,11]. Debessa et al. [12] assessed the coping strategies employed by households in the event of food insecurity shocks and the nexus between the types of coping strategies and resilience to food insecurity. The novelty of this paper lies in the incorporation of the household food insecurity access scale (HFIAS) and structural equation modeling. In addition, this study adds to existing knowledge by pointing out the nexus between coping strategies and the agricultural drought resilience to food insecurity of smallholder livestock farming households. The findings of this study will aid policymakers to develop appropriate policies to enhance the resilience of smallholder livestock farmers when faced with the effects of drought, which threatens food security.

2. Nexus between Coping Strategies and Households' Agricultural Drought Resilience to Food Insecurity/Conceptual Framework

According to Sassi [6], households employ four coping mechanisms when faced with food shortages. As part of these measures, they consider the quality of food consumed, increase food supply, receive assistance from neighbors, and ration food. Lehmann-Uschner and Kraehnert [8] stated that the presence of institutions (such as access to credit), drought relief, engagement in farm and non-farm activities, and consumption reduction will all influence household responses to agricultural drought. To put it another way, reducing consumption may be perceived negatively because it has a number of negative consequences, including immediate hunger as well as long-term effects on children's health and development [16].

Adverse events (agricultural drought) are hypothesized to lead to a short-term decline in assets and incomes and to long-term negative impacts on the livelihoods of smallholder livestock farmers [17]. The severe effects depend on the severity of the shocks, asset dynamics, and coping strategies. Lehmann-Uschner and Kraehnert [8] state that shocks, both directly and indirectly, affect households' resilience.

While resilience has various definitions, all these definitions share certain characteristics [18–21]. Most definitions of resilience emphasize the following characteristics: ability, mitigation, adaptation, coping, recovery, withstanding shocks, resistance, and bouncing back from shocks. In this study, resilience was defined as a household's ability to "bounce back" after being exposed to threats to its livelihood and to shocks (such as agricultural drought and food insecurity). Household resilience to food insecurity in response to agricultural drought was defined as the ability to maintain a certain level of income and well-being (food security). This was determined by the household's options for making a living and their ability to cope with agricultural drought. It refers, therefore, to both ex ante and ex post measures used in the reduction or mitigation of agricultural drought. The ability of a household to deal with agricultural drought was determined by the available options [22].

Agricultural drought was characterized by the dynamic nature of resilience, which can be divided into three categories: absorptive, adaptive, and transformative. Absorptive capacity emphasizes the ability to respond to agricultural drought with an initial, "persistent" response. Adaptive capacity deals with the ability to remain as functional as before, despite small but continuous changes in climate change shocks, such as agricultural drought. Transformational capacity refers to responding to challenges, such as droughts and prolonged disturbances, through a significant change in value, regimes, and financial, technological, and biological systems [23,24].

The FAO [25] acknowledges that food security is a highly flexible concept and that, generally, food insecurity manifests whenever people do not have adequate physical, social, and/or economic access to food. Guided by the above, food insecurity was therefore defined as a household's inability to access adequate food to meet its target consumption levels, in the face of shocks such as agricultural drought [26]. This scope was motivated by the dominant dimensions of food security in the study area, as informed by the Department of Agriculture, Forestry, and Fisheries (DAFF) and humanitarian organizations. However, the adopted definition still recognizes the central role of behavior patterns across coping strategies, as exhibited by vulnerable (and potentially vulnerable) individuals and the wider community affected, as is the case in the drought-prone Northern Cape Province of South Africa. To support this viewpoint, the study utilized the HFIAS, which has a broader measurement range of food insecurity severity status conditions, while accounting for a large spectrum of generic food security indicators. As informed by Coates et al. [27], in order to improve its performance, the standard HFIAS was modified and adapted in terms of the core food insecurity indicators to suit the context in which it was applied.

In this paper, the term "resilience to food insecurity" referred to the adaptive capacity of smallholder livestock farmers in South Africa's Northern Cape Province. According to Javadinejad et al. [28], the key mechanisms required for household resilience are social, economic, situational, and institutional preparedness. Furthermore, numerous studies have documented a wide range of factors influencing the methods and processes for achieving household resilience [16,29–32].

Fan et al. [33] proposed several frameworks for resilience analysis. However, the plethora of resilience analysis frameworks all has similar components [34]. This includes assessing the larger environment (or individual or other units of observation) in which a household resides; assessing the resources to which that household has access; determining how the shocks experienced by the household affect the household's economic returns on those uses; and assessing how the consequences of those uses may result in food and other goods and services consumption, savings, health, nutritional status, and other outcomes. As a result, resilience frameworks were frequently used to guide studies on household resilience to food insecurity [26,35,36]. In this study, a framework adapted from the work of Alinovi et al. [37] was used.

Figure 1 provides an overview of the conceptual framework for this study. The framework was chosen because it was originally proposed for analyzing households' resilience to food insecurity shocks, such as agricultural drought, in relation to coping strategies and adaptive capacities (Equations (1) and (2)). By using this framework, the study can determine the extent of variation in resilience-building among households, along with several factors influencing this variation. Assets, non-agricultural assets, adaptive capacity, social safety nets, and climate change are among the factors to be considered. These variables are regressed on the agricultural drought resilience index's (ADRI) outcome variable.

Frankenberger et al. [38] and Pasture [39] explored the relationship between coping strategies and resilience, proposing that maintaining a specific strategy can negatively impact resilience in the long run. To put it another way, negative coping strategies make it difficult to cope with future shocks. Therefore, it is possible that the level of coping strategies previously used by a household can play a part in its resilience status at a specific time (resilience to future food insecurity shocks).

The ADRI was calculated using principal component analysis (PCA) and variables related to livestock production and consumption, with and without drought seasons. Similarly, in the structural equation model, ADRI was used as an outcome variable against independent variables such as assets, adaptive capacity, social safety nets, and climate change indicators, as shown in Figure 1 and Equations (1) and (3).

Based on the literature findings and the resulting framework shown in Figure 1, the coping strategies used by households in response to the shock of food insecurity were assumed to influence household resilience. Therefore, households' coping mechanisms in response to agricultural droughts that threaten their food security were investigated using surveys, as discussed in Section 3.3. This study aimed to determine the relationship between the types of coping mechanisms used and household resilience status, which is a proxy for a household's ability to deal with food insecurity shocks.

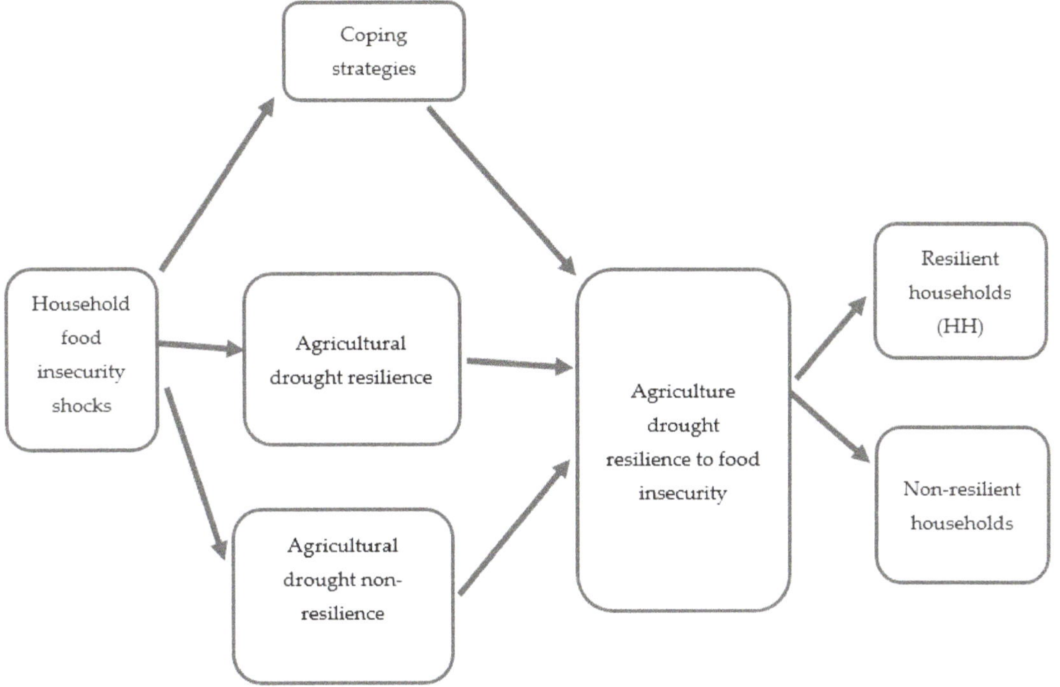

Figure 1. Conceptual framework representing the food insecurity shock–coping strategies–resilience nexus. Source: Author's adoption from observations of various studies.

3. Materials and Methods

3.1. Study Area Description

The Northern Cape is the largest and most sparsely populated province of South Africa, with Kimberley as the capital. The distances between towns are enormous, due to its sparse population and its size, which is just shy of that of the American state of Montana and slightly larger than Germany. The province is dominated by the Karoo Basin and consists mostly of sedimentary rocks and some dolerite intrusions. The south and south-east of the province is high-lying, situated 1200–1900 meters above sea level, in the Roggeveld and Nuweveld districts. The west coast is dominated by the Namaqualand region, famous for its spring flowers. The terrain is hilly to mountainous and consists of granites and other metamorphic rocks. The central areas are generally flat, with interspersed salt pans. Kimberlite (igneous rock) intrusions punctuate the Karoo rocks, giving the province its most precious natural resource, diamonds. The north is primarily in the Kalahari Desert, which is characterized by parallel red sand dunes and acacia trees on dry savanna [40].

The study was conducted in the Northern Cape Province of South Africa (Figure 2). The Northern Cape Province is situated in the northwest region of South Africa, shares international borders with Botswana and Namibia, and shares local borders with the Western and Eastern Cape Provinces in the south and the Free State and North West Provinces in the east [41]. The province's land area is 372,889 km^2, accounting for 30.5% of South Africa's total land area, with a population of 1.2 million people [42]. Frances Baard (12,800 km^2), John Taolo Gaetsewe (27,300 km^2), Namakwa (126,900 km^2), Pixley Ka Seme (103,500 km^2), and ZF Mgcawu (102,500 km^2) are the five district municipalities in the Northern Cape Province. The current research was carried out in the Frances Baard District Municipality (FBDM), which is divided into four local municipalities: Dikgatlong (2377.6 km^2), Magareng (1541.6 km^2), Phokwane (833.9 km^2), and Sol Plaatje (1877.1 km^2) [42] (Figure 2).

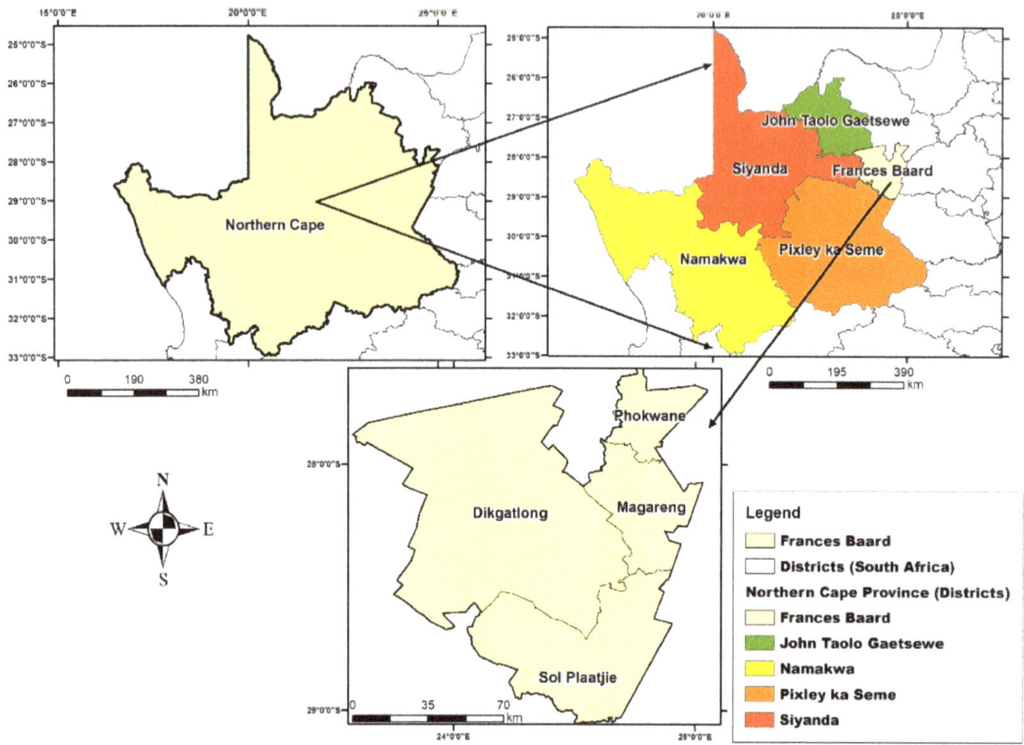

Figure 2. Maps of Northern Cape Province, the district municipalities of the Northern Cape, and the four local municipalities of Frances Baard District Municipality (FBDM). Source: FBDM [42].

The climate in the Northern Cape is arid and semi-arid. It is a large, dry area with a wide range of temperatures and topographical features. Rainfall is infrequent, ranging from 50 to 400 mm per year. The provinces' average annual rainfall/precipitation is 202 mm and is variable; for example, Kimberley experiences 497 mm of rainfall, Springbok, 195 mm, and Sutherland, 237 mm. Summer temperatures frequently rise above 40 °C. During winter, the average daytime temperatures are mild and may drop below 0 °C at night. Winter is usually frosty, with the southern area becoming bitterly cold, often experiencing snow and temperatures below −10 °C [40,43]. Evaporation levels exceed the annual average rainfall, which varies from 66 mm at Port Nolloth on the west coast to 414 mm at Kimberley, and 457 mm at Kuruman. The western areas, including Namaqualand, receive rainfall during the winter from April to September. The central, northern, and eastern parts of the province receive rain, mostly in the summer, from October to February [40].

3.2. Sample Size Determination and Sampling Procedure

The study employed a multistage sampling technique. In the first stage, the Northern Cape Province was chosen purposively because it represents the main livestock-producing province in South Africa. Additionally, the South African government declared the province a disaster area during this study [44,45]. The second stage of sampling involved the simple random selection of the FBDM using balloting. Within FBDM, four municipalities, namely, Phokwane, Magareng, Sol Plaatjie, and Dikgatlong, were purposively selected for sampling as the main livestock-producing municipalities.

A random sampling formula developed by Cochran [46], as well as Bartlett et al. [47], was used to determine the appropriate sample size for this study. Following a simple random formula, from a pool of 878 smallholder livestock farmers who applied for assistance from the national and local governments during the worst drought season ever recorded in South Africa, during the 2015/2016 crop season, 217 smallholder livestock farmers were selected. As part of the government's drought resilience activities, the government provided livestock feed, medication, improved access to resources, and provided training and information to help smallholder farmers become more resilient to drought.

3.3. Collection of Data

This study used both quantitative and qualitative methods. In addition to structured questionnaires, face-to-face interviews were conducted to collect data. Data were collected in both continuous and categorical forms on livestock production, assets, adaptive capacity, climate change, and social safety-net indicators. Face-to-face interviews were conducted, using a structured questionnaire, from October to December 2020. Ethical clearance was obtained from the University of the Free State.

3.4. Analytical Techniques

3.4.1. Agricultural Drought Resilience Index (ADRI)

The ADRI was calculated using PCA by aggregating livestock production in a normal year (WnPrn), livestock production during agricultural drought (WdPrd), the number of months in which a household consumes food produced by the household in a normal year (WcnMn), and the number of months in which a household consumes food produced by the household during an agricultural drought ($W_{cd}M_d$)(Table 1). Equation (1) is the formula for ADRI:

$$ADRI = Wt_nPr_n + Wt_dPr_d + Wt_{cn}Mo_n + Wt_{cd}Mo_d \tag{1}$$

where Wt represents each component, which is a weighted linear combination of variables determined by component loadings from the principal components, with a zero mean and unit variance.

Table 1. Principal component analysis of the agricultural drought resilience index (ADRI).

Variables	Communalities		Component Factors	Corr.ADRI
	Initial	extraction	1	
Mo_n	1	0.280	0.963	0.890
Mo_d	1	0.955	0.977	0.984
Pr_n	1	0.935	0.967	0.894
Pr_d	1	0.958	0.979	0.995

Total = 3.776. Chi-square = 2224.837; Bartlett's test of sphericity is significant at $p = 0.0000$; the Kaiser–Meyer–Olkin (KMO) test of sampling adequacy = 0.636; cumulative (%) = 94.402 and eigenvalue variances (%) = 94.402.

The SPSS software, Bartlett's test of sphericity, and the Kaiser–Meyer–Olkin (KMO) test were used to analyze the data. Four variables were examined: livestock production in a normal year (Prn), livestock production during an agricultural drought (Prd), the number of months in which a household consumes food during a drought year (Mod), and the number of months in which a household consumes food produced by the household in a normal year (Mon).

Based on these conceptual underpinnings, the study determined how agricultural drought resilience to food insecurity was related to coping strategies. As shown in Table 1, there was a high correlation among variables because they were measuring the same construct. There was no doubt that both the commonalities and the initial commonalities were all greater than 0.30, a positive sign.

Based on the eigenvalue analysis, one factor was extracted. In terms of the total variance explained, 94.402% of the components account for it. According to Bartlett's sphericity test, the null hypothesis that the inter-correlation matrix is an identity matrix is true. On the other hand, as the inter-correlation matrix was not derived from a population, the variable reduction is rejected. In terms of KMO statistics, the model had a KMO value of 0.636, while the Bartlett test of sphericity showed a significant result ($p = 0.000$, chi-square = 2224.837).

The ADRI can be written as follows (Equation (2)):

$$\text{ADRI} = \times 0.979 + \text{Prn} \times 0.967 + \text{M0d} \times 0.977 + \text{Mon} \times 0.963 \tag{2}$$

where:

ADRI: agricultural drought resilience index,

Prn: production of livestock in a normal year,

Prd: production of livestock in a drought year,

Mon: the period (number of months) in which the household consumed food in a normal year,

M0d: the period (number of months) in which the household consumed food in a drought year,

Numerical value: weights derived using PCA (component factors).

The ADRI of the study area was calculated using Equations (1) and (2). An ADRI value of greater than zero represents agricultural drought-resilient households, whereas an ADRI value of less than zero represents households that are not resilient (vulnerable) when faced with agricultural drought. According to the ADRI values, 79% (172) of livestock-farming households were not resilient to agricultural drought, while the remaining 21% (45) were resilient. Although agricultural drought is a frequent occurrence in the Northern Cape Province, it can have a significant impact on smallholder livestock farmers, and a lack of or delay in rainfall can lead to a drop in livestock production, resulting in food insecurity.

3.4.2. Household Food Insecurity Access Scale (HFIAS)

The Food and Nutrition Technical Assistance Project (FANTA) developed the HFIAS, which was used to link agricultural drought resilience with food insecurity [27]. The HFIAS score is a tool for determining food insecurity in households over the previous months. Therefore, to calculate each household's HFIAS score, nine "frequency of occurrence" questions are posed (Table 2).

Table 2. The nine "frequency of occurrence" questions.

No.	Frequency of Occurrence Questions
1	Concern about insufficient food
2	Unable to consume preferred foods
3	Consume a restricted variety of foods
4	Compelled to eat certain foods
5	Eat smaller meals
6	Eat fewer meals in a day
7	The household does not have any food of any kind
8	Go to bed hungry
9	Eat nothing for a whole day and night

Source: Author compilation, based on Coates et al. [27].

The answers to the nine questions above determine the household's food security. Higher-scoring households are more likely to be food-insecure. Based on the HFIAS scores, households are also classified into four categories: strongly food-secure, mildly food-insecure, moderately food-insecure, and severely food-insecure.

3.4.3. Structural Equation Modeling

A structural equation model was used to investigate the determinants and relationship between households' resilience to food insecurity and smallholder livestock farmers' coping strategies/adaptive capacity. The coping strategies included in the independent variables are sorted under each category of adaptive capacity, climate change, safety nets, and assets. Factor analysis models were used to measure the latent variables using observed variables, while regression models were used to model the relationship between latent variables [36,37,48]. Equation (3) depicts the structural equation model:

$$ADRI = f(ASS, ADC, SSF, CH) \tag{3}$$

where:

ADRI: agricultural drought resilience index,

ASS: assets including HFS (Herd/flock size- cattle, sheep, and goats), AA (agricultural assets- tractors, feeding equipment, livestock trailer, water tank, and corral system), and NAA (non-agricultural assets- house, television, chairs, radio, and bed),

ADC: adaptive capacity including perception, source of income (Incsource), migration and credit,

SSF: social safety nets including cash, training, food support, water rights, equipment, sanitary latrine, farm input,

CH: climate change including agricultural drought occurrence and intensity.

4. Results

4.1. Coping Strategies

When dealing with drought, farmers must have a coping strategy in place. Table 3 shows that the most common strategy for dealing with drought was selling livestock. Thus, selling livestock regulated the income fluctuations caused by drought. The smallholder livestock farming households utilized various coping strategies, ranging from selling livestock (21%) to leasing out their farms (1%). Smallholder livestock farmers used alternative land (20%), stored food (20%), requested food for their animals (16%), searched for other employment (6%), migrated (6%), raised drought-tolerant breeds (5%), received relief grants (3%), used their savings and investments (2%), and leased their farms (1%) as coping strategies (Table 3).

Table 3. Coping strategies adopted by smallholder livestock farmers in the Northern Cape Province of South Africa during a drought year.

Coping Strategies	%
selling livestock	21
alternative land	20
storing food	20
requested feed for their animals	16
searched for alternative employment	6
migrated	6
raised drought-tolerant breeds	5
received relief grants	3
savings and investments	2
leasing out their farms	1

Source: Author's compilation, based on the survey (2022).

4.2. Household Food Insecurity Access Scale (HFIAS)

The HFIAS was used to assess food insecurity. The respondents perceived food insecurity differently, depending on their level of spending power and financial well-being. The majority of respondents (71%) were concerned about not having enough food, while 62.7% ate limited amounts of food, and 60.4% ate smaller meals than they thought were necessary. More than half of the respondents (57.6%) ate fewer meals, and 55.3% ate what they did not want to eat. Less than half of the respondents (42.4%) reported not having food in the house, 36.4% went to bed hungry, and 34.1% went the entire day without eating. Food-secure households had higher resilience to food insecurity, whereas severely food-insecure households had lower resilience to food insecurity (Table 4).

Table 6. Results of the structural equation modeling.

Variables	Unstandardized Coefficient		Standardized Coefficient	Sig.
	ß	Std.error	ß	
Constant	11.37	2.09		
Assets (ASS)				
Herd/flock size (HFS)	3.44	0.68	0.33	0.00 ***
Agricultural assets (AA)	37.49	27.57	0.09	0.18
Non-agricultural assets (NAA)	−2.80	10.00	−0.02	0.78
Social safety nets (SSF)				
Cash	0.04	0.06	0.04	0.52
Training	0.10	0.06	0.12	0.09 *
Food support	0.06	0.06	0.08	0.30
Water rights	0.11	0.08	0.11	0.15
Garden equipment	0.27	0.11	0.20	0.012 **
Sanitary latrine	0.04	0.08	0.04	0.61
Farm input	0.12	0.06	0.15	0.032 **
Adaptive capacity (ADC)				
Perception	−0.15	0.06	−0.18	0.007 ***
Incsource	−0.24	0.13	−0.12	0.077 *
Credit	−0.54	0.16	−0.25	0.001 ***
Migration	0.06	0.11	0.04	0.60
Climate change (CH)				
Frequency	−0.05	0.03	−0.12	0.090 *
Intensity	−0.01	0.03	−0.02	0.83

*** 1%, ** 5%, * 10% significant. Source: Author's findings.

5. Discussion

According to the literature, households' responses to food insecurity include a variety of coping strategies. The Global Sustainable Development Report [49] asserts that people begin to rethink their consumption habits when anticipating a food shortage, rather than waiting until they are completely without food. Such situational changes in consumption habits are often viewed as short-term adjustments. Still, this can remain a normal habit, even when non-consumption-based strategies are employed. In particular, this is true when a community has faced long-term food insecurity, in terms of access and/or availability.

The majority of the respondents in this study indicated that they utilized livestock sales as coping strategies. This implied that smallholder farmers used livestock as a coping and adaptation mechanism because they sold livestock during agricultural droughts to enhance their resilience. Taking this into account, few respondents varied their livelihoods in any way, which left them vulnerable to drought issues. These findings are consistent with those of Acosta et al. [50], who investigated the role of livestock as a household coping strategy against climate shocks and discovered that livestock portfolios serve as a buffer against the effects of drought, supporting household income and consumption. To mitigate the impact of agricultural drought, it is necessary to diversify livelihood strategies through income-generating activities, both within and outside agriculture. As Kiani et al. [51] highlight, agricultural diversification raises farmers' adaptive capacity for the adoption of agricultural diversification and will enable them to generate tangible benefits by increasing their income through adopting sustainable agricultural livelihoods. These findings align with Bahta [10], who found that smallholder farmers sold livestock to cope with agricultural drought.

The ADRI found that most respondents were not able to cope with drought in agriculture. Therefore, governments and industry stakeholders should assist smallholder livestock farmers in improving their robustness. Assistance may include fodder, livestock medication, improving access to resources, and increasing the participation of smallholder farmers in drought-resilient agricultural activities, through training and information dissemination. This study concurs with Matlou and Bahta's [52] findings that most farming households in the Northern Cape Province were not drought-resistant. Furthermore, the findings are consistent with those of Adzawlaa et al. [53], who discovered that a lack of rainfall has a negative impact on farming household resilience.

The majority of respondents were concerned about not having enough food. This implied that most of the smallholder livestock farmers were vulnerable and needed assistance from governments and industry stakeholders to enhance their resilience. Hussein et al. [54] found similar results when households faced uncertainty about food availability. The authors found that 53.2% and 24.1% reported eating food that was insufficient in terms of quality and quantity, respectively. Half of the participants in the study (50.6%) also reported being unable to eat their preferred foods. Nearly a quarter (23.4%) reported eating smaller portions of meals, while 16.8% of households reported eating fewer meals. The overall food insecurity rate was 56.5%. Ansah et al. [55] reported similar findings, where high scores indicated that households had become more resilient, whereas low scores indicated that households had become more vulnerable and required additional assistance.

The structural equation modeling analysis revealed that assets, adaptive capacity, safety nets, and climate change indicators played a significant role in households' resilience to food insecurity. Consequently, a farming household's resilience to agricultural drought increased with the possession of additional assets. Households' safety nets increased when their assets increased; the more aware they were of climate change, the more resilient to agricultural drought they became. The results of this study are consistent with previous research indicating that more assets may make a household resilient to food insecurity [26,55–57]. The literature suggests that resilience is an essential step toward developing coping strategies and improving adaptive capacity [58].

Benefits that protect vulnerable households from food insecurity are social safety nets. The indicators of social safety nets positively impacted household resilience to food insecurity. Adzawlaa et al. [53], Dasgupta et al. [59], and Dejene and Cochrane [60] all agreed with this study's findings; however, Chakona and Shackleton [61] disagreed. Dasgupta et al. [59] highlighted the link between the social safety net and food insecurity. In the study by Adzawlaa et al. [53], social safety-net indicators significantly and positively impacted household resilience to food insecurity. Dejene and Cochrane [60] found that social safety nets are significant predictors of food insecurity. Chakona and Shackleton [61] found that there was no significant influence of social grants (social safety) on household food security as the funds were insufficient to fulfill all household members' needs.

There was a negative and significant impact of climate change (the frequency and severity of droughts) on household resilience to food insecurity. This implied that the dry and relentless climate, due to low annual precipitation, reduced livestock production in the province. The findings of this study concurred with those of Bahta and Myeki [62], who found that agricultural droughts impacted food production and food security.

This study only included smallholder livestock farmers in South Africa's Northern Cape Province and excluded smallholder crop farmers.

6. Conclusions

This study examined smallholder livestock farmers' coping strategies in the event of food insecurity shock, and the relationship between the types of coping strategies and their agricultural drought resilience to food insecurity in South Africa's Northern Cape Province. The most common strategy for dealing with drought was selling livestock, and the majority of smallholder livestock farmers (79%) were not drought-resilient. The indicators of assets, social safety nets, and adaptive capacity had a significant impact on household resilience to food insecurity. Climate change indicators had a negative and significant impact on households' abilities to cope with food insecurity. In other words, the greater the assets (such as livestock) of a farming household, the more resilient it would be to agricultural droughts. The findings also revealed that households benefited from social safety nets.

The Northern Cape Province has a hot summer climate combined with low rainfall (200 mm annually). Due to the dry and unforgiving climate, livestock production is reduced. In response to this issue, the government and stakeholders need to strengthen drought-relief programs to enhance the inhabitants' resilience to food insecurity, by focusing on less resilient smallholder farmers to increase their persistency, adaptability, and how they cope with agricultural drought.

The findings implied that firstly, to mitigate the impact of agricultural drought, it is necessary to diversify livelihood strategies via income-generating activities both within and outside agriculture. Secondly, the more assets a farming household owned, the higher their resilience to agricultural drought. The findings further indicated that benefiting from the social safety nets provided support for individual households. Therefore, governments and industry stakeholders should assist smallholder livestock farmers in improving their robustness. Assistance may include fodder, livestock medication, improving access to resources, and increasing the participation of smallholder farmers in drought-resilience agricultural activities through training and information dissemination.

The study recommends that improving policy is crucial to enhance the resilience of smallholder livestock farmers. The policy should not be limited to drought relief but should also improve various coping strategies. This includes encouraging smallholder farmers to raise drought-tolerant breeds, along with the acquisition of more resources and assets. The government needs to work with stakeholders to enhance the resilience of smallholder farmers by supporting the less resilient farmers. Improved access to agricultural credit and farm inputs and, subsequently, the accumulation of assets will reduce their vulnerability to food insecurity. In addition, the government should address off-farm employment as a source of income, and strengthen social safety nets, including providing training and disseminating information to smallholder farmers regarding drought preparation. As a result, these policies will aid smallholder farmers in being more resilient in times of climatic shock.

In general, this study's findings suggested that governments and non-governmental policymakers should focus on improving the resilience of smallholder farmers by expanding their access to resource bases, reducing food insecurity, and delivering timely drought relief.

This study used primary data collected through face-to-face interviews to assess the impact of agricultural drought on the resilience of smallholder farming households in the Northern Cape Province. The COVID-19 pandemic caused some data collection delays, and the language barrier was also a limitation. The most widely spoken languages in the Northern Cape Province are Afrikaans and Setswana (local South African languages), making communication between the researcher and the respondents difficult.

The study recommends that future research in developing countries should concentrate on the impact of agricultural drought on nutritional security for smallholder and commercial livestock and crop farmers, which was beyond the scope of this study.

Funding: The National Research Foundation (NRF) of South Africa funded this research; grant number TTK170510230380.

Institutional Review Board Statement: The study obtained an ethical clearance certificate from the University of the Free State General/Human Research Ethics Committee (GHREC), and the reference number is UFSHSD2020/0359/2704.

Informed Consent Statement: Informed consent was obtained from all subjects involved in the study.

Data Availability Statement: Data is available upon request from the corresponding author (Y.T.B).

Acknowledgments: I acknowledge and thank the National Research Foundation (NRF), Thuthuka funding instrument, for funding the project "Household resilience to agricultural drought in the Northern Cape province of South Africa" Contract Number/Project Number (TTK170510230380). I also acknowledge the contribution of Vuyiseka A. Myeki and L van der Westhuizen (Science Writer) for language-editing this manuscript.

Conflicts of Interest: The author declares no conflict of interest.

Glossary

AA	Agricultural assets
ADC	Adaptive capacity
ADRI	Agricultural Drought Resilience Index
ASS	Assets
Bartlett's test of sphericity	Compares an observed correlation matrix to the identity matrix
CH	Climate change
Chi-square	A statistical test used to examine the differences between categorical variables from a random sample in order to judge goodness of fit between expected and observed results.
$^{\circ}$C	Degrees Celsius
COVID-19	Coronavirus disease 2019
DAFF	Department of Agriculture, Forestry and Fisheries
FANTA	Food and Nutrition Technical Assistance Project
FAO	Food and Agriculture Organization of the United Nations
FBDM	Frances Baard District Municipality
GOAL	An international humanitarian response agency
HFIAS	Household Food Insecurity Access Scale
HFS	Herd/flock size
IFAD	International Fund for Agricultural Development
IFPRI	International Food Policy Research Institute
Km^2	Square kilometer or kilometer squared
KMO	Kaiser–Meyer–Olkin
Mod	Number of months that a household consumed food in a drought year
Mon	Number of months that a household consumed food in a normal year
Mm	Millimetre
NAA	Non-agricultural assets
NRF	National Research Foundation
PCA	Principal component analysis
Prd	Production of livestock in a drought year
Prn	Production of livestock in a normal year
p-value	Measure of the probability that an observed difference could have occurred just by random chance
SDGs	Sustainable Development Goals
SLF	Sustainable livelihood framework
SPSS	Statistical Package for the Social Sciences
SSF	Social safety nets
Stats SA	Statistics South Africa
UNDP	United Nations Development Programme
UNICEF	United Nations International Children's Emergency Fund
UNDESA	United Nations Department of Economic and Social Affairs
USAID	United States Agency for International Development
Wt	Weight—the loading of components of the first principal weights determined
WcnMn	Weight for the number of months during which the household consumed food in a normal year, multiplied by the actual amount of food produced
WcdMd	Weight for the number of months during which the household consumed food in a drought year, multiplied by the actual amount of food produced
WdPrd	Weight of livestock production in during drought year multiplied by the actual number of livestock produced
WFP	World Food Program
WHO	World Health Organization
WnPrn	Weight of livestock production in a normal year, multiplied by the actual number of livestock produced

References

1. Van Dijk, M.; Morley, T.; Rau, M.L.; Saghai, Y. A meta-analysis of projected global food demand and population at risk of hunger for the period 2010–2050. *Nat. Food* **2021**, *2*, 494–501. [CrossRef]
2. Ayala, A.; Meier, B.M. A human rights approach to the health implications of food and nutrition insecurity. *Public Health Rev.* **2017**, *38*, 10. [CrossRef] [PubMed]
3. Food and Agricultural Organization of the United Nations (FAO); International Fund for Agricultural Development (IFAD); UNICEF, World Food Programme (WFP); World health Organization (WHO). *The State of Food Security and Nutrition in the World 2021: Transforming Food Systems for Food Security, Improved Nutrition and Affordable Healthy Diets for All*; FAO: Rome, Italy, 2021.
4. Statistics South Africa (Stats SA). *Towards measuring the extent of food security in South. Africa: An. Examination of Hunger and Food Adequacy/Statistics South. Africa*; Report No. 03-00-14; Statistics South Africa: Pretoria, South Africa, 2017.
5. Manning, G. Hunger in South Africa and Solutions. 2021. Available online: https://borgenproject.org/hunger-in-south-africa/#:~{}:text=Hunger%20ratios%20in%20South%20Africa,the%20start%20of%20the%20pandemic (accessed on 20 February 2022).
6. Sassi, M. Coping Strategies of Food Insecure Households in Conflict Areas: The Case of South Sudan. *Sustainability* **2021**, *13*, 8615. [CrossRef]
7. Farzana, F.D.; Rahman, A.S.; Sultana, S.; Raihan, M.J.; Haque, M.A.; Waid, J.L.; Choudhury, N.; Ahmed, T. Coping strategies related to food insecurity at the household level in Bangladesh. *PLoS ONE* **2017**, *12*, e0171411.
8. Lehmann-Uschner, K.; Kraehnert, K. *When Shocks Become Persistent: Household-Level Asset Growth in the Aftermath of an Extreme Weather Event. Discussion Paper 1759*; DIW Berlin, German Institute for Economic Research: Berlin, Germany, 2018.
9. Masipa, T.S. The impact of climate change on food security in South Africa: Current realities and challenges ahead. *J. Disaster. Risk Stud.* **2017**, *9*, a411. [CrossRef]
10. Bahta, Y.T. Smallholder livestock farmers coping and adaptation strategies to agricultural drought. *AIMS Agric. Food* **2020**, *5*, 964–982. [CrossRef]
11. Myeki, V.A.; Bahta, Y.T. Determinants of Smallholder Livestock Farmers' Household Resilience to Food Insecurity in South Africa. *Climate* **2021**, *9*, 117. [CrossRef]
12. Debessa, A.A.; Tolossa, D.; Denu, B. Analysis of the Nexus Between Coping Strategies and Resilience to Food Insecurity Shocks: The Case of Rural Households in Boricha Woreda, Sidama National Regional State, Ethiopia. In *Food Systems Resilience*; Ribeiro-Barros, A.I., Tevera, D., Goulao, L.F., Daniel, L., Eds.; IntechOpen: London, UK, 2022; pp. 1–15.
13. Ansah, I.G.K.; Gardebroek, C.; Ihle, R. Shock interactions, coping strategy choices and household food security. *Clim. Dev.* **2021**, *13*, 414–426. [CrossRef]
14. Ado, A.M.; Leshan, J.; Savadogo, P.; Baba, A. Households' Coping Strategies to Food Insecurity: Insights from a Farming Community in Aguie District of Niger. *J. Environ. Earth Sci.* **2018**, *8*, 1–8.
15. Amoah, L.N.A.; Simatele, M.D. Food Security and Coping Strategies of Rural Household Livelihoods to Climate Change in the Eastern Cape of South Africa. *Front. Sustain. Food Syst.* **2021**, *5*, 692185. [CrossRef]
16. World Economic Forum. *The Global Risks Report, 16th Edition Insight Report*. 2021. —Published by the World Economic Forum. Available online: https://www3.weforum.org/docs/WEF_The_Global_Risks_Report_2021.pdf (accessed on 7 May 2022).
17. Islam, S.N.; Winkel, J. *Climate Change and Social Inequality*. UNDESA Working Paper, No.152 ST/ESA/2017/DWP/152. 2017. New York, NY, USA. Available online: https://www.un.org/esa/desa/papers/2017/wp152_2017.pdf (accessed on 3 May 2022).
18. Luthar, S.S.; Cicchetti, D. The construct of resilience: Implications for interventions and social policies. *Dev. Psychopathol.* **2000**, *12*, 857–885. [CrossRef] [PubMed]
19. Keil, A.; Zeller, M.; Wida, A.; Sanim, B.; Birner, R. What determines farmers' resilience towards ENSO-related drought? An empirical assessment in central Sulawesi, Indonesia. *Clim. Chang.* **2007**, *86*, 291–307. [CrossRef]
20. Gwaka, L.; Dubihlela, J. The resilience of smallholder livestock farmers in sub-saharan Africa and the risks imbedded in rural livestock systems. *Agriculture* **2020**, *10*, 270. [CrossRef]
21. Alinovi, L.; Mane, E.; Romano, D. Towards the measurement of household resilience to food insecurity: Applying a model to Palestinian household data. In *Deriving Food Security Information from National Household Budget Surveys*; Sibrian, R., Ed.; Experiences, Achievement, Challenges; FAO: Rome, Italy, 2008; pp. 137–152.
22. Grist, N.; Mosello, B.; Roberts, R.; Hilker, L. Resilience. *The Sahel: Building Better Lives With Children*. ODI. Available online: https://cdn.odi.org/media/documents/9369.pdf (accessed on 10 January 2022).
23. Bhandary, P.; Ringler, C.; de Pinto, A. The resilence landscape. In *Building Resilience to Climate Shocks in Ethiopia*; IFPRI and UNDP food policy report; Koo, J., Thurlow, J., Eldidi, H., Ringler, C., de pinto, A., Eds.; International Food Policy Research Institute: Washington, DC, USA, 2019; pp. 14–24.
24. McCaul, B.; Flores, G.C.; Mencía, S. Resilience for social systems: R4s approach user guidance manual. In *Developed by GOAL Resilience Innovation and Learning Hub*; GOAL Resilience Innovation and Learning Hub: Dublin, Ireland, 2019.
25. Food and Agricultural Organization of the United Nations (FAO). *The State of Food Insecurity in the World*; FAO: Rome, Italy, 2017.
26. Melketo, T.; Schmidt, M.; Bonatti, M.; Sieber, S.; Müller, K.; Lana, M. Determinants of pastoral household resilience to food insecurity in Afar region, northeast Ethiopia. *J. Arid Environ.* **2021**, *188*, 104454. [CrossRef]
27. Coates, J.; Swindale, A.; Bilinsky, P. Household Food Insecurity Access Scale (HFIAS) for Measurement of Food Access: Indicator Guide (Version 3). In *Food and Nutrition Technical Assistance Project (FANTA)*; Academy for Educational Development: Washington, DC, USA, 2007.

28. Javadinejad, S.; Dara, R.; Jafary, F. Analysis and Prioritization the Effective Factors on Increasing Farmers Resilience Under Climate Change and Drought. *Agric. Res.* **2021**, *10*, 497–513. [CrossRef]
29. United States Agency for International Development (USAID). Economics of Resilience to Drought in Ethiopia, Kenya and Somalia. In *Resilience Evidence Forum Report*; USAID Center for Resilience: Washington, DC, USA, 2018.
30. Food and Agricultural Organization of the United Nations (FAO). *El Nino Response Plan 2016*; FAO: Addis Ababa, Ethiopia, 2018.
31. Sunday, N.; Kahunde, R.; Atwine, B.; Adelaja, A.; Kappiaruparampil, J. How Specific Resilience Pillars Mitigate the Impact of Drought on Food Security: Evidence from Uganda. Department of Agricultural, Food, and Resource Economics. In *Research Paper #31 of USAID and IFPRI*; Michigan State University: East Lansing, MI, USA, 2021.
32. Birhanu, Z.; Ambelu, A.; Tesfaye, A.; Woldemichael, K. Understanding resilience dimensions and adaptive strategies to the impact of recurrent droughts in Borana zone, Oromia region, Ethiopia: A grounded theory approach. *Int. J. Environ. Res. Public Health* **2017**, *14*, 118. [CrossRef]
33. Fan, S.; Pandya-Lorch, R.; Yosef, S. *Resilience for Food and Nutrition Security*; IFPRI: Washington, DC, USA, 2014.
34. Hoddinott, J. Looking at development through a resilience lens. In *Resilience for Food and Nutrition Security; Part of an IFPRI, Book*; Fan, S., Pandya-Lorch, R., Yosef, S., Eds.; IFPRI: Washington, DC, USA, 2020; pp. 19–26.
35. Mulugeta, H. Green Famine in Ethiopia: Understanding the Cause of Increasing Vulnerability to Food Insecurity and Policy Response in the Southern Ethiopia Highlands. Ph.D. Thesis, University of Sussex, Brighton, UK, 2014.
36. Guyu, D.; Muluneh, W. Household resilience to seasonal food insecurity: Dimensions and magnitude in the "green famine" belt of Ethiopia. *Appl. Sci. Rep.* **2015**, *11*, 125–143.
37. Alinovi, L.; D'Errico, M.; Mane, E.; Romano, D. Livelihoods strategies and household resilience to food insecurity: An empirical analysis to Kenya. In *The European Report of Development in Dakar, Senegal, 28–30 June*; FAO: Rome, Italy, 2010; pp. 1–52.
38. Frankenberger, T.; Langworthy, M.; Spangler, T.; Nelson, S. Enhancing Resilience to Food Security Shocks. TANGO International, Inc. 2012. Available online: https://www.fsnnetwork.org/sites/default/files/revised_resilience_paper_may_28.pdf (accessed on 5 February 2022).
39. Pasteur, K. *From Vulnerability to Resilience: A Framework for Analysis and Action to Build Community Resilience*; Bourton on Dunsmore, Rugby: Practical Action; Practical Action Publishing Ltd: Warwickshire, UK, 2011.
40. uMoya-NILU Consulting (Pty) Ltd. *Air Quality Management Plan for the Northern Cape. Air Quality Baseline Assessment Report for Northern Cape Department of Environment and Nature Conservation*; Report No. uMN072-2017; uMoya-NILU Consulting: Kimberley, South Africa, 2017.
41. Bradshaw, D.; Nannan, N.; Laubscher, R.; Groenewald, P.; Joubert, J.; Nojilana, B.; Norman, R.; Pieterse, D.; Schneider, M. Mortality Estimates for Northern Cape Province. Available online: https://www.samrc.ac.za/sites/default/files/files/2017-07-03/northerncape.pdf (accessed on 14 February 2022).
42. Frances Baard District Municipality (FBDM). *Frances Baard Municipal District in the Northern Cape*; FBDM: Kimberly, South Africa, 2019; Available online: https://municipalities.co.za/map/134/frances-baard-district-municipality (accessed on 29 August 2021).
43. Statistics South Africa (Stats SA). Profile: Northern Cape. 2018. Available online: http://cs2016.statssa.gov.za/wp-content/uploads/2018/07/NorthernCape.pdf (accessed on 31 August 2021).
44. Statistics South Africa (Stats SA). *Community Survey 2016: Provincial Profile Northern Cape*; Report No. 03-01-14; Statistics South Africa: Pretoria, South Africa, 2017.
45. Department of Agriculture, Forestry and Fisheries (DAFF). Drought status in the agriculture sector. In *Portfolio Committee on Water and Sanitation*; DAFF: Pretoria, South Africa, 2018.
46. Cochran, W.G. *Sampling Techniques*, 3rd ed.; JohnWiley and Sons: New York, NY, USA, 1977.
47. Bartlett, J.E.; Kotrlik, J.W.; Higgins, C.C. Organizational research: Determining the appropriate sample size in survey research. *Inf. Techn. Learn. Perfor. J.* **2001**, *19*, 43–50.
48. Bollen, K.A. *Structural Equations with Latent Variables*; Wiley and Sons: New York, NY, USA, 1989.
49. Global Sustainable Development Report. *The Future is now-Science for Achieving Sustainable Development*; United Nations: New York, NY, USA, 2019.
50. Acosta, A.; Nicolli, F.; Karfakis, P. Coping with climate shocks: The complex role of livestock portfolios. *World Dev.* **2021**, *146*, 105546. [CrossRef]
51. Kiani, A.K.; Sardar, A.; Khan, W.U.; He, Y.; Bilgic, A.; Kuslu, Y.; Raja, M.A.Z. Role of Agricultural Diversification in Improving Resilience to Climate Change: An Empirical Analysis with Gaussian Paradigm. *Sustainability* **2021**, *13*, 9539. [CrossRef]
52. Matlou, R.C.; Bahta, Y.T. Factors influencing the resilience of smallholder livestock farmers to agricultural drought in South Africa: Implication for adaptive capabilities. *J. Disaster. Risk Stud.* **2019**, *11*, a805.
53. Adzawlaa, W.; Azumahb, S.B.; Anani, P.Y.; Donkoh, S.A. Analysis of farm households' perceived climate change impacts, vulnerability and resilience in Ghana. *Sci. Afr.* **2021**, *8*, e00397. [CrossRef]
54. Hussein, F.; Ahmed, A.; Muhammed, O. Household food insecurity access scale and dietary diversity score as a proxy indicator of nutritional status among people living with HIV/AIDS, Bahir Dar, Ethiopia. *PLoS ONE* **2018**, *13*, e0199511. [CrossRef] [PubMed]
55. Ansah, I.G.; Gardebroek, C.; Ihle, R. Resilience and household food security: A review of concepts, methodological approaches and empirical evidence. *Food Secur.* **2019**, *11*, 1187–1203. [CrossRef]
56. Matlou, R.; Bahta, Y.T.; Owusu-Sekyere, E.; Jordaan, H. Impact of Agricultural Drought Resilience on the Welfare of Smallholder Livestock Farming Households in the Northern Cape Province of South Africa. *Land* **2021**, *10*, 562. [CrossRef]

57. Cheteni, P.; Mokhele, X. Small-scale livestock farmers' participation in markets: Evidence from the land reform beneficiaries in the central Karoo, western Cape, South Africa. *S Afr. J. Agric. Ext.* **2019**, *47*, 118–136. [CrossRef]
58. Karman, A. Flexibility, coping capacity and resilience of organizations: Between synergy and support. *J. Organ. Change Manag.* **2020**, *33*, 883–907. [CrossRef]
59. Dasgupta, S.; Robinson, E.J.Z. Food Insecurity, Safety Nets, and Coping Strategies during the COVID-19 Pandemic: Multi-Country Evidence from Sub-Saharan Africa. *Int. J. Environ. Res. Public Health* **2021**, *18*, 9997. [CrossRef]
60. Dejene, M.; Logan Cochrane, L. Safety nets as a means of tackling chronic food insecurity in rural southern Ethiopia: What is constraining programme contributions? *Can. J. Dev. Stud.* **2022**, *43*, 157–175. [CrossRef]
61. Chakona, G.; Shackleton, C.M. Food insecurity in South Africa: To what extent can social grants and consumption of wild foods eradicate hunger? *World Dev. Perspect.* **2019**, *13*, 87–94. [CrossRef]
62. Bahta, Y.T.; Myeki, V.A. The Impact of Agricultural Drought on Smallholder Livestock Farmers: Empirical Evidence Insights from Northern Cape, South Africa. *Agriculture* **2022**, *12*, 442. [CrossRef]

Article

Income Change and Inter-Farmer Relations through Conservation Agriculture in Ishikawa Prefecture, Japan: Empirical Analysis of Economic and Behavioral Factors

Yoshitaka Miyake [1], Shota Kimoto [2], Yuta Uchiyama [3] and Ryo Kohsaka [1,*]

[1] Graduate School of Environmental Studies, Nagoya University, Nagoya 464-8601, Japan; miyake.yoshitaka.m8@f.mail.nagoya-u.ac.jp
[2] Ishikawa Prefectural Government, Kanazawa 920-8580, Japan; kshota@pref.ishikawa.lg.jp
[3] Graduate School of Human Development and Environment, Kobe University, Kobe 657-8501, Japan; yuta.uchiyama@landscape.kobe-u.ac.jp
* Correspondence: kohsaka@hotmail.com

Abstract: Conservation agriculture, also known as environment-friendly agriculture, is expected to contribute to global climate change mitigation and biodiversity conservation. To understand the effect of conservation agriculture on farmers and identify those factors, such as farmers' income change, that might affect practices of conservation agriculture, perceptions, and output, this study examined farmers' economic and behavioral factors, motivation, and satisfaction. We surveyed 51 farmers who are receiving subsidies to practice conservation agriculture in Ishikawa Prefecture, Japan. The survey is one of the first prefectural-scale studies that combines unique quantitative analysis of motivation and satisfaction levels (e.g., behaviors) in temporal sequence from the initial to current time to practice conservation agriculture. Our results showed that years of experience, trade with a retail shop, and the farmer's age can affect income change. With regard to social factors, the satisfaction of their fellowship with other farmers practicing conservation agriculture was also significantly correlated with income change. Simultaneously, this category of satisfaction was difficult to attain compared to the other categories. Thus, greater effort is needed to enhance support networking among conservation farmers. Furthermore, the work presented here also provides the opportunity for future research on temporal and spatial questions surveying economic and behavioral effects with consideration of the heightened policy promotion and entrance of large retail industries.

Keywords: organic agriculture; environment-friendly agriculture; regression tree; attitude; motivation; satisfaction

Citation: Miyake, Y.; Kimoto, S.; Uchiyama, Y.; Kohsaka, R. Income Change and Inter-Farmer Relations through Conservation Agriculture in Ishikawa Prefecture, Japan: Empirical Analysis of Economic and Behavioral Factors. *Land* **2022**, *11*, 245. https://doi.org/10.3390/land11020245

Academic Editor: Elisa Marraccini

Received: 17 January 2022
Accepted: 3 February 2022
Published: 6 February 2022

1. Introduction

Conservation agriculture, also known as environment-friendly agriculture, is regarded as a farm-level attempt to use agricultural practices that contribute to the protection of the environment via biodiversity conservation, water quality enhancement, and climate change mitigation and adaptation [1–3]. Thus, this study utilizes a broader-focus definition of conservation agriculture than the definition used by the Food and Agriculture Organization of the United Nations, which primarily emphasizes the control of soil erosion [4]. The intention to use a definition with a broader focus is closely related to Direct Payment for Conservation Oriented Agriculture (*Kankyō Hozengata Nōgyō Chokusetsu Shiharai Kōfukin* in Japanese), the Japanese policy of direct payment focusing on the reduction of chemical fertilizer and pesticide while supporting agricultural practice to conserve biodiversity with the recent addition of climate change mitigation measures [5]. Thus, the use of the term, conservation agriculture, is appropriate to analyze agriculture to preserve the environment in Japan, especially biodiversity conservation. The agricultural methods also include low-input agriculture to decrease agriculture's energy dependence [6–8]. The environmental benefit of integrated agriculture is relevant in this study, while the practice to integrate crop and livestock production for the cyclical exchange of input and output is not necessarily the focus of this inquiry [9,10]. As such, conservation agriculture is gaining salience in policy priorities as well. As a recent example, the European Union (EU) announced the "Farm to Fork Strategy" to promote a sustainable agri-food system, which includes conservation agriculture [11]. Similarly, the Japanese government is renewing its interest in conservation agriculture through a strategy announced in May 2021 entitled "Strategy for Sustainable Food Systems" (referred hereafter as MeaDRI: *Midori no Shokuryō Sisutemu Senryaku*), which aims to balance economic production and environment sustainability, pursue climate change mitigation, and decrease the dependence on fossil fuels in agriculture [12]. Via this strategy, Japan aims to promote conservation agriculture as an agricultural practice from individual to community levels to counter climate change and preserve biodiversity [5,12,13]. The strategy promotes this type of agriculture including various numerical targets (i.e., the extent of organic agriculture to reach 25% by 2050). In this study, conservation agriculture is introduced as an agricultural practice that includes reduction of chemical pesticides and fertilizers, promotion of organic agriculture, and winter-flooded rice fields, which ultimately lead to biodiversity conservation. It is regarded as a synergistic measure with multiple benefits for climate change mitigation, water pollution, and biodiversity conservation. However, the effects of the strategy and the development of conservation agriculture are still at an early stage in Japan. Besides the national and international strategies to mitigate carbon emission, factors on the farmers' side should be studied for the expansion of conservation agriculture if the strategies are to be translated into actions. The promotion of the current strategies with lofty goals demands a comprehensive quantitative study of farmers engaged in conservation agriculture as the foundation to conduct the new policies.

While the economic benefit of conservation farmers has always attracted the interest of scholars [14–16], recent studies have focused more on the social and cognitive behavioral changes in farmers who have adopted conservation agriculture [17]. Previous studies surveying farmers have mainly focused on the identification of factors that drive the adoption of methods for conservation agriculture [18–22]. According to the literature review conducted by Mozzato et al. [1], these studies initially analyzed the socio-economic variables of farmers and farm structures. They further aimed to capture the impact of behavioral factors, including social factors and attitudes such as motivation [18]. Meanwhile, the work of Bouttes et al. [13] and Dessart et al. [17] has focused more on additional behavioral factors, such as learning and the perception of cost and benefit. It is also critical to understand the satisfaction of practicing farmers at a micro level in local practices to understand the balance of environment, economy, and chemical or energy use. However, this type of study is very rare in the area of the so-called Monsson Asia, with the climate condition containing high humidity. To expand conservation agriculture in the area has been shadowed by the struggle of high cost resulting in low economic benefits.

Reflecting these previous findings, we focused more on the processes used by farmers to develop conservation agriculture. Besides the economic benefit, this study examined the factors underlying and/or related to satisfaction levels, which can further analyze the attitudinal and behavioral factors of farmers starting and continuing conservation agriculture. The study focused on the analysis of conservation farmers in Ishikawa Prefecture, Japan, especially the ones who benefited from policies on conservation agriculture. While there is undoubtedly a public interest for such practices to advance environmental and social goals, the continuity and expansion of these practices hinge on the motivations of "individual" farmers. As for the policy implications, this study aims to fill the gap among the policy's guiding principles, the farmers' motivations, and the economic influence on the ground through intensive surveys of farmers in a Japanese prefecture. In the following sections, we provide the results of review of existing studies and show the methods used, including sample size and statistical analysis. After an analysis of the results, we discuss the factors affecting farmers' income change and the topics related to the satisfaction of farmers, before providing a conclusion.

2. Existing Studies on Farmers and Conservation Agriculture

Recent studies on conservation agriculture have focused on those factors that led to the start of this agricultural practice by surveying farmers in specific locations and exploring the applicability of the results on a larger scale [1,17,23]. These studies analyzed factors such as the farmers' demographic characteristics, farm types, and local farm structure [1]. These become the base to predict the development of conservation agriculture [24]. One of the frontiers in this research area is the analysis of spatial attributes and interpersonal relationships, such as social capital [1,23,25]. The behavioral approach specifically focused on socio-psychological characteristics and learning [17]. Dessart et al. [17] classified these behavioral parameters into: (1) dispositional factors, which are stable personal characteristics (e.g., beliefs and basic preferences); (2) social factors, which are linked to social relationships (e.g., peer pressure) that affect behaviors; and (3) cognitive factors, which are concerned with learning and analysis to consider decision-making components, such as factors, results, and benefits. The systematic categorization of behavioral factors could encourage a detailed discussion and application of these factors, which are frequently interpreted as farmers' general preferences. Besides the categorical attributes of farmers and their farms, their will for social development and environmental conservation can affect the adoption of conservation agriculture [26]. In summary, the status of these personal, farm, and regional factors should be further explored to facilitate the adoption and sustenance of conservation agriculture.

Conservation agriculture has also been studied for biodiversity conservation and the enhancement of ecosystem services in addition to lower input and reduced tillage [27,28]. To explore the factors that facilitated the adoption of these methods and the status of the farmers, Japanese scholars have mostly conducted qualitative studies with minimal quantitative application, surveying farmers engaged in conservation and organic agriculture [29–31]. Fujita and Hatano [30] conducted a survey with comprehensive questions to capture the status of organic farmers in Japan. The study analyzed farmers' personal and family characteristics, farm types, income change, sales channels, and motivations to start organic agriculture. Similarly, Oda and Kiminami [29] surveyed farmers conducting conservation agriculture in Sado Island, Niigata Prefecture, which is a critical location for conserving the Japanese crested ibis (*Nipponia nippon*). Besides questions about socio-demographics, management, and the motivation explained above, they included questions about the farmers' engagement, joy of agriculture, and care for creatures. Uenishi [31] focused on the role of extension in promoting the adoption and further development of conservation agriculture by comparing two known cases in Japan: one in Sado Island and the other in Toyooka City, Hyogo Prefecture, which is known for the preservation of storks. Uenishi [31] claimed that enhancing farmers' understanding of cultivation standards, support, and marketing promoted adoption in the initial stage. Ensuring a sufficient price premium could further encourage conservation agriculture [31]. In summary, Japanese scholars have attempted to capture the adoption of conservation agriculture in cultural, environmental, and academic settings. From the empirical data, the economic benefit and satisfaction that farmers derive from conservation agricultural practices appears to be the major challenge for the adoption and development of these practices in Japan.

To date, academic discussions on farmers practicing conservation agriculture have mainly concentrated on the factors that led to the expansion of this agricultural method [18–22]. Less attention has been paid to the effects of involvement in conservation agriculture, such as income change and satisfaction. A choice of market channel can affect income [16], although farmer satisfaction has not been discussed. Alternatively, the so-called conventionalization in organic agricultural markets can lead to the marginalization of small farmers due to decreased prices and income caused by the involvement of agribusinesses and large supermarket chains of developed countries in the development of organic agriculture [32,33].

Fujita and Hatano [30] interviewed organic farmers in Japan regarding their income increase and the stability of management. They found that new farmers generally experienced income growth as, at the time of the survey, their median income had increased from a range of JPY 0.5 to 1 million to JPY 2 to 4 million from the time they started organic agriculture. Simple regression analysis found a weak correlation with years of experience. Emotional aspects, such as the joy of and care for the environment, were also covered by Oda and Kiminami [29], as mentioned earlier. Additionally, the socio-cultural situation of the development of conservation and organic agriculture has also been documented in descriptive books, including Suzuki [34] and Arai et al. [35]. Arai et al. identified limited sales channels and low income from organic agriculture as a bottleneck for Shirakawa Village, which is located in a mountainous area in Japan. Thus, the studies identifying the factors that contribute to the spread and sustenance of conservation agriculture are needed.

Along with industrialization and urbanization, agriculture and its related land use are associated with an increase in CO2 emissions [12,36]. Although agriculture is regarded as a contributing factor for CO2 emissions in developed countries [37,38], Anwar et al. [36] noted the opposite effect in developing countries and included conservation agriculture in his policy recommendations. Mitigation measures for CO2 emissions, which include renewable energy use, forest maintenance [36,39], and the topic of our study, conservation agriculture, are also now gaining salience.

As for farmers practicing conservation agriculture, studies have mainly focused on factors underlying the adoption of this agricultural method. In contrast, studies focusing on the effects of conservation agriculture on farmers are relatively scarce. Thus, this study focuses on the economic and behavioral effects of conservation agriculture by examining income change and satisfaction levels, among others. Although a range of items related to farmer satisfaction exists, the level of satisfaction can be further studied in relation to farm economy or efficiency [40]. The findings of this work will reinforce our understanding of practices of conservation agriculture, as well as provide policy implications from a broader perspective.

3. Methods

This study used a questionnaire survey of farmers. Because the prefecture provided the list of participants in relative policy schemes for the sampling frame, the results are robust enough to analyze farmers with evidence showing that they are implementing conservation agriculture in the prefecture. The response datasets were analyzed with a regression tree on income change and a multiple comparison procedure regarding their relationship with satisfaction items.

3.1. Study Site and Sample Selection

Our study site is Ishikawa Prefecture, which is located in the central northern part of Honshu Island. Kanazawa, the capital city, is known for its fine crafts and rich feudal history. The population of the prefecture and the city were 1.1 million and 463,583, respectively, in 2020 [41]. The number of farm managements in the prefecture was 13,636 in 2015 [42]. Farm managements cultivated 32,367 hectares, with an average size of 2.3 hectares. Traditional vegetables produced around the city and its northern region, the Noto Peninsula, are widely available in the prefecture at various venues including restaurants and morning markets [43]. The Noto Peninsula is registered as a Globally Important Agricultural Heritage System for its historically sustainable socio-ecological system of agriculture and fisheries, Satoumi and Satoyama, which are symbolized by rice terraces [44–46]. According to the 2015 Census of Agriculture and Forestry, there were 4017 conservation farmers in the prefecture, comprising 29.5% of all managements in the prefecture. This percentage is the 32nd highest in Japan [42]. However, the census data is based on the answers of individual farmers, who were not required to provide evidence to show that they were implementing conservation agriculture. Furthermore, the prefecture has also displayed promising trends in conservation agriculture. The share of organic dry fields of the prefecture's total amount of dry fields was 3.0% in 2018, the highest in Japan [47]. With such relative advantages of conservation agriculture within past and recent agricultural development in the area, this study can contribute to the development of conservation agriculture at the regional and national levels.

This study initially identified potential respondents through comprehensive documents retrieved from Ishikawa Prefecture's official bodies that list those farmers who are beneficiaries of conservation agriculture policies. To receive the subsidies attached to the policies, farmers need to submit evidence to verify their practice of conservation agriculture. These policies include: Specially Cultivated Agricultural Products (*Tokubetsu Saibai* in Japanese), Direct Payment for Conservation Oriented Agriculture, JAS (Japanese Agricultural Standards) Organic Standards certified by the prefecture, and *Ikimono Genki Mai Ninshō* (Certificate of Rice with Active Creatures). The four policies have 38, 98, 19, and 9 people or entities involved, respectively. Because farmers can form a group to apply for a policy and obtain certificates, all names on the lists may not clearly indicate whether they represent an individual farmer, a farmer group, or a corporate. As respondents from farmer groups and corporations basically represent an opinion of their organizations, this study regarded all respondents as individual farmers. The study assumes a respondent reflects the general characteristics and opinions of the group to answer the questionnaire.

After correcting the overlaps of farmers registering for multiple schemes and removing the farmers whose addresses were not on the beneficiary conservation farmers list, the names and addresses of 73 farmers practicing conservation agriculture in Ishikawa Prefecture were identified. In August 2020, we mailed each a questionnaire, clearly stating the research purpose while highlighting the voluntariness of participation. A total of 51 responses were collected before the December 2020 deadline, with 44 responses containing complete attributes to conduct statistical inference between income change and the items on satisfaction through conservation agriculture. Additionally, we interviewed three farmers by telephone in order to understand the reasons for their answers on their satisfaction with direct payment policies.

In this way, the study obtained a sampling frame of 73 conservation farmers. It assumed these farmers consisted of the population of conservation farmers in Ishikawa Prefecture. Their practice is rather certain as they can provide evidence of conservation agriculture. They are also suitable to analyze and discuss the adoption of policies to promote conservation agriculture. The sampling frame was possibly the best in practice to represent conservation farmers in Ishikawa Prefecture to discuss policy promotion. Furthermore, 44 responses were above the appropriate sample size of 40.70 calculated using Cochran's formula [48,49]. The calculation was performed using a confidence level of 95% and sampling error of 10%. The response rate to answer both the question on income change and those on satisfaction was derived from 44/73 (about 0.603). The population size for the analysis was 73.

3.2. Survey Questionnaire

The questionnaire used in this study included questions on the profiles and characteristics of the farmers, agricultural types, practices of conservation agriculture, sales and income, certificate types, motivations to start and satisfaction achieved through conservation agriculture, and satisfaction with policies at different times (Table 1).

3.2.1. Sales Channels and Income Change

Farmers were asked whether there were any changes in their sales channels before and after they started conservation agriculture. The choice of sales channels included: (1) direct consumer sales; (2) direct sales stores; (3) schools; (4) agricultural cooperatives; (5) consumer cooperatives; (6) processors; (7) wholesalers, except for agricultural and consumer cooperatives; (8) retail shops; (9) restaurants; (10) internet sales; and (11) others. Direct sales stores usually sell the products that a farmer supplies in person. The stores can be owned by local governments, agricultural cooperatives, or farmers themselves. These options were based on items listed in Fujita and Hatano [30] and MAFF [50] to reflect the situation of the agricultural and food markets in Japan and to enable comparisons. If many farmers chose retail shops as a sales channel and showed a low level of income increase, it would indicate that the market and distribution could be affected by conventionalization [32,33]. A high level of market intrusion can thus marginalize the production of small farmers and decrease their income.

Table 1. Ishikawa Prefecture conservation agriculture questionnaire items.

Category	Item	Answer Method
Individual	Age	Years
	Years of practicing conservation agriculture	Years
	Prospect of a successor	Multiple choice
Agriculture	Five best-sold crops and the sizes of the area on which each crop is produced	Crop name and land area by hectare
	Practices of conservation agriculture	Multiple answers possible
Sales and income	Annual sales	JPY 10,000
	Sales channels before starting conservation agriculture	Multiple answers possible
	Sales channels after starting conservation agriculture	Multiple answers possible
	Income change with conservation agriculture on land with certification and without certification	Multiple choice
Certification	Certificate type	Multiple answers possible
Motivation	Motivation to start conservation agriculture	List three items in order of relevance
Satisfaction	Satisfaction from conservation agriculture	For each item, a five-point scale ranging from very satisfied, satisfied, neither satisfied nor dissatisfied, somewhat dissatisfied, to dissatisfied
Policy	Satisfaction with policy at different times	For policy, a scale ranging from very satisfied to dissatisfied. Satisfaction in Year 1 and Year 4 or longer in conservation agriculture

Income changes were classified as slight or large changes (increase or decrease). The intervals for increase were less than 5%, 5% to less than 10%, 10% to less than 15%, and 15% or more. The intervals for decrease were less than 10%, 10% to less than 20%, 20% to less than 30%, 30% to less than 50%, and 50% or more. To analyze the causes and effects of income change, the intervals were simplified to increase, same, and decrease, which were denoted as 1, 0, and −1, respectively.

3.2.2. Certification

We covered the types of certificates that these farmers possess. These included Rice with Active Creatures, which is a certificate for specially cultivated agricultural products, JAS Organic Standards, Good Agricultural Practice (GAP), Hazard Analysis Critical Control Point (HACCP), Geographical Indications (GI), and Regional Collective Trademark (RCT). A farmer qualifies for a certificate for specially cultivated agricultural products if they decrease the frequency of applying certain chemicals to 50% or less compared with conventional agriculture and the nitrogen content of chemical fertilizers to 50% or less [51]. GAP certifies a farmer for sustainable attempts in production processes [52]. HACCP is a guideline for hygiene control in the handling of food [53]. GI registers locally or regionally unique products for intellectual property protection [52]. The RCT system allows the registration of trademarks linking a place name and a general commodity or service name [54].

3.2.3. Motivation to Start Conservation Agriculture

Farmers were asked to rank three items from a given list (adopted from Fujita and Hatano [30]) that would best describe their motivations to start conservation agriculture. The list included: (1) production of safe and secure foods; (2) production of high-quality foods; (3) interest in environmental protection and biodiversity; (4) disliking pesticides; (5) recommendation or advice from other practitioners; (6) inspiration from books, magazines, TV, or radio; (7) taking over a parent's agricultural methods; and (8) others.

3.2.4. Satisfaction from Conservation Agriculture

Farmers were asked about the satisfaction level for multiple aspects of conservation agriculture. The farmers' choices ranged from 5—very satisfied to 1—dissatisfied. Satisfaction levels were asked in relation to: (1) production of safe and secure foods; (2) production of high-quality foods; (3) protection of the environment and biodiversity; (4) fellowship with farmers practicing conservation agriculture; (5) interest and ambition in agriculture and agricultural methods; and (6) family relationship. The first three items were adjusted to correspond with the items provided in the motivation question.

3.2.5. Satisfaction with Policy at Different Times

We further analyzed farmers' satisfaction with three different supporting schemes at different periods (first to fourth year after the introduction of conservation agriculture). Farmers were asked to choose a level of satisfaction with each policy across each year of conservation agriculture. Their level of satisfaction was chosen based on a five-point scale: 1—dissatisfied to 5—very satisfied. Three direct payment schemes were involved, namely the Multifunctional Payment, which supports cooperative activities contributing to multifunctionality [5]; Direct Payment for Hilly and Mountainous Area, which attempts to pay the cost difference between the maintenance of a flat agricultural environment and a mountainous environment [5,55]; and Direct Payment for Conservation Oriented Agriculture, which supports farmers with the cost increase associated with agricultural activities that contribute to the protection of the environment [5].

3.3. Quantitative Analysis Methods

3.3.1. Regression Tree

This study estimated the factors affecting income change using a regression tree. The dependent variable was income increase, income decrease, or no change. The independent variables included farmers' individual information, the information on agriculture (including the practice of rice agriculture and the practices of conservation agriculture), annual sales, marketing channels before and after starting conservation agriculture, and whether they had a certification.

To identify the key factors, a regression tree calculates the entropy for each independent variable. The smallest entropy determines the value of a certain variable as a threshold to divide the samples. The calculation continues until the comparison to find a smaller entropy ends. Limiting the number of iterations to decide a threshold could simplify a tree. In the same way, the Gini impurity could decide the threshold.

3.3.2. Bonferroni Adjustment

This study conducted a multiple comparison procedure on three average scores of a satisfaction item. The respondents were divided into three groups based on the difference in income change: increase, same, and decrease. This enabled the study to conduct a statistical test on the three average values of the groups for each item of the satisfaction question.

The test included a Bonferroni adjustment to avoid a Type 1 error resulting from multiple testing. A Bonferroni adjustment lowers an evaluation criterion by dividing it by the number of tests. Compared with other statistical methods, a Bonferroni adjustment was appropriate in this study because the number of respondents in a divided group was too small to evaluate whether these samples followed a normal distribution and whether the population variances of the divided groups were equal. The test was performed using SPSS 27.

4. Results

4.1. Characteristics of Farmers and Their Conservation Agriculture

The majority of farmers in this study were more than 50 years old (76.5% of 51 valid answers), with an average age of 60.5. The average number of workers on a farm was 7.2, although this value might reflect the existence of the farmer groups for policy or certification purposes (50 valid answers). There were 28 farmers (57.1%) who indicated some prospect of a successor (49 valid answers). Of the five crops with the highest sales, rice was the dominant crop, as indicated by 37 (77.0%) farmers. The majority of farmers were engaged in rice agriculture (83.3% of 48 valid answers). The average number of years practicing conservation agriculture was 14.7 (48 valid answers). A total of 28 (58.3%) of farmers had been practicing it for more than 10 years (48 valid answers). The average length of time was 14.7 years.

According to the results of the question on practices of conservation agriculture, farmers were more likely to apply these methods to reduce chemical use than for habitat preservation. The three largest values for selected choices were related to the use of agricultural chemicals (72% for item (1), 46% for (2), and 42% for (3), among 50 valid responses in Table 2). Just over a third of farmers were certified organic (34%). The choices of methods of conservation agriculture with the three smallest values were related to habitat preservation (20% for item (8), 12% for (9), and 0% for (10)).

Table 2. Practicing methods of conservation agriculture (n = 50, multiple answers possible).

(1) Reducing chemical fertilizer and chemically synthesized fertilizer to less than a half of those for conventional agriculture	72%
(2) No application of herbicide over ridges between farmland plots	46%
(3) No application of systematic insecticide such as neonicotinoid pesticides	42%
(4) Organic certified with JAS	34%
(5) Practicing a winter-flooded rice field	24%
(6) Applying neither chemical fertilizer nor chemically synthesized pesticide without a JAS certificate	22%
(7) Rationalizing land use of conservation agriculture in neighboring fields	20%
(8) Leaving the edge of rice fields as a swale	12%
(9) Creating a biotope on an uncultivated rice field	6%
(10) Installing a fish waterway	0%
(11) Other methods	8%

4.1.1. Motivation to Start Conservation Agriculture

The three main motivation types to start conservation agriculture were the production of safe and secure foods, the production of high-quality foods, and interest in environmental conservation and biodiversity (80.9%, 68.1%, and 59.6%, respectively, with 47 valid respondents; Figure 1). However, few farmers indicated an interest in environmental protection and biodiversity as their single most important motivation (4.3%).

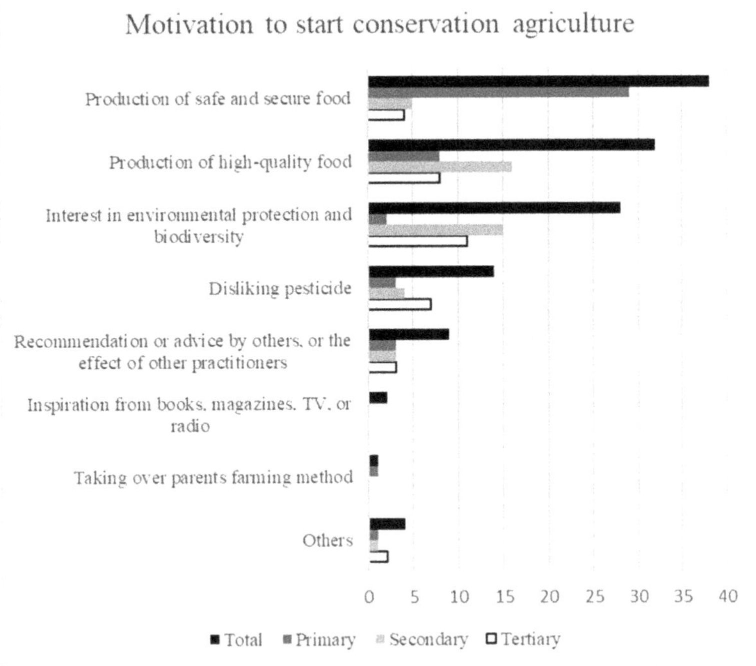

Figure 1. Ishikawa Prefecture farmers' motivation to start conservation agriculture (*n* = 47).

4.1.2. Sales and Sales Channels

The mean of farmers' annual sales was JPY 32.8 million and the median was JPY 9 million (49 valid answers). About half of the farmers had sales exceeding JPY 10 million (49.0%). A major sales channel was direct consumer sales (78.0% of 50 valid respondents), with the second-largest proportion being direct sales stores (42.0%). Certain farmers also sold their products to wholesalers, agricultural cooperatives, restaurants, and retail shops (40.0%, 36.0%, 32.0%, and 26.0%, respectively). They also made internet sales (26.0%).

Compared with before practicing conservation agriculture, farmers increased the number of sales channels from 2.3 to 3.1 on average (49 valid respondents). About half of the farmers increased their number of sales channel types (54.2%), while a few experienced a decrease (10.4%). More farmers traded with restaurants, direct consumer sales, wholesalers, retail shops, and made internet sales (19.8%, 18.9%, 13.5%, 11.8%, and 11.7% increase, respectively). Although the decrease was small, fewer farmers traded with agricultural cooperatives and consumer coops (6.9% and 4.2% decrease, respectively).

4.1.3. Certificates

More than half of the farmers earned certification for specially cultivated agricultural products (56.9% of 51 respondents). As mentioned above, 33.3% of the farmers were certified as organic under the JAS Organic Standards. The same proportion was certified with GAP (33.3%).

4.2. Factors That Affect Income Increase

This study separated the question about income change into two agricultural situations: one with a certification and the other without any certification. Four farmers provided the same responses in these two situations, while three answered differently. This study took account of answers in each category. Of the 44 responses with the complete attributes for a series of the analyses in this study, 22 experienced an income increase, 12 an income decrease, and 20 were neutral. The study generated a regression tree to analyze the factors of income increase among the farmers (Figure 2). The results showed that years of experience, trade with a retail shop, and age tended to cause an income increase. If a farmer had been engaged in conservation agriculture for fewer than six years, income tended to decrease. If they had been doing so for six or more years, their income tended to increase. Furthermore, among the farmers who neither engaged in conservation agriculture for fewer than six years nor traded with a retail shop to sell their products, those who were 62 years old or older tended to increase their income. Conversely, those under 62 years old were less likely to increase their income under the same conditions. Selling conservation agriculture products in retail shops is relatively new in Japan, and elderly farmers might not need to adapt to it very much to increase their income.

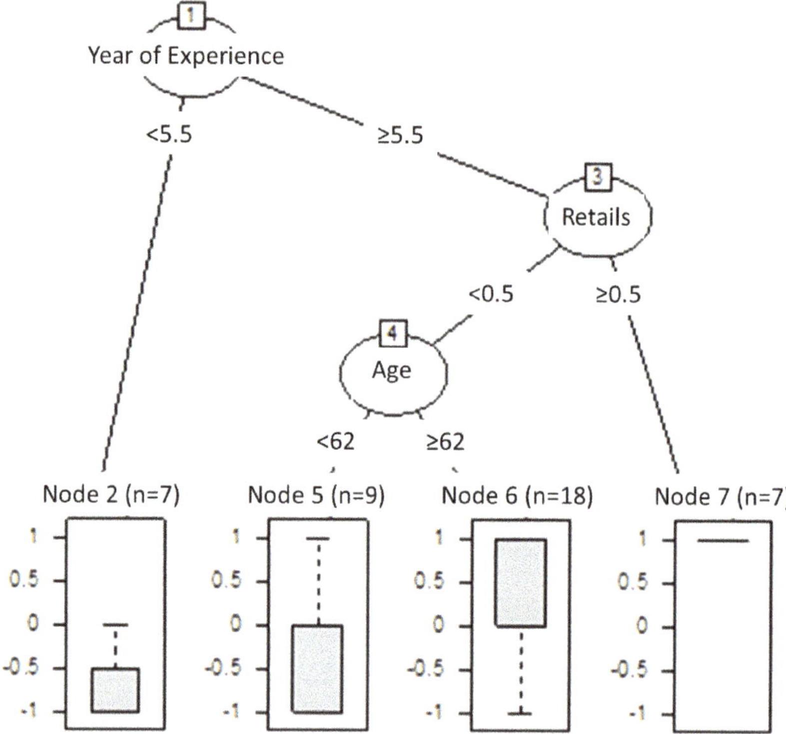

Figure 2. Regression tree of income increase among farmers practicing conservation agriculture. (Note: The nodes of years of experience, age, and retail indicate the number of years of conducting conservation agriculture, age of farmers, and trade with a retail shop where farmers sell their products, respectively).

4.3. Satisfaction from Conservation Agriculture

Satisfaction from conservation agriculture was first calculated and summarized by combining the five possible answers into the following three groups: (1) satisfied or very satisfied; (2) neither satisfied nor dissatisfied; and (3) somewhat satisfied or dissatisfied (Figure 3). Many farmers are satisfied or very satisfied with the production of safe and secure food, protection of the environment and biodiversity, and production of high-quality foods (86.7%, 80.4%, and 78.3%, respectively). On the contrary, the item with the lowest level of satisfaction was fellowship with farmers practicing conservation agriculture (only 39.1% were satisfied or very satisfied).

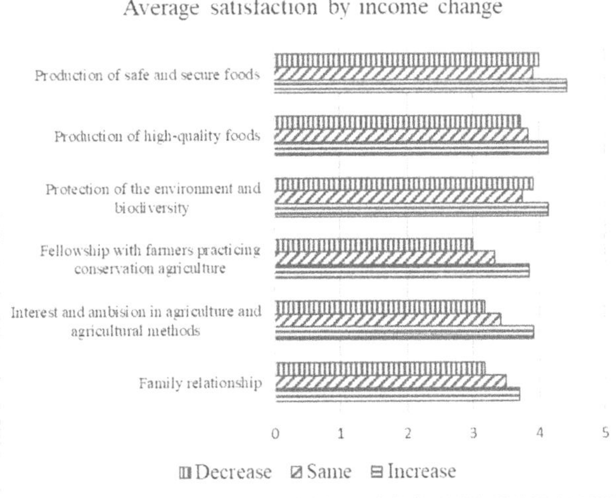

Figure 3. Ishikawa Prefecture farmers' satisfaction through conservation agriculture.

Additionally, Figure 4 shows the average scores of farmers' satisfaction according to income change: increase, same, and decrease (excluding those responses of farmers answering differently about income change versus the existence of certification). The average satisfaction with income increase was the largest for all items. Four items had an average satisfaction increase as income increased: production of high-quality foods, fellowship with farmers practicing conservation agriculture, interest and ambition in agriculture and agricultural methods, and family relationships. On the other hand, farmers who had the same level of income were the least satisfied with the production of safe and secure foods and protection of the environment and biodiversity.

Figure 4. Average satisfaction of farmers practicing conservation agriculture by income change.

Table 3 shows the primary motivations to start conservation agriculture and the average score of the corresponding satisfaction items. This could indicate that some initial motivations were more likely to be rewarding. The motivation to produce high-quality foods resulted in a high level of satisfaction (4.5). Two other motivations that resulted in a high level of satisfaction were the production of safe and secure foods (4.25) and the recommendation or advice of others, or the effect of other practitioners (4). The latter item corresponded to fellowship with farmers practicing conservation agriculture. Interest in environmental protection and biodiversity and taking over a parent's agricultural method had low initial motivation levels (2 and 1 farmers, respectively), while the average levels of satisfaction were neutral (both 3).

Table 3. Ishikawa Prefecture farmers' initial motivation and current satisfaction.

Motivation	Number of Farmers	Average Scores in Corresponding Satisfaction
Production of safe and secure foods	31	4.25
Production of high-quality foods	6	4.5
Interest in environmental protection and biodiversity	2	3
Recommendation or advice by others, or the effect of other practitioners	3	4
Books, magazines, TV, or radio	0	n/a
Taking over parents' agricultural method	1	3

Based on the result of multiple comparison of the averages of satisfaction level of income-change groups, Table 4 shows the result of a test with a Bonferroni adjustment. It indicates that the average satisfaction in fellowship with farmers practicing conservation agriculture was significantly different between the farmers with an income increase and those with an income decrease ($p < 0.05$) (Table 4). The result suggests that income change was related to satisfaction in the relationships with other farmers interested in biodiversity conservation measures.

Table 4. Results of multiple comparison procedure with a Bonferroni adjustment.

Variable	Group	Mean Difference	Standard Error	*p* Value	95% CI of the Difference Lower	Upper
Level of satisfaction in fellowship with farmers practicing conservation agriculture (1–5)	Income increase					
	Income decrease	−0.857 *	0.335	0.043	−1.693	−0.020

Note: * The mean difference is significant at the 0.05 level.

4.4. Satisfaction with Payment Policies

This study asked about satisfaction with three types of direct payment schemes in Japan: Multifunctional Payment, Direct Payment for Hilly and Mountainous Area, and Direct Payment for Conservation Oriented Agriculture. The first two payment types satisfied more farmers though most were not dissatisfied. Regarding Direct Payment for Conservation Oriented Agriculture, more farmers were dissatisfied than farmers evaluating the other subsidy schemes, whether they were in their initial year or had been practicing conservation agriculture for four or more years. Dissatisfaction could decrease as farmers practice conservation agriculture for many years.

4.4.1. Multifunctional Payment

Of the 20 farmers who answered the question about satisfaction with Multifunctional Payment in their first year of conservation agriculture, 9 (45%) were satisfied, 6 (30%)

were dissatisfied, and 5 (22.7%) were neutral. Of the 22 farmers who had been practicing conservation agriculture for four or more years, 12 (54.4%) were satisfied, 5 (22.7%) were dissatisfied, and 5 (22.7%) were neutral.

4.4.2. Direct Payment for Hilly and Mountainous Area

Of the 17 farmers who answered the question about satisfaction with Direct Payment for Hilly and Mountainous Area in their first year of conservation agriculture, 7 (41.2%) were satisfied, 5 (29.4%) were dissatisfied, and 5 (29.4%) were neutral. Of the 18 farmers who had been practicing conservation agriculture for four years or longer, 9 (50.0%) were satisfied, 4 (22.2%) were dissatisfied, and 5 (27.8%) were neutral.

4.4.3. Direct Payment for Conservation-Oriented Agriculture

Of the 23 respondents who answered the question about their satisfaction with Direct Payment for Conservation Oriented Agriculture in their first year of conservation agriculture, 7 (30.4%) were satisfied, 11 (47.8%) dissatisfied, and 5 (21.7%) neutral. Of the 24 participants who had been practicing conservation agriculture for four years or longer, 8 (33.3%) were satisfied, 9 (37.5%) were dissatisfied, and 7 (29.2%) were neutral.

5. Discussion
5.1. Factors Affecting Income Change

Years of experience, trade with a retail shop, and age can affect income change. Although Fujita and Hatano [30] documented a weak correlation between income increase and years of experience among new organic farmers in Japan, the findings of the current study provide a more in-depth understanding of how income increases in conservation agriculture. For instance, years of experience and age significantly affected income change; as the number of years increases, agricultural skills become more enhanced and the production more stabilized. Additionally, this study documented that trade with a retail shop would increase income, which is similar to the findings of Pham and Shively [16], who noted that market channel types can affect income. Although farmers in conservation agriculture may have several types of sales channels, the results showed that retail shops are the most effective channels for increasing income, particularly for younger farmers in conservation agriculture. One possible explanation is that farmers stand a better chance of increasing their income as retail industries become more interested in the products of conservation agriculture, especially organic products. Additionally, further development of the tea export market from conservation and organic production stimulates the conversion of tea producers [56]. Elderly farmers can increase their income by using more traditional sales channels for conservation agriculture, such as direct consumer sales and consumer cooperatives. By doing so, the available sales channels will change and farmers will have more time to adapt to these changes based on their respective ages. The dependency on retail industries to develop a conservation farm sector can be regarded as a sign of conventionalization, even though the concept was originally applied to organic agriculture [32,33,57]. According to studies on the political ecology of organic agriculture in California, exposure to market rule placed pressure on small organic farms and allowed the prioritization of large, specialized farms [32,33]. This phenomenon also lowered the interest in environmental protection among new farmers according to a study of German farmers [57]. Additionally, Oda and Kiminami [29] discussed the fact that elderly farmers were more interested in the environment than younger farmers, who tended to be more interested in subsidies. Oda and Kiminami [29] further highlighted that the feeling agriculture is a worthwhile pursuit can alleviate this issue. Resisting alienation by conventionalization could be a key when critical discourse is taken into consideration [58]. Fujita and Hatano [30] mentioned that organic farmers in Japan were less capable of negotiating prices with agricultural and consumer cooperatives than wholesalers. To lessen the negative effects of conventionalization in Japan, existing cooperative sales channels with a long supply chain might need to review their purchase and sales planning toward a shorter and more sustainable supply chain.

5.2. Characteristics of Satisfaction

This analysis showed that the enhancement of satisfaction by conservation agriculture was asymmetric through time. Most farmers surveyed were satisfied with personal and environmental attributes of motivations: the production of safe and secure food, protection of the environment and biodiversity, and production of high-quality food, which is similar to the findings of Mozzato et al. [1] and Morel and Léger [26]. The items of safe, secure, and high-quality food should be rewarding to pursue as a majority of farmers chose these personal attributes as motivations. The satisfaction that results from producing high-quality food demands more consideration as it was not observed in a previous study on barley production [59]. At the same time, farmers might not be completely sure about which motivations had to be fulfilled or were worth exploring when they started conservation agriculture. In this sense, motivation and satisfaction demonstrate time differentiation. A comparison of the various motivation and satisfaction items showed that environmental attributes could surpass personal attributes. The protection of the environment and bio-diversity eventually surpassed the production of high-quality foods. The protection of biodiversity including bees [60,61] and surrounding forest [62] may further the satisfaction. In the same way, interest and ambition in agriculture and agricultural methods might satisfy farmers who tend not to periodize them at an initial stage. A social factor, namely fellowship with farmers practicing conservation agriculture, might not be as satisfying as other items of satisfaction. This reflects that those organic farmers and farmers who reject chemical use are mainly in the minority. Practicing farmers tend to feel estranged or distance themselves from neighboring farmers during the initial period [30]. Although this feeling of loneliness subsides after a continued period, fellowship has the potential to be effective for increasing income. This issue demands more practical and academic attention.

5.3. Relation with Satisfaction

Furthermore, this study also showed that farmers' satisfaction could be related to other variables. Intensive information sharing is the key to improving the practice of conservation agriculture [63]. This study distilled the relationship between income change and social relations among farmers engaged in conservation agriculture. Referring to both motivation and satisfaction at the appropriate time can promote a balanced development of farmers. Our results indicated that more studies are required to understand both the characteristics of satisfaction and how they interact with interpersonal relationships for extracting the benefits of social factors on farm management. Farmers can share information to stabilize production in the same area and to manage better [2,17]. Farmers can effectively improve their costly or depleting processes when they have the information to compare their processes with those of nearby farmers [64]. Simultaneously, social factors might not work well in certain situations. Farmers might simply have no fellow farmers near them [17]. The differences in technologies, applications, and perceptions of agriculture and nature could potentially prevent them from networking with other farmers. Thus, a sensitive approach is required to evaluate whether a relationship with satisfaction is theoretical or practical.

6. Conclusions

This study gained unique insights into conservation agriculture, especially dynamic characters on perceptions of agricultural practice and the policy of farmers at the regional level. In summary, the study examined the effects of conservation agriculture on farmers (including factors influencing satisfaction) that contribute to the understanding of both the economic and behavioral factors affecting this practice. This study identified the relationship among income change, years of experience, trade with retail shops, and age. The relationship between income change and satisfaction with the farmer's social network was also identified. Because the level of satisfaction with the social network was lower than the other categories in this study, the promotion of the network is likely to be laborious. In addition, through discussions about the relationship between motivation and satisfaction, this study can infer that the development of conservation agriculture is dependent on the contexts of time and space, and careful considerations are necessary. Compared with the level of motivation on environmental protection, the satisfaction level grew through the years. Income changes can also reflect the changing market situation for organic agricultural products in Japan.

There are policy implications in three relevant domains, namely the change of sales channels, the inclusion of satisfaction, and networking policies, especially to improve farmers' income. First, this study identified that trade with a retail shop is a significant factor that increases income despite the concerns over the introduction of large retail industries in organic agricultural products [57]. Through the result of this study, the establishment of fairer transactions is more likely to happen with the introduction of retail industries in the sector. This possibility should be more certain when the government cooperates with farmers to calculate the cost gap between conventional and conservation agriculture. This will also reduce complaints about the direct payment scheme.

Next, the significant relationship between increasing income and the satisfaction with peer conservation that farmers observed in this study implies the potential for the government to apply network policies. With the ultimate goal of increased income, the application of network policies can help improve farmers' knowledge and facilitate information exchange among them and other stakeholders in the context of conservation agriculture [65,66]. Simultaneously, this study found that farmers tended to have difficulty in recognizing their satisfaction with networking. The government or other supporting agencies should provide well-coordinated networking among farmers.

Studies related to the satisfaction of farmers are still in the exploratory phase. Thus, further studies can explore the time- and space-specific characteristics of satisfaction and contribute to the development of conservation agriculture at both the initial and subsequent stages. The trend of satisfaction and income change in this study is focused on conservation agriculture in Ishikawa Prefecture. Future studies will examine this tendency in other Japanese regions. The result of this study will then be investigated in a relatively similar agronomical context with different policy promotion; for example, Taiwan and South Korea. South Korea conducts an extensive public procurement program to promote conservation and organic agriculture [67]. Furthermore, this study only focused on open-field agriculture. Future studies can target additional types of agriculture such as sustainable greenhouse agriculture, which is also a policy target in MeaDRI [12]. Additionally, knowledge transmission is also an issue to be examined in future studies [68,69]. Future international studies about knowledge transmission to promote conservation agriculture can help improve the balance between economic sustainability and environmental conservation. To understand the benefits and disadvantages of changes in the agricultural market, the historical context of organic and conservation agriculture needs to be considered (cf. [70]).

Author Contributions: Conceptualization, S.K., Y.M., Y.U. and R.K.; methodology, Y.M., S.K., Y.U. and R.K.; software, Y.M. and S.K.; validation, Y.M., S.K., Y.U. and R.K.; formal analysis, Y.M., S.K., Y.U. and R.K.; investigation, Y.M., S.K., Y.U. and R.K.; resources, R.K., S.K., Y.U. and Y.M.; data curation, Y.M., S.K., Y.U. and R.K.; writing—original draft preparation, Y.M., S.K., Y.U. and R.K.; writing—review and editing, Y.M., R.K., Y.U. and S.K.; visualization, Y.M., S.K., Y.U. and R.K.; supervision, R.K. and Y.U.; project administration, R.K., Y.U., Y.M. and S.K.; funding acquisition, R.K. All authors have read and agreed to the published version of the manuscript.

Funding: The study for this article was funded by Policy Research Institute MAFF of Japan as a Commissioned Project for the Policy Science of Agriculture, Forestry, and Fisheries between 2018 and 2020: Factors to Facilitate Production and Export of Organic Agricultural Products: Analysis of Status and Trends in EU and Implications for Japan. The JSPS KAKENHI Grant Numbers were JP16KK0053, JP17K02105, JP20K12398, and JP21K18456. Additionally, the study was supported by JST RISTEX Grant Number JPMJRX20B3, Japan, and JST Grant Number JPMJPF2110.

Institutional Review Board Statement: This study regarded and consulted the rules involving human subjects for research in the institutions of the authors.

Informed Consent Statement: Informed consent was obtained from all subjects involved in the study.

Data Availability Statement: Any request for the data of this study will be considered.

Conflicts of Interest: The authors declare no conflict of interest.

References

1. Mozzato, D.; Gatto, P.; Defrancesco, E.; Bortolini, L.; Pirotti, F.; Pisani, E.; Sartori, L. The role of factors affecting the adoption of environmentally friendly farming practices: Can geographical context and time explain the differences emerging from literature? *Sustainability* **2018**, *10*, 3101. [CrossRef]
2. Wezel, A.; Brives, H.; Casagrande, M.; Clement, C.; Dufour, A.; Vandenbroucke, P. Agroecology territories: Places for sustainable agricultural and food systems and biodiversity conservation. *Agroecol. Sustain. Food* **2016**, *40*, 132–144. [CrossRef]
3. McGranahan, D.A. Ecologies of scale: Multifunctionality connects conservation and agriculture across fields, farms, and landscapes. *Land* **2014**, *3*, 739–769. [CrossRef]
4. Food and Agriculture Organization of the United Nations. Conservation Agriculture. 2022. Available online: https://www.fao.org/conservation-agriculture/en/ (accessed on 1 February 2022).
5. Ministry of Agriculture, Forestry and Fisheries (MAFF). *About Direct Payment for Conservation Oriented Agriculture*; Ministry of Agriculture, Forestry and Fisheries: Tokyo, Japan, 2020. (In Japanese)
6. Atlin, G.N.; Frey, K.J. Breeding crop varieties for low-input agriculture. *Am. J. Altern. Agric.* **1989**, *4*, 53–58. [CrossRef]
7. Schaller, N. Mainstreaming low-input agriculture. *J. Soil Water Conserv.* **1990**, *45*, 9–12.
8. Fess, T.L.; Kotcon, J.B.; Benedito, V.A. Crop Breeding for low input agriculture: A sustainable response to feed a growing world population. *Sustainability* **2011**, *3*, 1742–1772. [CrossRef]
9. Liebig, M.A.; Herrick, J.E.; Archer, D.W.; Dobrowolski, J.; Duiker, S.W.; Franzluebbers, A.J.; Hendrickson, J.R.; Mitchell, R.; Mohamed, A.; Russell, J. Aligning land use with land potential: The role of integrated agriculture. *Agric. Environ. Lett.* **2017**, *2*, 170007. [CrossRef]
10. Hendrickson, J.R.; Hanson, J.D.; Tanaka, D.; Sassenrath, G. Principles of integrated agricultural systems: Introduction to processes and definition. *Renew. Agric. Food Syst.* **2008**, *23*, 265–271. [CrossRef]
11. European Commission. *Factsheet: From Farm to Fork: Our Food, Our Health, Our Planet, Our Future*; European Commission: Brussels, Belgium, 2020.
12. Ministry of Agriculture, Forestry and Fisheries (MAFF). *Strategy for Green Food System*; Ministry of Agriculture, Forestry and Fisheries: Tokyo, Japan, 2021.
13. Bouttes, M.; Darnhofer, I.; Martin, G. Converting to organic farming as a way to enhance adaptive capacity. *Org. Agric.* **2019**, *9*, 235–247. [CrossRef]
14. Ayuya, O.I. Organic certified production systems and household income: Micro level evidence of heterogeneous treatment effects. *Org. Agric.* **2019**, *9*, 417–433. [CrossRef]
15. Hansen, B.G.; Haga, H.; Lindblad, K.B. Revenue efficiency, profitability, and profitability potential on organic versus conventional dairy farms—Results from comparable groups of farms. *Org. Agric.* **2021**, *11*, 351–365. [CrossRef]
16. Pham, L.; Shively, G. Profitability of organic vegetable production in northwest Vietnam: Evidence from Tan Lac District, Hoa Binh Province. *Org. Agric.* **2019**, *9*, 211–223. [CrossRef]
17. Dessart, F.J.; Barreiro-Hurlé, J.; van Bavel, R. Behavioural factors affecting the adoption of sustainable farming practices: A policy-oriented review. *Eur. Rev. Agric. Econ.* **2019**, *46*, 417–471. [CrossRef]
18. Badu-Gyan, F.; Henning, J.I.; Grové, B.; Owusu-Sekyere, E. Examining the social, physical and institutional determinants of pineapple farmers' choice of production systems in Central Ghana. *Org. Agric.* **2019**, *9*, 315–329. [CrossRef]
19. Bui, H.T.M.; Nguyen, H.T.T. Factors influencing farmers' decision to convert to organic tea cultivation in the mountainous areas of northern Vietnam. *Org. Agric.* **2021**, *11*, 51–61. [CrossRef]
20. Digal, L.N.; Placencia, S.G.P. Factors affecting the adoption of organic rice farming: The case of farmers in M'lang, North Cotabato, Philippines. *Org. Agric.* **2019**, *9*, 199–210. [CrossRef]
21. Sharma, M.; Pudasaini, A. What motivates producers and consumers towards organic vegetables? A case of Nepal. *Org. Agric.* **2021**, *11*, 477–488. [CrossRef]
22. Väre, M.; Mattila, T.E.A.; Rikkonen, P.; Hirvonen, M.; Rautiainen, R.H. Farmers' perceptions of farm management practices and development plans on organic farms in Finland. *Org. Agric.* **2021**, *11*, 457–467. [CrossRef]
23. Chan, C.; Sipes, B.; Ayman, A.; Zhang, X.; LaPorte, P.; Fernandes, F.; Pradhan, A.; Chan-Dentoni, J.; Roul, P. Efficiency of conservation agriculture production systems for smallholders in Rain-Fed Uplands of India: A transformative approach to food security. *Land* **2017**, *6*, 58. [CrossRef]
24. Rudel, T.K.; Kwon, O.-J.; Paul, B.K.; Boval, M.; Rao, I.M.; Burbano, D.; McGroddy, M.; Lerner, A.M.; White, D.; Cuchillo, M. Do Smallholder, mixed crop-livestock livelihoods encourage sustainable agricultural practices? A meta-analysis. *Land* **2016**, *5*, 6. [CrossRef]

25. Bartolini, F.; Vergamini, D. Understanding the spatial agglomeration of participation in agri-environmental schemes: The case of the Tuscany region. *Sustainability* **2019**, *11*, 2753. [CrossRef]

26. Morel, K.; Léger, F. A Conceptual framework for alternative farmers' strategic choices: The case of French organic market gardening microfarms. *Agroecol. Sustain. Food* **2016**, *40*, 466–492. [CrossRef]

27. Uruma, H.; Kobayashi, R.; Nishijima, S.; Miyashita, T. Effectiveness of conservation-oriented agricultural practices on amphibians inhabiting Sado Island, Japan, with a consideration of spatial structure. *Jpn. J. Conserv. Ecol.* **2012**, *17*, 155–164. (In Japanese)

28. Kanazawa, S. No-till farming as sustainable and environment-friendly farming: Crop yeild in dry field and soil characteristics. *Jpn. Soc. Soil Sci. Plant Nutr.* **1995**, *66*, 286–297. (In Japanese)

29. Oda, M.; Kiminami, L. Study on the consciousness of the farm households introducing environment-friendly agriculture: Analysis based on the "certification system for the rice of living with Japanese crested ibis" in Sato City, Niigata Prefecture. *Niigata Daigaku Nōgakubu Kenkyū Hōkoku* **2014**, *66*, 85–104. (In Japanese)

30. Fujita, M.; Hatano, T. Opportunity and financial status in newly started and transformed organic farmers: Based on the questionnaire surveys from organic farmers. *Yūki Nōgyō Kenkyū* **2017**, *9*, 53–63. (In Japanese)

31. Uenishi, Y. Factors affecting the diffusion of conservation-oriented farming methods. *J. Rural Probl.* **2019**, *55*, 73–80. (In Japanese) [CrossRef]

32. Buck, D.; Getz, C.; Guthman, J. From farm to table: The organic vegetable commodity chain of northern California. *Sociol. Rural.* **1997**, *37*, 3–20. [CrossRef]

33. Guthman, J. The trouble with 'organic lite' in California: A rejoinder to the 'conventionalisation' debate. *Sociol. Rural.* **2004**, *44*, 301–316. [CrossRef]

34. Suzuki, Y. Sustainability and development of environmentally friendly agriculture in the hilly and mountainous farming areas—A case study of the south Aso and the mountainous region in Kyushu, Japan. *Ann. Jpn. Assoc. Econ. Geogr.* **1997**, *43*, 276–292. (In Japanese)

35. Arai, S.; Nishio, M.; Yoshino, T. *Connecting with Organic Farming and Living Close to a Community: Development of Yūki Hāto Netto in Shirakawa Town, Gifu Prefecture*; Tsukuba Shobō: Tokyo, Japan, 2021. (In Japanese)

36. Anwar, A.; Sarwar, S.; Amin, W.; Arshed, N. Agricultural practices and quality of environment: Evidence for global perspective. *Environ. Sci. Pollut. Res.* **2019**, *26*, 15617–15630. [CrossRef]

37. Smith, D.L.; Almaraz, J.J. Climate change and crop production: Contributions, impacts, and adaptations. *Can. J. Plant Pathol.* **2004**, *26*, 253–266. [CrossRef]

38. Oenema, O.; Wrage, N.; Velthof, G.L.; van Groenigen, J.W.; Dolfing, J.; Kuikman, P.J. Trends in global nitrous oxide emissions from animal production systems. *Nutr. Cycl. Agroecosyst.* **2005**, *72*, 51–65. [CrossRef]

39. Waheed, R.; Chang, D.; Sarwar, S.; Chen, W. Forest, agriculture, renewable energy, and CO_2 emission. *J. Clean. Prod.* **2018**, *172*, 4231–4238. [CrossRef]

40. Mena, Y.; Gutierrez-Peña, R.; Ruiz, F.A.; Delgado-Pertíñez, M. Can dairy goat farms in mountain areas reach a satisfactory level of profitability without intensification? A case study in Andalusia (Spain). *Agroecol. Sustain. Food* **2017**, *41*, 614–634. [CrossRef]

41. Ishikawa Prefectural Government. *Preliminary Report of the 2020 Census*; Ishikawa Prefectural Government: Kanazawa, Japan, 2021. (In Japanese)

42. Ministry of Agriculture, Forestry and Fisheries (MAFF). *Census of Agriculture and Forestry in Japan*; Ministry of Agriculture, Forestry and Fisheries: Tokyo, Japan, 2016. (In Japanese)

43. Kohsaka, R.; Tomiyoshi, M.; Matuoka, H. Tourist Perceptions of Traditional Japanese Vegetable Brands: A Quantitative Approach to Kaga Vegetable Brands and an Information Channel for Tourists at the Noto GIAHS Site. In *Aquatic Biodiversity Conservation and Ecosystem Services*; Nakano, S., Yahara, T., Nakashizuka, T., Eds.; Springer: Singapore, 2016; pp. 109–121.

44. Kamiyama, C. Food Provisioning Services via Homegardens and Communal Sharing in Satoyama Socio-Ecological Production Landscapes on Japan's Noto Peninsula. In *Sharing Ecosystem Services*; Saito, O., Ed.; Springer: Singapore, 2020; pp. 35–53.

45. Kohsaka, R.; Matsuoka, H. Analysis of Japanese municipalities with Geopark, MAB, and GIAHS certification: Quantitative approach to official records with text-mining methods. *SAGE Open* **2015**, *5*, 1–10. [CrossRef]

46. Kohsaka, R.; Ito, K.; Miyake, Y.; Uchiyama, Y. Cultural ecosystem services from the afforestation of rice terraces and farmland: Emerging services as an alternative to monoculturalization. *For. Ecol. Manag.* **2021**, *497*, 119481. [CrossRef]

47. Ministry of Agriculture, Forestry and Fisheries (MAFF). *About the Situation of Organic Agriculture*; Ministry of Agriculture, Forestry and Fisheries: Tokyo, Japan, 2020. (In Japanese)

48. Kotrlik, J.; Higgins, C. Organizational research: Determining appropriate sample size in survey research appropriate sample size in survey research. *Inf. Technol. Learn. Perform. J.* **2001**, *19*, 43.

49. Quevedo, J.M.D.; Uchiyama, Y.; Lukman, K.M.; Kohsaka, R. How blue carbon ecosystems are perceived by local communities in the coral triangle: Comparative and empirical examinations in the Philippines and Indonesia. *Sustainability* **2021**, *13*, 127. [CrossRef]

50. Ministry of Agriculture, Forestry and Fisheries (MAFF). *A National Survey through the Network Scheme to Communicate the Information of Agriculture, Forestry, and Fisheries in the Year Heisei 27: The Survey of Awareness and Intention About Environmentally Friendly Agricultural Products Including Organic Ones*; Ministry of Agriculture, Forestry and Fisheries: Tokyo, Japan, 2016. (In Japanese)

51. Ministry of Agriculture, Forestry and Fisheries (MAFF). *Label Guideline about Specially Cultivated Agricultural Products*; Ministry of Agriculture, Forestry and Fisheries: Tokyo, Japan, 2020. (In Japanese)

52. Ministry of Agriculture, Forestry and Fisheries (MAFF). *Annual Report on Food, Agriculture and Rural Areas in Japan FY 2019*; Ministry of Agriculture, Forestry and Fisheries: Tokyo, Japan, 2020. (In Japanese)
53. Ministry of Health, Labour and Welfare. HACCP, n.d. Available online: https://www.mhlw.go.jp/stf/seisakunitsuite/bunya/kenkou_iryou/shokuhin/haccp/index.html (accessed on 19 January 2021). (In Japanese)
54. Japan Patent Office. Regional Collective Trademark System, n.d. Available online: https://www.jpo.go.jp/e/system/trademark/gaiyo/chidan/ (accessed on 19 January 2021). (In Japanese)
55. Ito, J.; Feuer, H.N.; Kitano, S.; Asahi, H. Assessing the effectiveness of Japan's community-based direct payment scheme for hilly and mountainous areas. *Ecol. Econ.* **2019**, *160*, 62–75. [CrossRef]
56. Miyake, Y.; Kohsaka, R. Discourse of quality and place in geographical indications: Applying convention theory to Japanese tea. *Food Rev. Int.* **2022**, 1–26. [CrossRef]
57. Best, H. Organic Agriculture and the conventionalization hypothesis: A case study from west Germany. *Agric. Hum. Values* **2008**, *25*, 95–106. [CrossRef]
58. Giménez Cacho, M.M.T.; Giraldo, O.F.; Aldasoro, M.; Morales, H.; Ferguson, B.G.; Rosset, P.; Khadse, A.; Campos, C. Bringing agroecology to scale: Key drivers and emblematic cases. *Agroecol. Sustain. Food* **2018**, *42*, 637–665. [CrossRef]
59. Baker, B.P.; Meints, B.M.; Hayes, P.M. Organic barley producers' desired qualities for crop improvement. *Org. Agric.* **2020**, *10*, 35–42. [CrossRef]
60. Matsuzawa, T.; Kohsaka, R. Status and trends of urban beekeeping regulations: A global review. *Earth* **2021**, *2*, 933–942. [CrossRef]
61. Matsuzawa, T.; Kohsaka, R. Preliminary experimental trial of effects of lattice fence installation on honey bee flight height as implications for urban beekeeping regulations. *Land* **2022**, *11*, 19. [CrossRef]
62. Kohsaka, R.; Miyake, Y. The politics of quality and geographic indications for non-timber forest products: Applying convention theory beyond food contexts. *J. Rural Stud.* **2021**, *88*, 28–39. [CrossRef]
63. Bliss, K.; Padel, S.; Cullen, B.; Ducottet, C.; Mullender, S.; Rasmussen, I.A.; Moeskops, B. Exchanging knowledge to improve organic arable farming: An evaluation of knowledge exchange tools with farmer groups across Europe. *Org. Agric.* **2019**, *9*, 383–398. [CrossRef]
64. Chabe-Ferret, S.; Le Coent, P.; Reynaud, A.; Subervie, J.; Lepercq, D. Can we nudge farmers into saving water? Evidence from a randomised experiment. *Eur. Rev. Agric. Econ.* **2019**, *46*, 393–416. [CrossRef]
65. Kohsaka, R.; Tomiyoshi, M.; Saito, O.; Hashimoto, S.; Mohammend, L. Interactions of knowledge systems in shiitake mushroom production: A case study on the Noto Peninsula, Japan. *J. For. Res.-Jpn.* **2015**, *20*, 453–463. [CrossRef]
66. Miyake, Y.; Uchiyama, Y.; Kohsaka, R. Status and trends of urban organic agricultural policy in Japan: The survey on ordinance designated cities. *Org. Agric.* **2020**, *10*, 497–508. [CrossRef]
67. Willer, H.; Travnicek, J.; Meier, C.; Schlatter, B. (Eds.) *The World of Organic Agriculture: Statistics and Emerging Trends 2021*; Research Institute of Organic Agriculture: Frick, Switzerland; Organics International: Bonn, Germany, 2021.
68. Kizos, T.; Koshaka, R.; Penker, M.; Piatti, C.; Vogl, C.R.; Uchiyama, Y. The governance of geographical indications. *Br. Food J.* **2017**, *119*, 2863–2879. [CrossRef]
69. Tashiro, A.; Uchiyama, Y.; Kohsaka, R. Impact of geographical indication schemes on traditional knowledge in changing agricultural landscapes: An empirical analysis from Japan. *J. Rural Stud.* **2019**, *68*, 46–53. [CrossRef]
70. Miyake, Y.; Kohsaka, R. History, ethnicity, and policy analysis of organic farming in Japan: When "nature" was detached from organic. *J. Ethn. Foods* **2020**, *7*, 1–8. [CrossRef]

Article

Design Model and Management Plan of a Rice–Fish Mixed Farming Paddy for Urban Agriculture and Ecological Education

Jinkwan Son [1], Minjae Kong [2,*] and Hongshik Nam [3]

[1] Department of Agricultural Engineering, National Institute of Agricultural Sciences, Rural Development Administration, Jeonju-si 54875, Jeollabuk-do, Korea; son007005@korea.kr
[2] Department of Crop Protection, National Institute of Agricultural Sciences, Rural Development Administration, Wanju-gun 55365, Jeollabuk-do, Korea
[3] Department of Organic Agriculture, National Institute of Agricultural Sciences, Rural Development Administration, Wanju-gun 55365, Jeollabuk-do, Korea; namdalli@korea.kr
* Correspondence: alswogud@korea.kr; Tel.: +82-63-238-3288

Abstract: Imparting knowledge on agriculture and ecology is important for the preservation of nature. This study suggested the design of a rice–fish mixed farming (RFMF) paddy for urban agriculture and ecological education in Korea. This RFMF paddy supports the growth of rice as well as freshwater fish. ANOVA statistical analysis was conducted, and an RFMF paddy was necessary for urban agriculture/education and confirmed that biodiversity was high. To this aim, the design of a 10 m × 10 m RFMF paddy was suggested. Vegetation, insects, and aquatic invertebrates of the RFMF paddy constituted approximately 40 species more than a conventional paddy. The quality of an actual farm's soil and water was assessed, and techniques for the co-cultivation of rice and fish are proposed. The soil must comply with the standards of Korean paddy soil, and the water must be in the temperature range of 15 to 35 °C. In the proposed design, approximately 44.0 kg rice can be produced, and catfish can grow up to 30 cm. The study suggested many experiences using rice and freshwater fish. On the basis of our study design, a virtual model of an RFMF paddy was developed in consideration of the accessible space. The development of RFMF paddies in educational institutions can promote biodiversity in cities while providing ecological education regarding aquatic plants and insects.

Keywords: agricultural; urban; rice–fish farming; biodiversity; education; experience; museum

1. Introduction

Since the institutionalization of the five-day workweek in the 2000s, the demand for urban agriculture, vegetable gardens, and rural tourism has been increasing due to growing interest in health, leisure, and the environment [1–5]. Learning experiences related to nature and ecology are considered important educational tools to promote environmental conservation and to understand the harmony between humans and nature by allowing people to observe and touch living organisms in the field [6–9]. Among them, ecological education and agriculture experience studies have been conducted mostly in suburban farms, fields developed on institutional campuses (i.e., urban agriculture), or vegetable gardens within cities [9–11]. These agricultural education experiences are primarily limited to field and greenhouse crops, such as lettuce, Napa cabbage, and red pepper [12–15]. Although rice is the primary source of food in Korea, conducting educational studies on rice farming throughout the year is challenging [5,16,17] owing to the difficulties with periodically conducting studies that focus on experiential education in suburban fields, such as long-distance travel and inaccessibility [18]. Urban locations, such as science museums, exhibition halls, and public-relations halls, provide an attractive alternative for hosting rice farming education and experiences [19].

Citation: Son, J.; Kong, M.; Nam, H. Design Model and Management Plan of a Rice–Fish Mixed Farming Paddy for Urban Agriculture and Ecological Education. *Land* **2022**, *11*, 1218. https://doi.org/10.3390/land11081218

Academic Editors: Jay Mar D. Quevedo, Norie Tamura, Yuta Uchiyama and Ryo Kohsaka

Received: 11 July 2022
Accepted: 31 July 2022
Published: 2 August 2022

Publisher's Note: MDPI stays neutral with regard to jurisdictional claims in published maps and institutional affiliations.

The number of visitors to Korea's science museums has surpassed 4 million annually; therefore, customized education programs for visitors are considered essential [20]. However, most science museum programs focus on physics, chemistry, and astronomy [21]. Programs on natural sciences are available but are limited to Earth, earthquakes, and insect observation. Rice fields in Korea, the high ratio of cultivation land use, and the observation of the biodiversity in rice cultivation spaces may be interesting for visitors [22–24].

A rice–fish mixed farming (RFMF) paddy is a system used to add value to conventional farming and involves the culturing of catfish, loach, carp, and shrimp in paddies [25,26]. In Korea, rice farming is a 10,000-year-old tradition, and the wisdom underlying RFMF has been passed on from ancient times (206 BC) [25]. In a mixed farming paddy, a portion of the paddy is designated as a habitat for growing freshwater fish; this is traditionally called a *dumbung*, which corresponds to a pond-type wetland [27–31]. The wetland space outside a paddy has various environmental and ecological benefits [25].

The primary benefits of rice farming include the development of public interest regarding wetlands (e.g., biodiversity, flood control, and water quality protection), as well as its market value, food security, environmental conservation, employment maintenance, regional development, and social and cultural value enhancement [32–36]. Among them, organic farming can improve soil fertility, mitigate GHG and carbon, and preserve biodiversity and various ecosystem services [37–42]. For this reason, the RFMF paddy is found to have higher economic values than conventional paddy (CP) ecosystem services [43,44]. If rice paddies are developed in educational spaces, such as science museums, performing rice farming throughout the year will be possible. Additionally, these locations can be used as places to provide ecological education to visitors, enabling them to collect and observe animals and plants in this space [31].

At present, the utilization of ecological spaces to provide rice farming experiential education is possible only in suburban paddies [45]. Therefore, in this study, we propose a design plan to simultaneously offer rice agriculture education and an ecological experience through the development of an RFMF paddy where rice and freshwater fish co-exist in an easily accessible space of a science museum, a public-relations hall, or an exhibition hall. This study was conducted with the aim of contributing to the preservation of urban biodiversity, the enhancement of the environment (i.e., climate mitigation), and the provision of related education.

2. Materials and Methods

The purpose of the present study is to judge the possibility of using an RFMF paddy for ecological education for the ecological welfare of city dwellers and develop a detailed plan for its construction. This study comprises the following stages: First, we examined the ecosystem services of an RFMF paddy and conducted an expert survey to understand its value for eco-educational use (Section 2.1). Second, a field survey was conducted to identify the difference between the biodiversity of the RFMF paddy and that of the conventional paddy (Section 2.2). Third, soil and water quality was measured at a farm to determine the appropriate conditions for creating an RFMF paddy (Section 2.3), and the paddy design was developed while ensuring both that it could be built in a city and a science center and that it provides an accurate experience in terms of its contents in the space (Section 2.4). A detailed description of the materials and methods is provided below.

2.1. Evaluating the Ecosystem Service Function of the RFMF Paddy

We compared the development of public interest regarding conventional paddies, which are commonly found in Korea, with that regarding RFMF paddies, which have different structural characteristics for growing freshwater fish, based on the outcomes of a survey conducted among experts from various related domains (i.e., environmental, biology engineering, and agricultural fields). The evaluation of ecosystem services must jointly consider the opinions of experts and relevant knowledge. To design the expert survey questionnaire, we considered 17 functions based on the ecosystem services in rural areas introduced by Son et al. [28,46], who reviewed 11 previous studies related to ecosystem services and selected these 17 functions to be considered when implementing developmental projects, such as land-use changes in agricultural ecosystems. The primary functions are groundwater recharge, water storage, water purification, flood control, aquatic insect habitats, amphibian and reptile habitats, vegetation diversity, landscape creation, experience/education, avian habitats, climate regulation, air quality regulation, fishery habitats, rest areas, biological control, genetic diversity maintenance, and mammalian habitats.

The occupations of experts who responded to the questionnaire included 13 (23.6%) business officers, 18 (32.7%) institute researchers, and 19 (34.5%) university professors. Regarding educational degrees, 43 (78.2%) held a doctoral degree, and 3 (5.5%) were enrolled in doctoral courses. Furthermore, 16 experts from biological, 20 from environmental, 14 from engineering, and 6 from agricultural major fields participated in the questionnaire (see Table 1). Regarding these functions, the 56 experts were requested to score the weaknesses or required improvements in each function, using a 7-point Likert scale (+3, +2, +1, 0, −1, −2, or −3), to allow the analysis of the mean value (importance score) of each function. (+) indicates a positive value, and (−) indicates a negative value. Expert opinions on each function were obtained and analyzed. The collected data were analyzed using the SPSS software ver. 19.0 (IBM SPSS Statistics Institute, Chicago, IL, USA).

Table 1. General information of 56 expert respondents.

Category Classification (*n* = 56)		Respondence	
		Number	%
	University Professor	19	33.9
	Institute Researcher	18	32.1
Work	Business Officer	13	23.2
Fields	Public Official	4	7.1
	Graduate Student	2	3.6
	Doctor	43	76.8
Education	Doctoral Course	3	5.4
Degree	Master's Degree	6	10.7
	University Student	4	7.1
	Environmental	20	35.7
Major	Biological	16	28.6
Fields	Engineering	14	25.0
	Agricultural	6	10.7

2.2. Biodiversity of the RFMF Paddy

On the basis of the survey outcomes, the extent of biodiversity that can be improved through RFMF paddies was measured and analyzed on farms in Mundang-ri, Hongdong-myeon, Hongseong-gun, and Chungcheongnam-do, Korea. These farms implemented the RFMF method five times a year (May to September 2019). In Korea, there are very few RFMF paddies. This is because it is difficult to manage and control weeds, diseases, and pests. For this reason, organic rice paddies represent under 1.0% of all paddies in Korea [47–49]. Figure 1 shows pictures of the space created for research purposes.

The vegetation flora was examined by installing three $2 \times 2 \text{ m}^2$ sub-plots per study site, according to the Braun-Blanquet method, and classification and identification were confirmed by Lee [50]. The biodiversity of insects and aquatic invertebrates was investigated by sweeping, which was conducted three times at the waterside edge. The captured individuals were identified and counted in the field using the Korean Animal Name List and the Korean Insect List [51]. Unidentified species were fixed in ethyl alcohol and transported to the laboratory for identification. The community analysis of insects and aquatic invertebrates was conducted using the dominance index (DI), the diversity index (H'), the richness index (RI), and the evenness index (J') [52–55]. Analysis of variance (ANOVA) of the data was performed using SPSS software ver. 19.0. A combined ANOVA was performed using a cultivar as a fixed variable according to [56]. Based on the level of significance calculated from the F-value of the ANOVA, Duncan's multiple range tests were applied at $p \leq 0.05$ for mean comparisons among the various treatments.

Figure 1. Survey site of RFMF paddy (**left**) and conventional rice paddy (**right**) in Hongseong, Korea.

2.3. Soil and Water Quality Analyses of the RFMF Paddy

To determine suitable conditions for developing an RFMF paddy, the soil chemistry of the farms was analyzed using soil and plant analysis methods [57], at the National Institute of Agricultural Sciences, the Rural Development Administration. The considered analysis items were pH, electrical conductivity (EC), organic matter (OM), Av.P_2O_5, potassium (K), calcium (Ca), magnesium (Mg), and Av.SiO_2. Soil pH and EC were measured using a pH/EC meter (Orion StarTM A215 pH meter, Thermo-Scientific, Calsbad, CA, USA), after extraction, by mixing the pretreated soil sample and deionized water in a ratio of 1:5. The OM content was analyzed through a dry continuous method using an elemental analyzer (VarioMAX CN, Elementar, Langenselbold, Germany). The available phosphate (Av.P_2O_5) was analyzed using the Lancaster method, namely, by measuring the absorbance at 720 nm (UV-2600, Shimadzu, Kyoto, Japan). Exchangeable cations (potassium, Ex. K; calcium, Ex. Ca; and magnesium, Ex. Mg) were extracted with 1 M NH_4OAc (pH 7.0) buffer solution and analyzed using ICP (Integra XL, GBC Scientific Equipment Ltd., Braeside, VIC, Australia). Available silicate (Av. SiO_4) was analyzed by measuring the difference in color developed by redox reaction through measuring the absorbance at 700 nm (UV-2600, Shimadzu, Kyoto, Japan).

In this study, the temperature and pH of paddy water were analyzed to determine the conditions for establishing a suitable water environment for aquatic organisms. Species commonly found in a paddy and *dumbung* were presented as collectible and observable species for experiential education. Reflecting on the structural characteristics of the real farm, we propose a design plan and provide data on the necessary soil environment characteristics for developing a functional RFMF paddy. We also present a method for future rice farming and freshwater fish management, including details for ensuring suitable water quality for paddy management.

2.4. Composition Design and Educational Use Plan of the RFMF Paddy

On the basis of the study results, a virtual model was developed using a three-dimensional design of the accessible green space in front of the Agricultural Science Museum of the Rural Development Administration, which conducts various agricultural studies in Korea.

The design that was tested incorporates two separate regions for the growth of rice and fish. The experimental area was the experimental paddy field in the National Institute of Agricultural Sciences, the Rural Development Administration. It was installed exactly according to the envisioned design, and the growth of rice and fish was observed for 1 year (Figure 2).

Figure 2. Creation ((**left**) April) and management ((**right**) June) of a rice–fish mixed farming paddy.

The educational application of the RFMF paddy was presented in terms of rice and fish farming. The experience content refers to the analysis results of Han, Son, Choi, and Yoon [58], who analyzed 3007 types of experiences in 168 rural tourism villages in Korea.

3. Results

3.1. Public Interest Regarding RFMF Paddies

On the basis of the expert evaluation results (Table 2, Appendix A), we assessed the increase in ecological and environmental as well as experience and educational functions resulting from the introduction of an RFMF paddy compared with that for a conventional paddy. The mean scores of the 17 wetland functions ranged from 0.89 to 2.39, where (+) indicates a positive value, and (−) indicates a negative value; all function scores were positive, and the mean score of the amphibian and reptile habitat function was the highest (2.39 ± 0.69). Thus, species diversity and population size can be increased with the development of a waterway-type *dumbung*, through its function in providing a habitat for amphibians and reptiles. The results also indicated that functions related to biodiversity, such as aquatic insect habitats (2.36 ± 0.66), fishery habitats (2.34 ± 0.78), vegetation diversity (2.13 ± 0.78), and avian habitats (2.05 ± 0.94), will show large improvements. The experience and education (2.29 ± 0.64) function would also increase, considering the expert opinion that the waterway-type *dumbung* can be used for ecological experiences and education, such as fishing and organism collection (Table 2). Paddies have various functions, including rice production for food [59,60]. Some other agricultural functions include food security, the maintenance of the viability of rural communities, land conservation, the sustainable management of renewable natural resources, and environmental protection (through, e.g., biodiversity conservation and aesthetic landscape development) [61]. Moreover, waterway-type *dumbungs* act as wetlands, increasing the ecological function of paddies [25,26]. Therefore, irrigation ponds and canals are important elements in rice farming [60,62]. Several studies have focused on ponds in terms of their important role in the biodiversity conservation of agricultural lands [63–65]. The combination of these aspects will increase the efficiency of environmental and ecological functions.

The outcomes resulting from the analysis of expert opinions regarding the effects of combined agriculture on ecological service enhancement indicate that all functions will be improved. Many previous studies have considered paddies as spaces that benefit various environments [66–70]. Regarding their value as habitats for amphibians and reptiles, positive opinions were obtained, considering that they can act as an ecological spawning ground and hiding place for various amphibians (i.e., Seoul pond frogs and salamanders) and reptiles, due to the greater water depth and extended open surface owing to the ecological farming method. However, it was expressed, as a negative opinion, that the movement of amphibians and reptiles will be restricted by artificial insect screens and partitions installed for fish farming. The movement patterns of amphibians can be affected by anthropogenic obstacles [71], and artificial structures limit the movement of amphibians and reptiles in agricultural ecosystems [72]. Nevertheless, positive opinions suggested that such screens can be used by aquatic insects for shelter during the midsummer drainage period and vulnerable winter season. In addition, the artificial introduction of fish could reduce the number of aquatic insects and sources, causing a disturbance. Although some experts have argued that fish species diversity will increase as a result of building a *dumbung* to grow freshwater fish, others have highlighted that fish species other than artificially introduced species cannot enter due to the blocked and isolated structure. Although small-scale fish farms have a modest impact on water quality [73], fish activity has been reported to have a negative impact [74]. In terms of vegetation diversity, many experts have suggested that waterways play a bigger role as a habitat than the conventional paddy and, as a result, the influx of plants found in wetlands, such as submerged, merged, and floating plants, will present a large increase. However, the artificial management of the habitats of freshwater fish may introduce unnecessary species, such as naturalized plants and invasive species, or decrease diversity. This is consistent with the finding that the high distribution of naturalized plants is highly influenced by the presence of humans in rural areas due to associated land-use patterns [75–77].

Table 2. Expert assessment of RFMF ecosystem services.

Function	Mean [1]
Amphibian and reptile habitat	2.39 ± 0.69 [F]
Aquatic insect habitat	2.36 ± 0.66 [F]
Fishery habitat	2.34 ± 0.78 [F]
Experience and education	2.29 ± 0.64 [F]
Vegetation diversity	2.13 ± 0.78 [F]
Avian habitat	2.05 ± 0.94 [E,F]
Groundwater recharge	1.71 ± 1.12 [D,E]
Water storage	1.68 ± 1.13 [D,E]
Maintenance of genetic diversity	1.66 ± 1.22 [D,E]
Biological control	1.59 ± 1.07 [D]
Water purification	1.46 ± 1.28 [C,D]
Mammalian habitat	1.39 ± 1.00 [B,C,D]
Creating landscape	1.32 ± 0.98 [B,C,D]
Climate regulation	1.09 ± 1.01 [A,B,C]
Rest area	1.02 ± 1.13 [A,B]
Air quality regulation	0.98 ± 0.98 [A,B]
Flood control	0.89 ± 1.20 [A]

[1] Test result is statistically significant (Duncan): $^A < {}^B < {}^C < {}^D < {}^E < {}^F$.

In addition to rice farming, the fishing of freshwater fish, collection of organisms from waterways, and provision of ecological education are possible. These activities may enhance the organic safety of the produced rice in addition to diversifying income sources by expanding the freshwater fish-harvesting event into a village festival. Some studies have attempted to introduce storks through reducing the use of chemical fertilizers [78–80]. Growing rice in areas where there are Japanese and Korean storks contributes to the development of the rice brand, and tourism can also be improved, as the number of visiting tourists interested in storks will increase [79]. Such relationships among agricultural product production, brand development, education, and tourism form an essential industrial structure in rural areas [81–83].

3.2. Structural Composition Plan for the RFMF Paddy

We propose a plan for the development of an RFMF paddy, considering its various benefits, for urban students (Figure 3). The design was based on an empirical test at a farm actually implementing an RFMF paddy. The size of the RFMF paddy for rice agriculture and ecological education was 10 m × 10 m (=100 m^2 = 1 a). The bank was reinforced by digging the soil inside the paddy, and a 1.2 m *dumbung* was built around it.

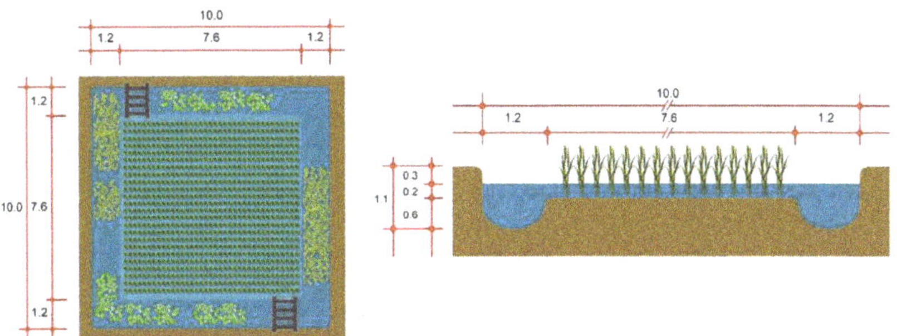

Figure 3. Proposed RFMF paddy design (unit: m).

A bridge-shaped 0.6 m wide passageway was installed to facilitate movement over the waterway to the paddy. The rainfall pattern in Korea is irregular [84,85], and temperature and water are the most important environmental factors in rice farming [86–88]. Therefore, installing an irrigation hole is necessary to periodically water the paddy, in order to counter the effects of irregular rainfall. Additionally, drainage holes must be installed to prevent flooding caused by heavy rain. Korea has historically lost large amounts of agricultural land every year to floods [89,90]. The height of a typical paddy bank is usually maintained at approximately 26 cm [91], and the depth should not exceed 80 cm to ensure the safety of the fish habitat. Moreover, a shallower water depth may be required if the space is to be used by children aged <6 years. Thus, it is recommended that the water depth should not exceed 30 cm, the threshold applied in standards for children's experience facilities [92]. Moreover, getting out of paddy soil is challenging, due to the high clay content [93]. Hence, installing safety bars is recommended when developing an experiential wetland [94].

3.3. Biodiversity of the RFMF Paddy and the Selection of Target Species

We investigated the biodiversity of the RFMF paddy. The field of investigation is vegetation flora, insects of land, and aquatic invertebrates of a hydrographic *dumbung*. Through this investigation, it was confirmed whether there was a species difference between a conventional paddy (CP) and an RFMF paddy (Table 3). In addition, it was possible to identify common species that can be used for experience and education.

The number of observed plant species (taxa) among the various paddy types was in the following order: RFMF$_A$-type (35.6 ± 8.9 species), RFMF$_B$-type (34.8 ± 6.3), and RFMF$_C$-type (34.6 ± 9.8) > CP (23.8 ± 5.3). This corresponds to approximately 10 more species found in the RFMF paddy than in the conventional paddy (CP). However, as a result of ANOVA analysis, this was not statistically verified. In general, in Korean rice paddies, there is a similar variety of plant taxa to the CP study site. Comparing the total number of taxa of the CP and RFMF sites, there is a difference of 26 compared to 32 types. The hydrographic dump is a space that develops wetland plants, which has contributed to improving plant diversity.

The results of examining the diversity of land insects are as follows: RFMF$_A$-type (27.4 ± 11.0 species), RFMF$_C$-type (34.4 ± 14.2), RFMF$_B$-type (19.8 ± 5.4), and CP (19.4 ± 11.7). The RFMF$_B$-type paddy has 15 more species than the conventional paddy (CP). However, as a result of ANOVA analysis, this has not been statistically verified. In addition, CP was found to have a low diversity index (H′) and a high dominance index (DI). This means that the RFMF paddy is more diverse in terms of land insects than the CP.

Table 3. Biodiversity analysis of RFMF paddies.

Classification	RFMF$_A$	RFMF$_B$	RFMF$_C$	CP	Significance	
					F-Value	Post Hoc
Vegetation						
Total taxa	75	85	91	59	-	-
Average	35.6 ± 8.9	34.8 ± 6.3	34.6 ± 9.8	23.8 ± 5.3	2.602	N.S
Insect						
Species	27.6 ± 11.3	34.4 ± 14.2	19.8 ± 5.4	19.4 ± 11.7	2.054	N.S
Individual	57.0 ± 17.8	83.8 ± 26.8	68.4 ± 45.1	61.4 ± 43.7	0.554	N.S
H'	4.41 ± 0.66	4.61 ± 0.80	3.70 ± 0.50	3.40 ± 0.86	3.189	N.S
J'	0.94 ± 0.01	0.92 ± 0.05	0.87 ± 0.10	0.85 ± 0.05	2.410	N.S
RI	6.54 ± 2.48	7.51 ± 2.85	4.64 ± 0.84	4.41 ± 2.28	2.230	N.S
DI	0.21 ± 0.06	0.22 ± 0.12	0.34 ± 0.14	0.44 ± 0.15	4.185 *	CP > C > A, B
Aquatic invertebrates						
Species	19.4 ± 3.1	19.4 ± 3.9	20.6 ± 4.7	4.2 ± 5.8	14.882 ***	C, B, A > CP
Individual	61.8 ± 16.7	87.8 ± 13.1	95.6 ± 39.7	16.2 ± 22.4	10.146 **	C, B, A > CP
H'	3.85 ± 0.22	3.61 ± 0.81	3.83 ± 0.36	1.11 ± 1.52	11.253 ***	A, C, B > CP
J'	0.90 ± 0.01	0.84 ± 0.14	0.88 ± 0.03	0.33 ± 0.45	6.683 **	A, C, B > CP
RI	4.47 ± 0.54	4.22 ± 0.95	4.32 ± 0.70	1.02 ± 1.42	14.814 ***	A, C, B > CP
DI	0.31 ± 0.04	0.39 ± 0.18	0.34 ± 0.06	0.21 ± 0.29	0.883	N.S

* RFMF$_A$: rice + catfish; RFMF$_B$: rice + loach; RFMF$_C$: rice + shrimp; CP: conventional paddy. Test result according to Duncan's multiple range statistically significant at the p = 0.5 level (*), 0.01 level (**), and 0.001 level (***); NS = nonsignificant result. H': diversity index, J': evenness index, RI: richness index and DI: dominance index.

The number of observed aquatic invertebrate species according to type among the various paddy types was in the following order: RFMF$_A$-type (19.4 ± 3.1 species) > RFMF$_B$-type (19.4 ± 3.9) > RFMF$_C$-type (20.6 ± 4.7) > CP (4.2 ± 5.8). Approximately about 15 species were more found in the RFMF paddy than in the conventional paddy. The RFMF paddy with a hydrographic *dumbung* showed a higher number of species, and it can be concluded that the development of a *dumbung* provided expanded space for various habitats. In indices such as diversity, richness, and the evenness index, results in the RFMF paddy were greater than those of the CP. It can be concluded that the *dumbung* created for RFMF contributes to the diversity of aquatic invertebrates.

RFMF paddies are typically managed without fertilizers and pesticides and are filled with freshwater. Jungle rice (*Echinochloa colona*), water foxtail (*Alopecurus geniculatus*), clover (*Trifolium* spp.), dandelion (*Taraxacum* spp.), spiny sowthistle (*Sonchus asper*), conyza (*Erigeron canadensis*), curly dock (*Rumex crispus*), giant chickweed (*Stellaria pubera*), common groundsel (*Senecio vulgaris*), wood sorrel (*Oxalidaceae* spp.), water pepper (*Persicaria hydropiper*), violet (*Viola* spp.), mugwort (*Artemisia* spp.), water parsley (*Oenanthe javanica*), green foxtail (*Setaria viridis*), common duckweed (*Spirodela polyrhiza*), and water hyacinth (*Eichhornia crassipes*) are species that are commonly found in conventional paddies and in banks [95,96].

Weeds growing around rice fields have been used as the main ingredients for oriental medicines for a long time, and, if an educational program was established based on this content, their utilization could be highly promoted [31]. Depending on the input, Far Eastern catfish (*Silurus asotus*), loaches (*Misgurnus anguillicaudatus*), Chinese weatherfish (*Misgurnus mizolepis*), crucian carp (*Carassius carassius*), Asiatic ricefish (*Oryzias latipes*), minnow (*Zacco platypus*), and lake prawns (*Palaemon paucidens*) may inhabit the waterway; it has also been considered that black-spotted pond frog (*Rana nigromaculata*) and tree frog (*Hyla japonica*) can be introduced to the paddy habitat [28,29,97–101]. Although the diversity will increase in the waterway space along with the number of years of maintenance, various common aquatic biota, such as leeches (Hirudinea), Korean muljara (*Appasus japonicus*), water scorpions (*Laccotrephes japonensis*), water beetle (*Asellus hilgendorfii*), diving beetle (*Cybister japonicus*), dark diving beetle (*Hydrophilus acuminatus*), ranatra (*Ranatra chinensis*), Chinese mystery snail (*Cipangopaludina chinensis*), and Radix auricularia (*Lymnaea auricularia*), can appear and inhabit the environment [102–109].

Moreover, bees visit and collect pollen during the rice flowering period [110]. Dragonflies, mantises, grasshoppers, ladybugs, stink bugs, bees, sulfur butterflies, and cabbage butterflies are insects that commonly appear in rice fields [50,108,110–114], and exploring the contribution of the science museum RFMF paddy toward the ecosystem is possible by investigating the species that emerge after its development. Figure 4 presents an illustration of an example RFMF paddy containing various organisms that can be used in such an investigation.

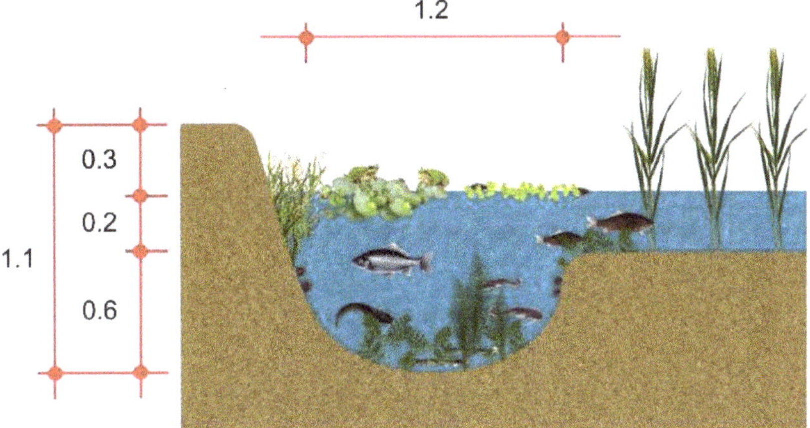

Figure 4. Illustration of an example RFMF paddy containing biological species that can be investigated (unit: m).

3.4. Soil and Water Quality Management Plan for the RFMF Paddy

The soil from a real farm with an RFMF paddy was analyzed to determine the optimal characteristics of the soil environment for developing our RFMF paddy. The soil chemistry results showed that the OM was 1.9%, the pH was 6.6, and available phosphate was 178.0 mg/kg (Table 4). These values are within the recommended chemical ranges for Korean paddy soil [115,116]. Korea's Rural Development Administration conducts soil analysis for the cultivation of rice paddies [117]. The analysis results of our study (Table 4) are similar to those of general rice paddies. However, it is necessary to add additional fertilizers according to the type of rice or the exploitation of organic materials.

Table 4. Topsoil environment of the RFMF paddy.

pH (1:5)	EC (ds/m)	OM (g/kg)	Av.P_2O_5 (mg/kg)	Ex.(cmolc/kg)			Av.SiO_2 (mg/kg)
				K	Ca	Mg	
6.6	0.4	19.0	178.0	0.1	6.3	1.7	307.0

When developing an RFMF paddy, soil with little water loss and the appropriate amount of OM should be prepared to ensure that the soil characteristics are suitable for rice farming [118–120]. When developing a wetland or paddy field within a science museum or city, collecting the topsoil from a nearby paddy and placing it at the target site can help stabilize the soil physicochemical properties for rice growth and easily and quickly establish biodiversity by introducing buried seeds and larvae. Many weed seeds have been reported to be buried in paddy soil [121,122], which are useful for rapidly developing vegetation diversity [123–125]. The introduction of soil (including many plant seeds) has provided ecological restoration in urban areas [126,127].

Tap water can be used as supply water to the paddies in a city. The temperature and quality of the paddy water affect rice growth and yield [86–88,128–131]. Several field experiments have been performed to determine the effects of irrigation conditions on paddy water temperature [132–135]. As a result, when the water temperature and quality differ from those in existing paddy fields, the growth of the rice and living organisms may be negatively affected. The examination of the characteristics of the water environment of the RFMF paddy showed that pH ranged between 7.1 and 8.0, and the water temperature ranged between 20.2 and 30.7 °C during the freshwater period (June–October) for rice production (Figure 5). In fact, the optimal water temperature for rice growth was approximately 15–35 °C, and best temperature was around 30 °C [27,62,136]. While the optimal possible pH for rice growth was 5.0–9.0, the best pH was approximately 6.5–8.5 [137–139]. The water supply for urban agriculture is bound to use public tap water. However, there may be a difference between paddy water and public tap water depending on the season. Rapid changes in pH and water temperature can affect rice production, freshwater fish, and other aquatic organisms [136,140]. Therefore, the water should be stored in a primary tank with a pH maintained between 6.5 and 8.5 before introduction into the paddy to avoid sudden changes in water temperature when irrigating for urban agriculture.

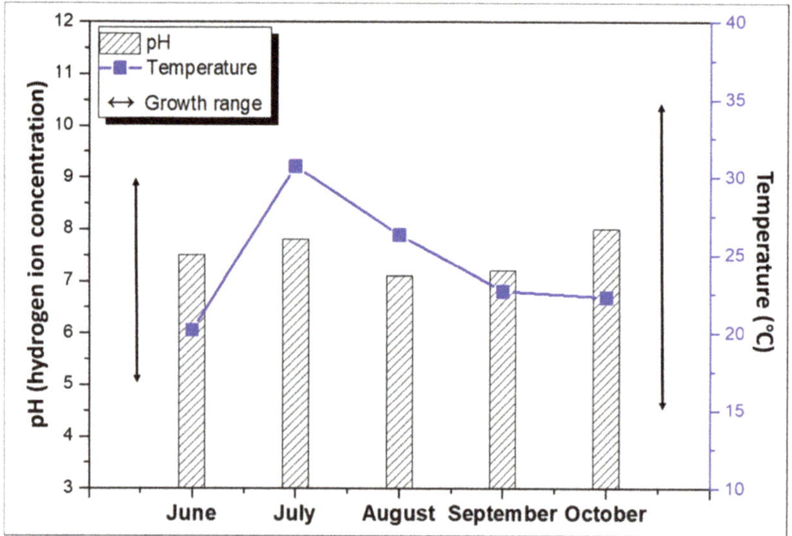

Figure 5. pH and temperature of RFMF paddy water; growth range was presented in [137–139] (pH) and [27,62,136] (temperature).

3.5. Rice and Freshwater Fish Management Methods and Related Educational Contents

Rice cultivation methods in Korea vary greatly depending on the production purpose, region, and rice variety [141–143]. Therefore, it is appropriate for operations to be conducted according to the given situation. The rice variety used in this study is 'Jopyeong'. It generally takes 6 months from sowing to produce rice in Korea [144] (Choe et al., 2003); however, 'Jopyeong' plants can produce rice in 4 months, which is faster than other varieties [145,146]. Shortening the period may be advantageous for the purpose of experience and education. Traditionally, Korea has been an agricultural country, with many rural tourism villages. The ecological experiences of rice and fish farming based on rice paddies in these villages were linked to the present study by analyzing the work of Han, Son, Choi, and Yoon [58].

For rice growth, sound rice seeds must first be selected by brining with saline water. Then, these seeds should be sterilized by soaking in hot water at 60 °C for 10 min. Afterward, they are sprouted, sown in a seedbed, and grown in a rice nursery (with the seedbed and rice nursery comprising an experience). After dividing the RFMF paddy around March, 100 kg/a of manure is applied as basic fertilizer in April, and the paddy is irrigated. The earlier it is irrigated, the more stable the paddy ecosystem, as aquatic organisms can emerge earlier in the paddy. In May, rice seedlings grown in the rice nursery are planted. After installing 30 cm × 20 cm grid lines, 3–5 seedlings are planted at a depth of 2–3 cm (comprising a rice planting experience). The water depth should be maintained between 7 and 10 cm for a week after planting, such that the roots can be quickly established. As the rice grows taller, the water depth needs to be increased. As the rice and weeds compete for nutrients and sunlight, effective weeding must be carried out to manage the weeds (comprising weeding experiences), which are managed through organic farming methods, in consideration of the freshwater fish that are present, and when the occurrence of diseases and pests are observed (based on visual inspection of the bank and inside the paddy and on organic farming education). Rice blooms from July–August and can be harvested, after approximately 40 days, in August and September (comprising rice harvesting and threshing experiences). After harvesting the rice, a new shoot—called an offshoot—emerges from the rice plant root stump. Following its harvest in August, rice is ripened from the new shoot that emerges from the stump. It is possible to recover the RFMF paddy used for rice production for a year by using the rice straw and maintaining the aquatic ecosystem through irrigation (with rope-, straw thatch-, and sandal-making experiences using rice straw). Freshwater rice cultivation is a representative agricultural tradition in East Asia [147,148]. The resulting variety of crafts is a cultural characteristic of many East Asian countries, including Korea, Japan, and China [149]. As mentioned above, various experiential learning experiences can be provided through rice and freshwater fish farming. However, caution is required to ensure measures are in place for preventing young and elderly people from falling into the freshwater fish habitat while visiting the paddy.

To inform our design, we analyzed the growth and yield of an actual RFMF paddy operating at a farm. The results indicated that a 10 a (1000 m^2) RFMF paddy can produce approximately 762.0 kg rice (Table 5). Organic fertilizer was added to Site 1, while nothing was applied to Site 2. At Site 2, the OM present in the soil and the nitrogen and phosphorus contained in the fish feed helped in the production of rice. Using eco-friendly organic products as fertilizers for paddy fields, it is possible to match the rice production yield of Site 1. According to Table 5, the rice cultivation area of 57.76 m^2 in the developed 10 m × 10 m wide RFMF paddy can produce 44.0 kg rice per year, organic fertilizer added to the paddy can produce 47.7 kg (Site 1), and adding nothing to the paddy can produce 40.3 kg (Site 2). The produced rice can be branded as being representative of the food culture of East Asia, which is associated with the traditional practices of making rice cakes and wine [150,151].

Table 5. Results of the rice yield survey of an actual RFMF paddy in a farm.

Type	Plant (cm)	Culm Length (cm)	Panicle		Yield (kg/10a)
			(cm)	(No./m^2)	
Site 1	99.1	78.5	20.6	321	826
Site 2	95.2	73.9	21.3	287	698
Average	97.2	76.2	21.0	304.0	762.0

Catfish, carp, oriental weatherfish, loaches, and shrimp have been introduced into RFMF paddies in Korea [26,152]. When a habitat is established in a paddy, the feed cost is expected to decrease, as the fish will feed on aquatic insects and pests in the water [153,154] which, in turn, positively influences rice growth [155]. This reduces the amount of pesticides needed for rice cultivation and ensures safe production. Freshwater fish were supplied by nurseries and aquafarms two weeks after planting the rice seeds, which is necessary for root establishment, and the fish were introduced to the RFMF paddy from the end of May to early June (freshwater fish introduction). Measuring the amount of dissolved oxygen, ammonia, nitrous acid, and pH of the habitat of the freshwater fish is necessary to assess the changes in water quality. The amount of feed supply is controlled with respect to the growth of fish, which is determined through measuring their length and weight (comprising freshwater fish feeding and size measurement experiences). The freshwater fish can be partially harvested by fishing or with fish traps, nets, etc., when the paddy is irrigated, or fully harvested after the completion of rice production and drainage of water in October/November using scoop and landing nets (comprising freshwater fish fishing and harvesting experiences). The total number, size (length and weight), and growth status of the harvested freshwater fish are examined, and the harvested fish can be stored in a tank over the winter, sold, or used as food (comprising a freshwater fish cooking experience). Catfish grown for approximately 4 months in the waterway-type dumbung of an RFMF paddy can grow up to 30 cm in size and be used as samples for the provision of various experiences and/or as food. The freshwater organisms grown at the study site were catfish, loaches, and shrimp, confirming the possibility of using this setup to grow these three freshwater fish species. However, if juvenile fish are used, it will be difficult to grow fish big enough to be used for food within a few months. Therefore, we suggest using older fish for the experience, as the experience of catching and raising fish in the city is considerably different to that in an aquafarming context.

Table 6 presents the experiential education contents through rice and freshwater fish co-production in the RFMF paddy. A correct interpretation must accompany the environmental education of a mixed ecology paddy—the subject of this study. The number of visitors to Korea's science museums surpasses 4 million annually and, thus, customized education for visitors is essential [20]. The purpose of this study was to implement an environment–ecology–agriculture program for science museum visitors. Understanding information correctly [156,157] and interpretation are important [158] for understanding and educating about the environment. In addition, the roles of marketers and site operators can determine the success or failure of pilot operations such as research sites [159,160]. These processes can contribute to enhancing the image of the operating institution, promoting park activities, increasing local economic value, changing visitor behaviors, and conserving resources [161].

Table 6. RFMF paddy experiential education contents.

Type	Experiential Contents (Han et al., 2015 [58])
Rice farming	Making a seedbed and growing a rice nursery; rice planting; weeding; observation of paddy (bank and inside); organic farming education; rice harvesting and threshing; and rope-making, straw-thatching, and sandal-making using rice straw
Fish farming	Freshwater fish introduction, feeding, size measurement, fishing and harvesting, and cooking

3.6. Development of an RFMF Paddy in an Urban Education Space

We developed a virtual model of an RFMF paddy using a 3-dimensional model for the Agricultural Science Museum of the Rural Development Administration to provide experiences and educational materials for agricultural and ecological education (Figure 6). This education and experience program includes rice production, rice planting, weeding in the paddy, harvesting, threshing, and crafting using rice straw. It is also possible to provide fish feeding, fishing and harvesting, and cooking experiences using freshwater fish. Ecological survey education and experience may include surveys on land and aquatic animals, plants in the paddy, bees during the rice flowering period, and the ecological food chain. The RFMF paddy model can be used as an educational and experiential space for various people, especially urban residents and children. Cities lack natural spaces [162,163] or sufficient space to practice agriculture [163–167]; therefore, several studies have considered the introduction and development of various green areas in urban cities. The RFMF paddy presented in the virtual model can serve as such a space, with great value due to its beneficial impacts on agriculture, green areas, ecology, education, and the environment.

Figure 6. Virtual model (**below**) of an RFMF paddy at the Agricultural Science Museum (**above**).

4. Discussion

The rural areas and agricultural sector of Korea have made diverse attempts to improve agricultural management. Among them, organic agricultural products have produced a higher income than conventional agriculture, and eco-friendly agricultural products are being produced nationwide using various farming methods. This study investigated a new type of complex farming that produces freshwater fish as well as organic rice for the environmental and ecological functions of rural areas. This study proposed a design plan to simultaneously provide paddy-farming education and ecological experience by developing an RFMF paddy in an educational space, such as a science museum or an experience hall.

This study analyzed expert opinions regarding the effects of complex agriculture on the enhancement of ecosystem services. The expert opinion is that the RFMF paddy can be used for ecological experiences and education (2.29 ± 0.64, first grade), such as fishing and organism collection.

This study suggested the size of an RFMF paddy for agriculture and ecological education in the Science Center of approximately 1a (10 × 10 m = 100 m^2), which is achievable during its development.

The RFMF-type with a hydrographic *dumbung* presented a high number of species. Vegetation in the RFMF-type comprised approximately 34.6~35.6 species, and the CP comprised 23.8 species. A similar result was observed for insects, the RFMF-type comprised approximately 19.8~34.4 species, and the CP, 19.4 species. Additionally, aquatic invertebrates in the RFMF-type were made up of 19.4~20.6 species, and in the CP, only 4.2 species. It can be concluded that the development of a *dumbung* provided expanded space for various habitats. In indices such as diversity, richness and the evenness index, results in the RFMF paddy were greater than those in the CP. It can be concluded that the *dumbung* created for RFMF contributes to the diversity of the species. The target plant species for the RFMF paddy were rice, water foxtail, clover, dandelion, spiny sowthistle, conyza, curly dock, giant chickweed, common groundsel, wood sorrel, water pepper, violet, mugwort, water parsley, green foxtail, common duckweed, and water hyacinth, which are commonly found in paddies and banks. The aquatic organisms found in the waterways included Far Eastern catfish; loaches; Chinese weatherfish; crucian carp; Asiatic ricefish; minnow; lake prawns; and amphibians, such as black-spotted pond frog and tree frogs. During the flowering period of rice, it will be possible to observe bees, dragonflies, mantises, grasshoppers, long-headed grasshoppers, ladybugs, sulfur butterflies, and cabbage butterflies, which are commonly observed insects in rice fields. According to the survey results, the vegetation and aquatic invertebrates of the RFMF paddy comprised 40 species more than a conventional paddy. However, this study did not identify biodiversity after the construction of the RFMF paddy in urban areas.

To develop an RFMF paddy in a green space, collecting the topsoil from a nearby paddy (Table 4) and adding it to the target site is advantageous because it can stabilize soil physicochemical properties. Caution is needed when using groundwater and tap water so as not to change the temperature (15–35 °C) of the RFMF paddy water to a large extent. Rice and freshwater fish farming is possible if the pH is maintained between 6.5 and 8.5.

In the proposed design, approximately 44.0 kg of rice can be produced, and catfish can grow up to 30 cm. If organic fertilizers are added, the production and quality of rice can be increased. The educational/experiential aspects of rice and freshwater fish production methods include rice production, rice planting, paddy weeding, rice harvesting, threshing, crafting using rice straw, feeding freshwater fish, freshwater fish fishing, harvesting, and cooking.

5. Conclusions

In this study, designs for application in urban areas, the prediction of biodiversity effects, target species selection, water and soil composition and management, rice production prediction, and experience programs were presented. The RFMF paddy proposed in this study can improve the ecosystem service function of the urban area.

During RFMF operations, care should be taken to prevent young and elderly people from falling into the deep *dumbung* during the experience. Ecological education and experience may include surveys on paddy land and aquatic animals, plants in paddy fields, bees during the rice flowering period, and the ecological food chain. However, the limitations of this study were that the study design was not actually applied and the effects of biodiversity and the environment on the surrounding areas could not be quantitively evaluated while developing an RFMF paddy. Therefore, we will install an RFMF paddy in urban areas through follow-up research. Through that, we will scientifically evaluate the biodiversity and ecosystem service function for RFMF paddies.

The results of this study can be used for agricultural–environmental–ecological education in urban areas, such as science museums, exhibition halls, and public-relations halls. The role of the science museum manager is important for creation, management, and operation. The RFMF paddy can be helpful in discovering the diverse contents of science museums and urban green designs. To date, ecological education and urban agriculture studies have been separate from biology, ecology, and agriculture. However, RFMF rice paddies can be used as an experience space for urban residents for ecological and urban agriculture education. Furthermore, in addition to the educational benefits, mixed farming paddies may contribute to the preservation of biodiversity in the urban area and improve the environment in terms of climate mitigation.

Author Contributions: Conceptualization, J.S. and H.N.; methodology, J.S. and M.K.; software, J.S. and H.N.; validation, H.N. and J.S.; data curation, J.S.; writing—original draft preparation, J.S. and M.K.; writing—review and editing, J.S.; project administration, H.N.; funding acquisition, H.N. All authors have read and agreed to the published version of the manuscript.

Funding: This work was supported by the Research and Development Project, National Institute of Agricultural Sciences, Rural Development Administration (Project Number: PJ01493904), in 2021 and 2022.

Institutional Review Board Statement: Not applicable.

Informed Consent Statement: Not applicable.

Data Availability Statement: Not applicable.

Acknowledgments: This study was supported by the RDA Fellowship Program (J.S. and M.K.) of National Institute of Agricultural Sciences, Rural Development Administration, Korea (2022).

Conflicts of Interest: The authors declare no conflict of interest.

Appendix A

Table A1. Expert assessment result according to the ecosystem function effects of rice–fish mixed farming.

Functions	Major field				Mean (n = 56)	F-Test [1]
	Environmental (n = 20)	Biological (n = 16)	Engineering (n = 14)	Agricultural (n = 6)		
Amphibian and reptile habitat	2.50 ± 0.59 [a,b,D]	2.50 ± 0.71 [a,b,F]	2.00 ± 0.76 [a,E,F]	2.67 ± 0.47 [b,C]	2.39 ± 0.69 [F]	NS
Aquatic insect habitat	2.40 ± 0.58 [a,b,D]	2.56 ± 0.61 [b,F]	1.93 ± 0.70 [a,D,E,F]	2.67 ± 0.47 [b,C]	2.36 ± 0.66 [F]	3.184 *
Fishery habitat	2.35 ± 0.73 [a,C,D]	2.31 ± 0.85 [a,D,E,F]	2.21 ± 0.86 [a,F]	2.67 ± 0.47 [a,C]	2.34 ± 0.78 [F]	NS
Experience and education	2.20 ± 0.60 [a,C,D]	2.38 ± 0.70 [a,E,F]	2.21 ± 0.67 [a,F]	2.50 ± 0.50 [a,B,C]	2.29 ± 0.64 [F]	NS
Vegetation diversity	2.15 ± 0.73 [a,b,C,D]	2.31 ± 0.68 [a,b,D,E,F]	1.71 ± 0.80 [a,B,C,D,E,F]	2.50 ± 0.76 [b,B,C]	2.13 ± 0.78 [F]	NS
Avian habitat	2.10 ± 0.54 [a,C,D]	2.25 ± 0.90 [a,C,D,E,F]	1.79 ± 1.15 [a,C,D,E,F]	2.00 ± 1.41 [a,A,B,C]	2.05 ± 0.94 [E,F]	NS
Groundwater recharge	1.70 ± 1.10 [a,B,C]	1.50 ± 1.00 [a,A,B,C,D]	1.64 ± 1.34 [a,B,C,D,E,F]	2.50 ± 0.50 [a,B,C]	1.71 ± 1.12 [D,E]	NS
Water storage	1.70 ± 1.05 [a,B,C]	1.44 ± 1.12 [a,A,B,C]	1.50 ± 1.24 [a,B,C,D,E,F]	2.67 ± 0.47 [b,C]	1.68 ± 1.13 [D,F]	NS
Maintenance of genetic diversity	1.35 ± 1.28 [b,A,B]	1.81 ± 1.07 [a,b,B,C,D,E,F]	1.57 ± 1.35 [a,b,B,C,D,E,F]	2.50 ± 0.50 [a,B,C]	1.66 ± 1.22 [D,E]	NS
Biological control	1.45 ± 1.02 [a,A,B]	1.75 ± 1.03 [a,B,C,D,E,F]	1.50 ± 1.18 [a,B,C,D,E,F]	1.83 ± 1.07 [a,A,B,C]	1.59 ± 1.07 [D]	NS
Water purification	1.30 ± 1.27 [a,A,B]	1.38 ± 1.62 [a,A,B]	1.64 ± 1.04 [a,B,C,D,E,F]	1.83 ± 0.69 [a,A,B,C]	1.46 ± 1.28 [C,D]	NS

Land **2022**, 11, 1218

Table A1. *Cont.*

Functions	Major field				Mean (n = 56)	F-Test [1]
	Environmental (n = 20)	Biological (n = 16)	Engineering (n = 14)	Agricultural (n = 6)		
Mammalian habitat	1.25 ± 0.89 [a,A,B]	1.63 ± 1.11 [a,B,C,D,E]	1.21 ± 1.01 [a,A,B,C,D,E,F]	1.67 ± 0.94 [a,A,B,C]	1.39 ± 1.00 [B,C,D]	NS
Creating landscape	1.45 ± 0.80 [a,A,B]	1.25 ± 1.09 [a,A,B]	1.14 ± 1.12 [a,A,B,C,D,E]	1.50 ± 0.76 [a,A,B]	1.32 ± 0.98 [B,C,D]	NS
Climate regulation	1.10 ± 0.83 [a,A,B]	1.25 ± 1.03 [a,A,B]	0.86 ± 1.30 [a,A,B,C]	1.17 ± 0.69 [a,A]	1.09 ± 1.01 [A,B,C]	NS
Rest area	1.00 ± 1.00 [a,A]	1.00 ± 1.27 [a,A,B]	0.93 ± 1.22 [a,A,B,C,D]	1.33 ± 0.94 [a,A]	1.02 ± 1.13 [A,B]	NS
Air quality regulation	1.25 ± 0.94 [a,A,B]	0.75 ± 0.75 [a,A]	0.71 ± 1.28 [a,A,B]	1.33 ± 0.47 [a,A]	0.98 ± 0.98 [A,B]	NS
Flood control	1.20 ± 0.93 [a,b,A,B]	0.69 ± 1.04 [a,A]	0.29 ± 1.44 [a,A]	1.83 ± 0.90 [b,A,B,C]	0.90 ± 1.20 [A]	3.305 *
F-test [2]	5.916 ***	5.519 ***	3.053 ***	2.590 **	14.503 ***	-

* Test result is statistically significant at the $p = 0.5$ level (*), 0.01 level (**), and 0.001 level (***); NS = nonsignificant result. [1] Results according to the major field types; lowercase letters indicate the four major fields (Duncan): width [a] < [b] < [c]. [2] Result according to the ecosystem service functions; uppercase letters indicate the 17 functions (Duncan): length [A] < [B] < [C] < [D] < [E] < [F].

References

1. Na, Y.E. Driving projects of urban agriculture for the energy independence. *Korean J. Environ. Agric.* **2010**, *29*, 304–308. [CrossRef]
2. Lee, I.H.; Lee, H.J.; Lee, S.B.; Jeon, I.C.; Kim, Y.G. The effect on participating in the urban farming in the farm village experience tourism of urbanite. *J. Korean Inst. Lands Archit.* **2012**, *40*, 79–88. [CrossRef]
3. Kim, E.J.; Lee, Y.K.; Kim, S.B.; Lim, C.S.; Park, M.J. A study on the Development Course of Guideline for Fostering the Rural Village Roads. *J. Korean Soc. Rural Plan.* **2013**, *19*, 189–207. Available online: https://kiss15.kstudy.com/kiss5/viewer.asp (accessed on 15 September 2021). [CrossRef]
4. Jang, D.H.; Soh, S.Y. A Study on Management of Urban Agriculture—The case of agriculture in Seoul. In *Bulletin of the Agricultural College*; Chonbuk National University: Jeonju, Korea, 2005; Volume 36, pp. 86–102. Available online: https://kiss18.kstudy.com/kiss10/download_viewer.asp (accessed on 15 September 2021).
5. Seo, H.S.; Hwang, J.H. A study on attribution of purchasing environment-friendly agricultural products in villages for rural tourism. *Korean J. Org. Agric.* **2014**, *22*, 47–65. [CrossRef]
6. Getz, D.A.; Karow, J.J. Inner City Preferences for Trees and Urban Forestry Programs. *Technol. Arboricult.* **1982**, *8*, 258–263. Available online: https://agris.fao.org/agris-search/search.do?recordID=US19830950742 (accessed on 15 September 2021).
7. Lee, Y.H.; Lee, S.D.; Oh, C.H.; Kong, J.Y. Activation Of Agricultural Awareness Education through "Short-Term Agricultural Experiential Learning Program"—A Case Study in Bahl-Ahn Agricultural High School, Jpoaugrensa. *J. Agric. Educ. Hum. Resour Dev.* **1998**, *30*, 25–38. Available online: https://kiss.kstudy.com/thesis/thesis-view.asp?key=1589018 (accessed on 15 September 2021).
8. Choi, E.Y.; Jeong, Y.N.; Kim, S.Y. Analysis of the Relationship between the Importance of Urban Farming Ordinance Factors and Participation Satisfaction as well as Sustainability for Vitalizing of the Urban Farming—Focused on Seoul City Urban Farming Participants. *J. Urban Des.* **2014**, *15*, 173–188. Available online: https://www.kci.go.kr/kciportal/ci/sereArticleSearch/ciSereArtiView.kci?sereArticleSearchBean.artiId=ART001945897 (accessed on 15 September 2021).
9. Jang, G.S.; Kim, Y.H.; Choi, Y.S.; Kim, S.H.; Kim, J.M.; Bae, S.J.; Cho, Y.G.; Koo, T.H. A research of soil environmental health in urban garden, Gwangju. *Korean J. Environ. Agric.* **2016**, *35*, 87–96. [CrossRef]
10. Jang, B.G.; Choi, Y.J.; Hwang, J.I. Needs assessment for urban agricultural program. *J. Agric. Extension Commun. Dev.* **2011**, *18*, 511–529. [CrossRef]
11. Choi, J.S. Amendments of urban planning related regulations and policies for securing the space available for urban agriculture. *Jepa* **2014**, *22*, 131–166. [CrossRef]
12. Jang, J.; Oh, C.H. Needs analysis of experiential learning Program for Eco-Friendly School Farm Activation -Target of teachers in elementary school, Korean. *Org. Agric.* **2012**, *20*, 283–296. Available online: https://kiss16.kstudy.com/kiss6/download_viewer.asp (accessed on 15 September 2021).
13. Kong, M.J.; Park, K.L.; Son, J.K.; Shin, J.H. Evaluation on Actual Condition and Image Analysis of Roof Garden in Seoul, Korea. *J. Korea Soc. Environ. Restor. Technol.* **2012**, *15*, 69–83. Available online: https://www.koreascience.or.kr/article/JAKO201216536729393.pdf (accessed on 15 September 2021). [CrossRef]
14. Won, S.Y. Farming Types and Plant Resources Proper to Vegetable Gardening of Rooftop for Urban Agriculture Activists, The Korean Society of Floral. *Art Des.* **2013**, *28*, 181–221. Available online: https://kiss.kstudy.com/thesis/thesis-view.asp?key=3150847 (accessed on 15 September 2021).

15. An, P.G.; Kong, M.J.; Lee, S.M.; Kim, S.B.; Jo, J.L.; Kim, N.C.; Shin, J.H. A study on the morphological management of major landscape elements in organic farming. *Korean J. Rural Plan.* **2020**, *26*, 107–116. [CrossRef]

16. Kim, S.E. The Effect of Rice Co-Brand Assets, Trust, and Attachment on Loyalty. *J. Digital Convergence.* **2022**, *20*, 401–410.

17. Kim, T.K. Proper Space and Its Conditions for Ecology-Culture(connected)-Environmental Education. *Environ. Educ.* **2010**, *23*, 62–81. Available online: https://www.dbpia.co.kr/pdf/pdfView.do?nodeId=NODE01532582&mark=0&useDate=&ipRange=N&accessgl=Y&language=ko_KR&hasTopBanner=false (accessed on 15 September 2021).

18. Han, J.S. Regional Activities to attract Affiliated Population through Agricultural Education and Experimental Activities: Focused on Exchange Programs between Seoul and Other Cities/Counties. *J. Korean Geogr. Soc.* **2019**, *54*, 435–448. Available online: https://www.dbpia.co.kr/pdf/pdfView.do?nodeId=NODE08768123&mark=0&useDate=&ipRange=N&accessgl=Y&language=ko_KR&hasTopBanner=false (accessed on 15 September 2021).

19. Kim, M.J. Reinventing Vacant Lands for Urban Agriculture: Evaluating Milwaukee's Vacant Land Programs and Initiatives. *J. Land Hous. Urban Affairs.* **2020**, *11*, 59–68.

20. Moon, J.W.; Na, J.Y. Elementary school teachers' perception and the status of education program on science museum field trips. *J. Elem. Sci. Educ.* **2019**, *38*, 87–101. [CrossRef]

21. Park, S.M. Analysis of Scientific Exhibits in the Gwacheon National Science Museum and Their Utilization for Science Education. Master's Thesis, The Graduate School of Education Hanyang University, Seoul, Korea, 2011. Available online: http://www.riss.kr/link?id=T12507214&outLink=K (accessed on 15 September 2021).

22. Hwang, Y.M.; Shin, D.H.; Bae, G.K. Farmhouse Management Strategy and Yield of Farmhouses. *Korean Rural. Sociol. Soc.* **2016**, *26*, 87–121. Available online: https://www.dbpia.co.kr/pdf/pdfView.do?nodeId=NODE07180420&mark=0&useDate=&ipRange=N&accessgl=Y&language=ko_KR&hasTopBanner=false (accessed on 15 September 2021).

23. Kim, H.; Kim, K.J.; Sun, Y.; Jo, Y.J.; Kim, T.Y.; Moon, M.J. Change in Biodiversity and Community Structures in Agricultural Fields depending on Different Farming Methods. *Korean J. Org. Agric.* **2018**, *26*, 687–706. [CrossRef]

24. Myeong, S.J. The Influence of Preadulthood Experiences in Rural Areas on Korean Adults' Perceptions on the Ecosystem Services of Rice-Paddies. *Korean Soc. Environ. Eucation.* **2019**, *32*, 488–497.

25. Kong, M.J.; Kim, C.H.; Lee, S.M.; Park, K.L.; An, N.H.; Cho, J.L.; Lim, J.A.; Kim, H.S.; Nam, H.S.; Son, J.K. The Effect Analysis of Vegetation Diversity on Ricefish Mixed Farming System in Paddy Wetland. *J. Wet Res.* **2018**, *20*, 398–409. Available online: https://www.earticle.net/Article/A343965 (accessed on 15 September 2021).

26. Nam, H.S.; Park, K.L.; An, N.H.; Lee, S.M.; Cho, J.L.; Kim, B.R.; Lim, J.A.; Lee, C.W.; Kong, M.J.; Son, J.K. Wetland function evaluation and expert assessment of organic rice-freshwater fish mixed farming. *J. Wet Res.* **2018**, *20*, 161–172.

27. Halwart, M.; Gupta, M.V. *Culture of Fish in Rice Fields*; Food and Agriculture Organization and the World Fish Center: Rome, Italy, 2004. Available online: https://ideas.repec.org/b/wfi/wfbook/16334.html (accessed on 15 September 2021).

28. Kim, J.O.; Shin, H.S.; Yoo, J.H.; Lee, S.H.; Jang, K.S.; Kim, B.C. Functional evaluation of small-scale pond at paddy field as a shelter for mudfish during midsummer drainage period, Korean. *J. Environ. Agric.* **2011**, *30*, 37–42.

29. Choe, L.J.; Han, M.S.; Kim, M.R.; Cho, K.J.; Kang, K.K.; Na, Y.E.; Kim, M.H. Characteristics communities structure of benthic macroinvertebrates at irrigation ponds, within paddy field, Korean. *J. Environ. Agric.* **2013**, *32*, 304–314.

30. Son, J.K.; Kong, M.J.; Kang, D.H.; Nam, H.S.; Kim, N.C. The comparative studies on the urban and rural landscape for the plant diversity improvement in pond wetland. *J. Wet Res.* **2015**, *17*, 62–74. [CrossRef]

31. Son, J.K.; Kong, M.J.; Kang, B.H.; Kim, M.H.; Kang, D.H.; Lee, S.Y.; Han, S.H. An analysis on use patterns of oriental medicine of pond wetland plants for the ecological experience in rural tourism village. *J. Wet Res.* **2017**, *19*, 230–239.

32. Bradley, G.A. *Urban Forest Landscape*; University of Washington Press: Washington, DC, USA, 1995. Available online: https://www.hfsbooks.com/books/urban-forest-landscapes-bradley (accessed on 15 September 2021).

33. Lee, G.B.; Kim, C.H.; Kim, J.G.; Lee, D.B.; Lee, S.B.; Na, S.Y. How Soil Characteristics and Vegetation Influence the Inflow of Sewage in a Tributary of the Mankyeong River. *Inst. Environ. Sci.* **2003**, *12*, 9–21. Available online: https://kiss.kstudy.com/thesis/thesis-view.asp?key=2221039 (accessed on 15 September 2021).

34. Reid, W.V.; Mooney, H.A.; Cropper, A.; Capistrano, D.; Carpenter., S.R.; Chopra, K.; Zurek, M.B. Ecosystems and Human Well-being-Synthesis: A Report of the Millennium Ecosystem Assessment. Island Press: Washington, DC, USA, 2005.

35. Díaz, S.; Fargione, J.; Chapin, F.S., III; Tilman, D. Biodiversity loss threatens human well-being. *PLoS Biol.* **2006**, *4*, e277. [CrossRef] [PubMed]

36. Ahn, I.; Park, J.S.; Kim, S.H.; Maeng, W.Y.; Lee, I.E. Comparative analysis of technical system by six organic rice cultivation type in the Southern Provinces. *Org. Agric.* **2012**, *20*, 535–542. [CrossRef]

37. Avasthe, R.; Babu, S.; Singh, R.; Das, S. Impact of Organic Food Production on Soil Quality. *Conserv. Agric. Adv. Food Secur. Chaning Clim.* **2017**, *1*, 387–396. Available online: https://www.researchgate.net/publication/326827479_Impact_of_Organic_Food_Production_on_Soil_Quality (accessed on 15 September 2021).

38. Ok, J.H.; Cho, J.L.; Lee, B.M.; An, N.H.; Shin, J.H. Monitoring for Change of Soil Characteristics by repeated Organic Supply of Comport and Green Manures in Newly reclaimed Organic Upland Field. *Korea J. Org. Agric.* **2015**, *23*, 813–827. [CrossRef]

39. Venkat, K. Comparison of Twelve Organic and Conventional Farming Systems: A Life Cycle Greenhouse Gas Emissions Perspective. *J. Sustain. Agric.* **2012**, *36*, 620–649. [CrossRef]

40. Kim, C.G.; Kyun, J.H.; Kim, Y.G. Effects of Organic Farming on Greenhouse Gas Emission Reduction. *J. Clim. Chang. Reser.* **2016**, *7*, 335–339. [CrossRef]

41. Han, M.S.; Nam, H.K.; Kang, K.K.; Kim, M.R.; Na, Y.E.; Kim, H.R.; Kim, M.H. Characteristics of Benthic Invertebrates in Organic and Conventional Paddy Field. *Korean J. Environ. Agric.* **2013**, *32*, 17–23. [CrossRef]
42. Kremen, C.; Miles, A. Ecosystem Services in Biologically Diversified versus Conventional Farming Systems: Benefits, Externalities, and Trade-Offs. *Ecol. Soc.* **2012**, *17*, 40. [CrossRef]
43. Arunrat, N.; Sereenonchai, S. Assessing Ecosystem Services of Rice–Fish Co-Culture and Rice Monoculture in Thailand. *Agronomy* **2022**, *12*, 1241. [CrossRef]
44. Liu, D.; Tang, R.; Xie, J.; Tian, J.; Shi, R.; Zhang, K. Valuation of ecosystem services of rice-Fish coculture systems in Ruyuan County. *China Ecosyst. Serv.* **2020**, *41*, 101054. [CrossRef]
45. Kim, S.Y.; Choi, J.Y.; Park, C.; Song, W.K.; Kim, S.Y.L. Changes of Ecosystem Services in Agricultural Area According to Urban Development Scenario? For the Namyangju Wangsuk District 1, the 3rd Phase New Town? *J. Environ. Impact Assess.* **2021**, *30*, 117–131. [CrossRef]
46. Son, J.K.; Kong, M.J.; Kang, D.H.; Lee, S.Y.; Han, S.H.; Kang, B.H.; Kim, N.C. A study on selection of butterfly and plant species for butterfly gardening. *J. Korean Soc. Rural. Plan.* **2015**, *21*, 1–12. [CrossRef]
47. Han, Y.S.; Nam, H.S.; Park, K.L.; Lee, Y.M.; Lee, B.M.; Park, K.C. Evaluation of Soil Carbon Storages in the Organic Farming Paddy Fields. *J. Korea Org. Res. Recy. Asso.* **2020**, *28*, 73–82.
48. Kim, H. Issues on Overcoming Present Crises of Organic Agriculture through its Philosophy and Principle. *Korean J. Org. Agric.* **2017**, *25*, 53–69.
49. Lee, Y.S.; Han, H.G.; Lim, S.G.; Kim, K.S.; Cheong, C.J. Physico-Chemical Characteristics in the Sedimnets in Loach Farms. *J. Korean Soc. Envir. Technol.* **2013**, *14*, 315–323. Available online: https://kiss16.kstudy.com/kiss61/download_viewer.asp (accessed on 15 September 2021).
50. Lee, T.B. *Coloured Flora of Korea*; Hyangmunsa: Seoul, Korea, 2006.
51. Baek, M.K.; Hwang, J.M.; Jeong, G.S.; Kim, T.W.; Kim, M.C.; Lee, Y.J.; Jo, Y.B.; Park, S.W.; Lee, H.S.; Goo, D.S.; et al. *Checklist of Korean Insects*; Nature Publishing & Ecology Academic: Seoul, Korea, 2010.
52. Margalef, R. Information theory in Ecology. *Gen. Syst.* **1958**, *3*, 36–71. Available online: https://scholar.google.com/scholar_lookup?title=Information%20theory%20in%20ecology&author=R.%20Margalef&journal=YearB%20Soc%20Gen%20Syst%20Res&volume=3&pages=36-71&publication_year=1958 (accessed on 15 September 2021).
53. Mcnaughton, S.J. Relationships among Functional Properties of Californian Grassland. *Nature* **1967**, *216*, 144–168. Available online: https://www.nature.com/articles/216168b0 (accessed on 15 September 2021). [CrossRef]
54. Pielou, E.C. Shannon's Formula as a Measure of Specific Diversity: Its Use and Misuse. *Amer. Nat.* **1969**, *100*, 463–465. Available online: https://www.journals.uchicago.edu/doi/abs/10.1086/282439 (accessed on 15 September 2021). [CrossRef]
55. Pielou, E.C. *Ecological Diversity*; John Wiley & Son: New York, NY, USA, 1975. Available online: https://www.abebooks.com/first-edition/ECOLOGICAL-DIVERSITY-Pielou-E-C-John/12152548852/bd (accessed on 15 September 2021).
56. Zar, J.H. *Biostatistical Analysis*, 3rd ed.; Prentice-Hall International Editions: Englewood Cliffs, NJ, USA, 1996.
57. Korea Forest Research Institute. Analysis Methods of Soil and Plant. 2014. Available online: https://www.google.co.kr/url?sa=t&rct=j&q=&esrc=s&source=web&cd=&ved=2ahUKEwjyjMGd65PzAhUZyosBHfUtA_UQFnoECAoQAQ&url=http%3A%2F%2Fknow.nifos.go.kr%2Fbook%2FMo.file%3Fcid%3D162927%26rid%3D5&usg=AOvVaw0AG9TQk_uJbyKhdTLL1qpZ (accessed on 15 September 2021).
58. Han, S.H.; Son, J.K.; Choi, Y.J.; Yoon, Y.S. A study on the types classification and analysis of experience activities in rural tourism village. *J. Agric. Ext. Commun. Dev.* **2015**, *22*, 31–41. [CrossRef]
59. Matsuno, Y.; Nakamura, K.; Masumoto, T.; Matsui, H.; Kato, T.; Sato, Y. Prospectsfor multifunctionality of paddy rice cultivation in Japan and other countries inmonsoon Asia. *Paddy Water Environ.* **2006**, *4*, 197. [CrossRef]
60. Yoon, C.G. Wise use of paddy fields to partially compensate for the loss of natural wetlands. *Paddy Water Environ.* **2009**, *7*, 366. [CrossRef]
61. Organization for Economic Co-Operation and Development. Multifuctionality towards an Analytical Framework. 2001. Available online: https://www.oecd-ilibrary.org/docserver/9789264192171-sum-en.pdf?expires=1627970124&id=id&accname=guest&checksum=A1AD894AC6784ACE5C38A7DD05A776E1 (accessed on 15 September 2021).
62. Natuhara, Y. Ecosystem Services by Paddy Fields as Substitutes of Natural Wetlands in Japan. *Ecol. Eng.* **2013**, *56*, 97–106. Available online: https://www.sciencedirect.com/science/article/pii/S092585741200153X (accessed on 15 September 2021). [CrossRef]
63. Hazell, D.; Hero, J.; Lindenmayer, D.; Cunningham, R. A comparison of constructed and natural habitat for frog conservation in an Australian agricultural landscape. *Biol. Conserv.* **2004**, *119*, 61–71. [CrossRef]
64. Williams, P.; Whitfield, M.; Biggs, J.; Bray, S.; Fox, G.; Nicolet, P.; Sear, D. Comparative biodiversity of rivers, streams, ditches and ponds in an agricultural landscape in Southern England. *Biol. Conserv.* **2004**, *115*, 329–341. [CrossRef]
65. Davies, B.R.; Biggs, J.; Williams, P.; Whitfield, M.; Nicolet, P.; Sear, D.; Bray, S.; Maund, S. Comparative biodiversity of aquatichabitats in the European agricultural landscape. *Agric. Ecosyst. Environ.* **2008**, *125*, 1–8. [CrossRef]
66. Sebastián-González, E.; Sánchez-Zapata, J.A.; Botella, F. Agricultural ponds as alternative habitat for waterbirds: Spatial and temporal patterns of abundance and management strategies. *Eur. J. Wildl. Res.* **2010**, *56*, 11–20. [CrossRef]
67. Tscharntke, T.; Klein, A.M.; Kruess, A.; Steffan-Dewenter, I.; Thies, C. Landscape perspectives on agricultural intensification and biodiversity–ecosystem service management. *Ecol. Lett.* **2005**, *8*, 857–874. [CrossRef]

68. Antle, J.M.; Stoorvogel, J.J. Predicting the Supply of Ecosystem Services from Agriculture. *Am. J. Agric. Econ.* **2006**, *88*, 1174–1180. Available online: https://www.jstor.org/stable/4123588 (accessed on 15 September 2021). [CrossRef]

69. Food and Agriculture Organization. *Scoping Agriculture–Etland Interactions. Towards a Sustainable Multiple-Response Strategy*; FAO Water Reports; FAO, Food and Agriculture Organization of the United Nations: Washington, DC, USA, 2008; Volume 33. Available online: http://www.fao.org/3/i0314e/i0314e.pdf (accessed on 15 September 2021).

70. Porter, J.; Costanza, R.; Sandhu, H.; Sigsgaard, L.; Wratten, S. The value of producing food, energy, and ecosystem services within an agro-ecosystem. *AMBIO J. Hum. Environ.* **2009**, *38*, 186–193. [CrossRef] [PubMed]

71. Ribaudo, M.; Greene, C.; Hansen, L.; Hellerstein, D. Ecosystem services from agriculture: Steps for expanding markets. *Ecol. Econ.* **2010**, *69*, 2085–2092. [CrossRef]

72. Raymond, L.R.; Hardy, L.M. Effects of a clearcut on a population of the mole salamander, Ambystoma talpoideum, in an adjacent unaltered forest. *J. Herpetol.* **1991**, *25*, 509–512. [CrossRef]

73. Marc, M.J. Amphibian Activity, Movement Patterns, and Body Size in Fragmented Peat Bogs. *J. Herpetol.* **2001**, *35*, 13–20. Available online: https://www.millenniumassessment.org/documents/document.356.aspx.pdf (accessed on 15 September 2021).

74. Pitta, P.; Karakassis, I.; Tsapakis, M.; Zivanovic, S. Natural versus Mariculture Induced Variability in Nutrients and Plankton in the Eastern Mediterranean. *Hydrobiologia* **1999**, *391*, 194. Available online: https://www.researchgate.net/publication/285829414_Natural_vs_mariculture_induced_variability_in_nutrients_and_plankton_in_the_eastern_Mediterranean (accessed on 15 September 2021).

75. Karakassis, I.; Tsapakis, M.; Hatziyanni, E.; Pitta, P. Diel variation of Nutrients and Chlorophyll in Sea Bream and Sea Bass Cages in the Mediterranean. *Fresenius Environ. Bull.* **2001**, *10*, 283. Available online: https://www.researchgate.net/publication/222092760_Diel_variation_of_nutrients_and_chlorophyll_in_sea_bream_and_sea_bass_cages_in_the_Mediterranean (accessed on 15 September 2021).

76. Vilà, M.; Pino, J.; Font, X. Regional assessment of plant invasions across different habitat types. *J. Veg. Sci.* **2007**, *18*, 35–42. [CrossRef]

77. Valladares, F.; Niinemets, Ü. Shade tolerance, a key plant feature of complex nature and consequences. *Annu. Rev. Ecol. Evol. Syst.* **2008**, *39*, 237–257. [CrossRef]

78. Milbau, A.; Stout, J.C.; Graae, B.J.; Nijs, I. A hierarchical framework for integrating invasibility experiments incorporating different factors and spatial scales. *Biol. Invas.* **2009**, *11*, 941–950. [CrossRef]

79. Hayashi, K.; Nishimiya, H. Restoring Rice Paddy Field Habitats to Reintroduce the Oriental White Stork in Toyooka City. *TEEB: Worldwide*. 2012. Available online: http://www.cbd.int/financial/pes/japan-pesstork.pdf (accessed on 15 September 2021).

80. Sawauchi, D.; Yamamoto, Y. Assessment of carbon dioxide emissions from biodiversity-conscious farming: A case of stork-friendly farming in Japan. *Low Carbon Econ.* **2015**, *6*, 7–12. [CrossRef]

81. Tanaka, W.; Itsukushima, R. Attempt to Develop High-Value rice In the Shimojin District, Mashiki Town, Kumamoto Prefecture: Transition Into Sustainable Local Community Using Disaster Recovery from the 2016 Kumamoto Earthquakes as a Branding Strategy. In *Decision Science for Future Earth*; Springer: Singapore, 2021; pp. 233–251. Available online: https://link.springer.com/chapter/10.1007/978-981-15-8632-3_12 (accessed on 15 September 2021).

82. Hyski, M.; Chudy-Hyski, D. Regional Tourism Brand—Need or Necessity in the Aspect of Socio-Economic Development of Polish Mountain Rural Areas. *SIE* **2018**, *6*, 200–211. Available online: http://journals.ru.lv/index.php/SIE/article/viewFile/3235/3248 (accessed on 15 September 2021). [CrossRef]

83. Ciodyk, T. Agroturystyka w Polsce—Znaczenie, Szanse i Bariery Rozwoju. *Bez Granic* **2000**, *6*, 6–8. Available online: http://agro.icm.edu.pl/agro/element/bwmeta1.element.agro-article-7bcf0397-6f36-4482-a380-f208f6e73aa2 (accessed on 15 September 2021).

84. Chudy-Hyski, D. Uwarunkowania Turystycznego Kierunku Rozwoju Gorskich Obszarow Wiejskich Polski. *Infrastrukt. I Ekol. Teren. Wiejskich. Ser. Monogr.* **2009**, *1*, 9–311. Available online: http://www.infraeco.pl/pl/art/a_15399.htm?plik=525 (accessed on 15 September 2021).

85. Kim, B.-J.; Kripalani, R.H.; Oh, J.-H.; Moon, S.-E. Summer monsoon rainfall patterns over South Korea and associated circulation features. *Theor. Appl. Climatol.* **2002**, *72*, 65–74. [CrossRef]

86. Lee, J.Y.; Kwon, M.; Yun, K.S.; Min, S.K.; Park, I.H.; Ham, Y.G.; Jin, E.K.; Kim, J.; Seo, K.; Kim, W.; et al. The long-term variability of Changma in the east Asian summer monsoon system: A review and revisit. *Asia Pac. J. Atmos. Sci.* **2017**, *53*, 257–272. [CrossRef]

87. Matsushima, S.; Tanaka, T.; Hoshino, T. Analysis of yield-determining process and its application to yield-prediction and culture improvement of lowland rice: LXX. Combined effects of air-temperatures and water-temperatures at different stages of growth on the grain yield and its components of lowland rice. *Jpn. J. Crop. Sci.* **1964**, *33*, 53–58, (In Japanese with English Summary).

88. Shimono, H.; Hasegawa, T.; Iwama, K. Response of growth and grain yield in paddy rice to cool water at different growth stages. *Field Crops Res.* **2002**, *73*, 67–79. [CrossRef]

89. Shimono, H.; Hasegawa, T.; Fujimura, S.; Iwama, K. Responses of leaf photosynthesis and plant water status in rice to low water temperature at different growth stages. *Field Crops Res.* **2004**, *89*, 71–83. [CrossRef]

90. Jung, W.K.; Lim, S.K.; Kim, M.S. Changes of Agricultural Land Use after Flooding Analyzed by Landsat-TM Data. *Korean J. Int. Agric.* **1999**, *11*, 155–160. Available online: http://db.koreascholar.com/article.aspx?code=286120 (accessed on 15 September 2021).

91. Lee, D.Y.; Kim, U.; Yoo, C.S. Climate change impacts on meteorological drought and flood. *J. Korea Water Resour. Assoc.* **2004**, *37*, 315–328. [CrossRef]

92. Jeon, J.H.; Yoon, C.G.; Ham, J.H.; Jung, K.W. Model Development for Nutrient Loading Estimates from Paddy Rice Fields in Korea. *J. Environ. Sci. Health B* **2004**, *39*, 845–860. Available online: https://pubmed.ncbi.nlm.nih.gov/15620091/ (accessed on 15 September 2021). [CrossRef]

93. Ministry of the Interior and Safety. *Children's Play Facility Safety Management Manual*; Ministry of the Interior and Safety: Sejong, Korea, 2018. Available online: https://www.mois.go.kr/synap/skin/doc.html?fn=BBS_201812170248321760&rs=/synapFile/202109/&synapUrl=%2Fsynap%2Fskin%2Fdoc.html%3Ffn%3DBBS_201812170248321760%26rs%3D%2FsynapFile%2F202109%2F&synapMessage=%EC%A0%95%EC%83%81 (accessed on 15 September 2021).

94. Manceau, A.; Tommaseo, C.; Rihs, S.; Geoffroy, N.; Chateigner, D.; Schlegel, M.; Tisserand, D.; Marcus, M.A.; Tamura, N.; Chen, Z.S. Natural speciation of Mn, Ni, and Zn at the micrometer scale in a clayey paddy soil using X-ray fluorescence, absorption, and diffraction. *Geochim. Cosmochim. Acta* **2005**, *69*, 4007–4034. [CrossRef]

95. Lee, S.; Kim, M.; Kang, B.; Son, J. The development of evaluation indicator for eco-experience in rural village. *J. Agric. Extension Commun. Dev.* **2014**, *21*, 1125–1147. [CrossRef]

96. Oh, Y.J.; Sohn, S.I.; Kim, C.S.; Kim, B.Y.; Kang, B.H. Phytosociological Classification of vegetation in paddy levee, Korean. *J. Environ. Agric.* **2008**, *27*, 413–420.

97. Kang, B.H.; Son, J.K. The study on the evaluation of environment function at small stream. In the case of Hongdong Stream in Hongsung gun. *J. Korean Soc. Environ. Eng.* **2011**, *14*, 81–101. [CrossRef]

98. Caton, B.P.; Foin, T.C.; Hill, J.E. A plant growth model for integrated weed management in direct-seeded rice: II. *Field Crops Res.* **1999**, *62*, 145–155. [CrossRef]

99. Cheong, S.W.; Sung, H.C.; Park, D.S.; Park, S.R. Population Viability Analysis of a Gold-Spotted Pond Frog (Rana chosenica) Population: Implications for Effective Conservation and Re-Introduction. *Korean J. Environ. Biol.* **2009**, *27*, 73–81. Available online: http://www.koreascience.kr/article/JAKO200920258463369.page (accessed on 15 September 2021).

100. Park, S.H.; Cho, K.H. Comparison of health status of Japanese tree frog (*Hyla japonica*) in a rural and an urban area, ecology and resilient infrastructure. *Ecol. Resil. Infrastruct.* **2017**, *4*, 71–74. [CrossRef]

101. Park, S.G.; Ra, N.Y.; Jang, Y.S.; Woo, S.H.; Koo, K.S.; Jang, M.H. Comparison of movement distance and home range size of gold-spotted pond frog (Pelophylax chosenicus) between rice paddy and ecological park—focus on the planning alternative habitat, ecology and resilient. *Infrastructure* **2019**, *6*, 200–207.

102. Sakurai, T.O. *River Restoration and Habitat Conservation*; Shinzansya Publishing Co, Ltd.: Tokyo, Japan, 2003; pp. 23–55.

103. Lambdon, P.W.; Pyšek, P.; Basnou, C.; Arianoutsou, M.; Essl, F.; Hejda, M.; Jarošík, V.; Pergl, J.; Winter, M.; Anastasiu, P.; et al. Alien Flora of Europe: Species Diversity, Temporal Trends, Geographical Patterns and Research Needs. *Preslia* **2008**, *80*, 149. Available online: http://www.preslia.cz/P082Lam.pdf (accessed on 15 September 2021).

104. Yoon, S.M. Taxonomic Study of the Genus Cercyon (Coleoptera, Hydrophilidae) in Korea. Master's Thesis, Hannam National University, Daejeon, Korea, 2008. Available online: http://dl.nanet.go.kr/law/SearchDetailView.do?cn=KDMT1200853467 (accessed on 26 September 2021).

105. Botham, M.S.; Rothery, P.; Hulme, P.E.; Hill, M.O.; Preston, C.D.; Roy, D.B. Do urban areas act as foci for the spread of alien plant species? An assessment of temporal trends in the UK. *Divers. Distrib.* **2009**, *15*, 338–345. [CrossRef]

106. Han, M.S.; Bang, H.S.; Kim, M.H.; Kang, K.K.; Jung, M.P.; Lee, D.B. Distribution characteristics of water scavenger beetles (Hydrophilidae) in Korean paddy field. *Korean J. Environ. Agric.* **2010**, *29*, 427–433. [CrossRef]

107. Han, M.S.; Kim, M.H.; Bang, H.S.; Na, Y.E.; Lee, D.B.; Kang, K.K. Geographical distribution of diving beetles (Dytiscidae) in Korean paddy ecosystem, Korean. *J. Environ. Agric.* **2011**, *30*, 209–215.

108. Cho, K.T.; Kim, H.W.; Kim, H.R.; Jeong, H.M.; Lee, K.M.; Kang, T.G.; You, Y.H. Landscape ecological characteristics of habitat of Nannophya pygmaea Rambur (Libellulidae, Odonata), an endangered species for conservation. *J. Wet Res.* **2012**, *14*, 667–674.

109. Jung, J.W.; Bae, Y.H.; Lee, H.N.; Kim, S.J.; Kim, H.S. A study on the benefit estimation by artificial wetland construction. *J. Wet Res.* **2020**, *22*, 39–48.

110. Pu, D.Q.; Shi, M.; Wu, Q.; Gao, M.Q.; Liu, J.F.; Ren, S.P.; Yang, F.; Tang, P.; Ye, G.Y.; Shen, Z.C.; et al. Flower-visiting insects and their potential impact on transgene flow in rice. *J. Appl. Ecol.* **2014**, *51*, 1357–1365. [CrossRef]

111. Lee, J.G.; Hong, S.S.; Kim, J.Y.; Park, K.Y.; Lim, J.W.; Lee, J.H. Occurrence of stink bugs and pecky rice damage by stink bugs in paddy fields in Gyeonggi-do, Korea. *Korean J. Appl. Entomol.* **2009**, *48*, 37–44. [CrossRef]

112. Son, J.K.; Kong, M.J.; Kang, D.H.; Kang, B.H.; Yun, S.W.; Lee, S.Y. The comparative studies on the terrestrial insect diversity in protected horticulture complex and paddy wetland. *J. Wet Res.* **2016**, *18*, 386–393. [CrossRef]

113. Boscutti, F.; Sigura, M.; De Simone, S.; Marini, L. Exotic plant invasion in agricultural landscapes: A matter of dispersal mode and disturbance intensity. *Appl. Veg. Sci.* **2018**, *21*, 250–257. [CrossRef]

114. Kim, N.H.; Choi, S.W.; Lee, J.S.; Lee, J.H.; Ahn, K.J. Spatiotemporal changes of beetles and moths by habitat types in agricultural landscapes, Korean. *Korean J. Environ. Biol.* **2018**, *36*, 180–189. [CrossRef]

115. Hur, B.K.; Rim, S.K.; Kim, Y.H.; Lee, K.Y. Physico-Chemical Properties on the Management Groups of Paddy Soils in Korea. *Korean J. Soil Sci. Fert.* **1997**, *30*, 62–66. Available online: https://www.koreascience.or.kr/article/JAKO199719262658018.page (accessed on 15 September 2021).

116. Kang, S.S.; Roh, A.S.; Choi, S.C.; Kim, Y.S.; Kim, H.J.; Choi, M.-T.; Ahn, B.K.; Kim, H.W.; Kim, H.K.; Park, J.H.; et al. Status and changes in chemical properties of paddy soil in Korea. *Korean J. Soil Sci. Fert.* **2012**, *45*, 968–972. [CrossRef]

117. Korean Soil Information System. Available online: http://soil.rda.go.kr/soil/sibi/sibiPrescript.jsp (accessed on 24 July 2022).

118. Shiratori, Y.; Watanabe, H.; Furukawa, Y.; Tsuruta, H.; Inubushi, K. Effectiveness of a subsurface drainage system in poorly drained paddy fields on reduction of methane emissions. *Soil Sci. Plant. Nutr.* **2007**, *53*, 387–400. [CrossRef]

119. Xie, J.; Wu, X.; Tang, J.J.; Zhang, J.E.; Chen, X. Chemical Fertilizer Reduction and Soil Fertility Maintenance in Ricefish-Coculture System. *Front. Agric. China* **2010**, *4*, 422–429. [CrossRef]

120. Xie, J.; Hu, L.; Tang, J.; Wu, X.; Li, N.; Yuan, Y.; Yang, H.; Zhang, J.; Luo, S.; Chen, X. Ecological mechanisms underlying the sustainability of the agricultural heritage rice? fish coculture system. *Proc. Natl Acad. Sci. USA* **2011**, *108*, E1381–E1387. [CrossRef] [PubMed]

121. Watanaba, H.; Miyahara, M. Seed Dormancy of Some Lowland Weeds of Scirpus Buried in Paddy Soil. In Proceedings of the 12th Asian-Pacific Weed Science Society Conference, Taipei, Tawan, 29 November–5 December 1989; pp. 309–315. Available online: https://www.cabdirect.org/cabdirect/abstract/19922316782 (accessed on 15 September 2021).

122. Wei, H.; Bai, W.; Zhang, J.; Chen, R.; Xiang, H.; Quan, G. Integrated rice-duck farming decreases soil seed bank and weed density in a paddy field. *Agronomy* **2019**, *9*, 259. [CrossRef]

123. Keddy, P.A.; Reznicek, A.A. Great Lakes vegetation dynamics: The role of fluctuating water levels and buried seeds. *J. Gr Lakes Res.* **1986**, *12*, 25–36. [CrossRef]

124. Van Der Valk, A.G.; Pedersen, R.L. Seed Banks and the Management and Restoration of Natural Vegetation. In *Ecology of Soil Seed Banks*; Leck, M.A., Parker, V.T., Simpson, R.L., Eds.; Academic Press: San Diego, CA, USA, 1989; pp. 329–346. Available online: https://eurekamag.com/research/033/322/033322489.php (accessed on 15 September 2021).

125. Tsuyuzaki, S. Survival characteristics of buried seeds 10 years after the eruption of the usu volcano in northern Japan. *Can. J. Bot.* **1991**, *69*, 2256. [CrossRef]

126. Doherty, K.D.; Butterfield, B.J.; Wood, T.E. Matching seed to site by climate similarity: Techniques to prioritize plant materials development and use in restoration. *Ecol. Appl.* **2017**, *27*, 1010–1023. [CrossRef]

127. Kim, N.C.; Kim, H.Y.; Choi, M.Y. The Study on the Utilization of Soil Seed Bank for the Restoration of Original Vegetation. *J. Korean Env. Res. Technol.* **2015**, *18*, 201–2014.

128. Whittle, C.A.; Duchesne, L.C.; Needham, T. The importance of buried seeds and vegetative propagation in the development of postfire plant communities. *Environ. Rev.* **1997**, *5*, 79–87. [CrossRef]

129. Shimono, H.; Hasegawa, T.; Moriyama, M.; Fujimura, S.; Nagata, T. Modeling spikelet sterility induced by low temperature in rice. *Agron. J.* **2005**, *97*, 1524–1536. [CrossRef]

130. Usui, Y.; Sakai, H.; Tokida, T.; Nakamura, H.; Nakagawa, H.; Hasegawa, T. Rice grain yield and quality responses to free-air CO_2 enrichment combined with soil and water warming. *Glob. Change Biol.* **2016**, *22*, 1256–1270. [CrossRef]

131. Li, D.; Nanseki, T.; Chomei, Y.; Yokota, S. Production efficiency and effect of water management on rice yield in Japan: Two-stage DEA model on 110 paddy fields of a large-scale farm. *Paddy Water Environ.* **2018**, *16*, 643–654. [CrossRef]

132. Sharifi, H.; Hijmans, R.J.; Hill, J.E.; Linquist, B.A. Water and Air Temperature Impacts on Rice (Oryza Sativa) Phenology. *Paddy Water Environ.* **2018**, *16*, 467–476. Available online: https://link.springer.com/article/10.1007%2Fs10333-018-0640-4 (accessed on 15 September 2021). [CrossRef]

133. Nishida, K.; Yoshida, S.; Shiozawa, S. Theoretical analysis of the effects of irrigation rate and paddy water depth on water and leaf temperatures in a paddy field continuously irrigated with running water. *Agric. Water Manag.* **2018**, *198*, 10–18. [CrossRef]

134. Nishida, K.; Uo, T.; Yoshida, S.; Tsukaguchi, T. Effect of water depth, amount of irrigation water, and irrigation timing on water temperature in paddy field during continuous irrigation with cool running-water. *Trans. Jpn. Soc. Irrig. Drain. Rural. Eng.* **2015**, *300*, 185–194, (In Japanese with English Abstract).

135. Nishida, K.; Ninomiya, Y.; Uo, T.; Yoshida, S.; Shiozawa, S. Relationship between irrigation conditions and distribution of paddy water temperature under continuous irrigation with running water. *Trans. Jpn. Soc. Irrig. Drain. Rural. Eng.* **2016**, *303*, 391–401, (In Japanese with English Abstract).

136. Nishida, K.; Uo, T.; Yoshida, S.; Tsukaguchi, T. Reduction of water temperature in paddy field and leaf temperature by continuous irrigation with running water during nighttime, water Land. *Environ. Eng.* **2013**, *81*, 297–300. (In Japanese)

137. Ahmed, N.; Bunting, S.W.; Rahman, S.; Garforth, C.J. Community-based climate change adaptation strategies for integrated prawn-fish-rice farming in Bangladesh to promote social-ecological resilience. *Rev. Aquacult.* **2014**, *6*, 20–35. [CrossRef]

138. Shamshuddin, J.; Panhwar, Q.; Shazana, M.; Elisa, A.; Fauziah, C.; Naher, U. Improving the Productivity of Acid Sulfate Soils for Rice Cultivation Using Limestone, Basalt, Organic Fertilizer and/or Their Combinations. *Sains Malays.* **2016**, *45*, 392. Available online: http://www.ukm.my/jsm/pdf_files/SM-PDF-45-3-2016/08%20J.%20Shamshudin.pdf (accessed on 15 September 2021).

139. Mishra, A.; Mohanty, R.K. Productivity enhancement through rice-fish farming using a two-stage rainwater conservation technique. *Agric. Water Manag.* **2004**, *67*, 119–131. [CrossRef]

140. Hill, J.E.; Roberts, S.R.; Brandon, D.M.; Scardaci, S.C.; William, J.F.; Wick, C.M.; Canevari, W.M.; Weir, B.L. Rice Production in California. In *Cooperative Extension*; University of California, Division of Agricultural and Natural Resources: Los Angeles, CA, USA, 1992; p. 12. Available online: http://hdl.handle.net/2027/coo.31924063112290 (accessed on 15 September 2021).

141. Rothuis, A.J.; Vromant, N.; Xuan, V.T.; Richter, C.J.J.; Ollevier, F. The effect of rice seeding rate on rice and fish production, and weed abundance in direct-seeded rice-fish culture. *Aquaculture* **1999**, *172*, 255–274. [CrossRef]

142. Fiehn, O. Metabolomics? The Link between Genotypes and Phenotypes. *Plant Mol. Biol.* **2002**, *48*, 155–171. Available online: https://link.springer.com/article/10.1023/A:1013713905833 (accessed on 15 September 2021). [CrossRef]

143. Cho, H.J.; Heo, K.; Umemoto, T.; Wang, M.H. Physiological Properties of two Japonica Rice (*Oryza sativa* L.) cultivars: Odae and Ilpum. *J. Appl. Biol. Chem.* **2007**, *50*, 127–131. Available online: https://www.koreascience.or.kr/article/JAKO200706717340149.page1ff8 (accessed on 15 September 2021).

144. Kang, Y.; Lee, B.M.; Lee, E.M.; Kim, C.H.; Seo, J.A.; Choi, H.K.; Kim, Y.S.; Lee, D.Y. Unique metabolic profiles of Korean rice according to polishing degree, variety, and geo-environmental factors. *Foods* **2021**, *10*, 711. [CrossRef]

145. Ku, B.I.; Kang, S.K.; Sang, W.G.; Lee, M.H.; Park, H.K.; Kim, Y.D.; Lee, J.H. Study of Rice Double Cropping Feasibility in Korea to Cope with Climate Change. *Theses J. Agric. Life Sci.* **2014**, *45*, 39–46. Available online: https://kiss.kstudy.com/thesis/thesis-view.asp?key=3281103 (accessed on 15 September 2021).

146. Nam, J.K.; Ko, J.K.; Kim, K.Y.; Kim, B.K.; Kim, W.J.; Ha, K.Y.; Shin, M.S.; Ko, J.C.; Kang, H.J.; Park, H.S.; et al. A new early-maturing rice with Blast, bacterial blight, rice stripe virus resistant and high grain quality, 'Jopyeong'. *Korean J. Breed Sci.* **2013**, *45*, 177–182. [CrossRef]

147. Kim, S.Y.; Seo, J.H.; Bae, H.K.; Hwang, C.D.; Ko, J.M. Rice cultivars adaptable for rice based cropping systems in a paddy field in the Yeongnam plain area of Korea. *Korean J. Agric. Sci.* **2018b**, *45*, 355–363.

148. Fernando, C.H.; Paggi, J.C. Cosmopolitanism and latitudinal distribution of freshwater zooplanktonic Rotifera and Crustacea, SIL. *Int. Ver. Für Theor. Und Angew. Limnol. Verh.* **1998**, *26*, 1916–1917.

149. Jin, G.Y.; Yan, S.D.; Udatsu, T.; Lan, Y.F.; Wang, C.Y.; Tong, P.H. Neolithic Rice Paddy from the Zhaojiazhuang Site, Shandong, China. *Chin. Sci. Bull.* **2007**, *52*, 3376–3384. Available online: https://link.springer.com/article/10.1007%2Fs11434-007-0449-9 (accessed on 15 September 2021). [CrossRef]

150. Crawford, G.W. East Asian plant domestication. In *Archaeology of Asia*; Stark, M.T., Ed.; Blackwell Publishing: Oxford, UK, 2006; pp. 77–95. [CrossRef]

151. Fuller, D.Q.; Castillo, C. Diversification and Cultural Construction of a Crop: The case of Glutinous Rice and Waxy Cereals in the Food Cultures of Eastern Asia. In *The Oxford Handbook of the Archaeology of Food and Diet*; Lee-Thorp, J., Ed.; Oxford University Press: Oxford, UK, 2016. Available online: https://discovery.ucl.ac.uk/id/eprint/1489806/1/FullerCastillo_Glutinous%20rice.pdf (accessed on 15 September 2021).

152. Fukami, H.; Higa, Y.; Hisano, T.; Asano, K.; Hirata, T.; Nishibe, S. A review of red yeast rice, a traditional fermented food in Japan and East Asia: Its characteristic ingredients and application in the maintenance and improvement of health in lipid metabolism and the circulatory system. *Molecules* **2021**, *26*, 1619. [CrossRef] [PubMed]

153. Nam, H.S.; Byeon, Y.W.; Park, K.C.; Park, K.L.; Lee, Y.M.; Han, E.J.; Kim, C.H.; Kong, M.J.; Son, J.K. Economic value evaluation of ecosystem services in organic ricefish mixed farming system in paddy Wetland. *J. Wet. Res.* **2020**, *22*, 286–294.

154. Ahmed, N.; Garnett, S.T. Integrated rice-fish farming in Bangladesh: Meeting the challenges of food security. *Food Sec.* **2011**, *3*, 81–92. [CrossRef]

155. Kim, Y.J.; Han, H.S.; Park, Y.G.; The Plan for Activation of Insect Industry. Korea Rural Economic Institute. 2015. Available online: https://scienceon.kisti.re.kr/commons/util/originalView.do?cn=TRKO201600001475&dbt=TRKO&rn (accessed on 15 September 2021).

156. Frei, M.; Becker, K. Integrated rice-fish culture: Coupled production saves resources. *Nat. Resour. Forum* **2005**, *29*, 135–143. [CrossRef]

157. Chen, J.S. Travel Motivation of Heritage Tourists. *Tour. Anal.* **1998**, *2*, 213–215. Available online: https://www.ingentaconnect.com/content/cog/ta/1998/00000002/F0020003/art00006#Refs (accessed on 15 September 2021).

158. Alderson, W.T.; Low, S.P. *Interpretation of Historic Sites*; American Association for State and Local History: Nashville, TN, USA, 1976. Available online: https://www.abebooks.com/servlet/BookDetailsPL?bi=30255552310&searchurl=an%3Dwilliam%2Balderson%2Bshirley%2Bpayne%2Blow%26sortby%3D17%26tn%3Dinterpretation%2Bhistoric%2Bsites&cm_sp=snippet-_-srp1-_-title1 (accessed on 15 September 2021).

159. Beckmann, E.A. Evaluating visitors' reactions to interpretation in Australian national parks. *J. Interpret. Res.* **1999**, *4*, 5–19. [CrossRef]

160. Poria, Y.; Biran, A.; Reichel, A. Visitors' preferences for interpretation at heritage sites. *J. Travel Res.* **2009**, *48*, 92–105. [CrossRef]

161. Tilden, F. *Interpreting Our Heritage*; University of North Carolina Press: Chapel Hill, NC, USA, 2009. Available online: https://uncpress.org/book/9780807858677/interpreting-our-heritage/ (accessed on 15 September 2021).

162. Stone, D.; Hanna, J. Health and Nature: The Sustainable Option for Healthy Cities, August 2003. Available online: https://www.researchgate.net/publication/253758814_Health_and_Nature_the_sustainable_option_for_healthy_cities (accessed on 15 September 2021).

163. Bahriny, F.; Bell, S. Patterns of urban Park use and their relationship to factors of quality: A case study of Tehran, Iran. *Sustainability* **2020**, *12*, 1560. [CrossRef]

164. Corbould, C. Feeding the Cities: Is Urban Agriculture the Future of Food Security. *Future Directions International*; Pty Ltd.: WA, Austria, 2013; pp. 1–7. Available online: https://www.futuredirections.org.au/wp-content/uploads/2013/11/Urban_Agriculture-_Feeding_the_Cities_1Nov.pdf (accessed on 15 September 2021).

165. Palmer, L. Urban agriculture growth in US cities. *Nat. Sustain.* **2018**, *1*, 5–7. [CrossRef]

166. Multilateral Environmental Agreement. Ecosystems and Human Wellbeing: Multiscale Assessment. In *Millennium Ecosystem Assessment*; Google Books; Island Press: Washington, DC, USA, 2005; p. 4. Available online: https://islandpress.org/books/ecosystems-and-human-well-being-multiscale-assessments (accessed on 15 September 2021).

167. National Research Institute of Agricultural Economics. The Result of Evaluation of Public Benefit Accompanying Agriculture and Rural Area by a Replacement Cost Method. *J. Agric. Econ.* **1998**, *52*, 113–138. Available online: https://www.maff.go.jp/primaff/kanko/nosoken/attach/pdf/199810_nsk52_4_04.pdf (accessed on 15 September 2021). (In Japanese).

Article

Urban Gardening and Wellbeing in Pandemic Era: Preliminary Results from a Socio-Environmental Factors Approach

Diana Harding [1], Kevin Muhamad Lukman [2], Matheus Jingga [3], Yuta Uchiyama [4], Jay Mar D. Quevedo [5] and Ryo Kohsaka [5,*]

[1] Faculty of Psychology, Padjadjaran University, Jl. Raya Bandung Sumedang KM. 21, Sumedang 45363, Indonesia; diana.harding@unpad.ac.id
[2] LAMINA, Yayasan Lamun, Depok 16422, Indonesia; kevin.muhamad.lukman@gmail.com
[3] PT Nojorono Tobacco International, Kudus 58311, Indonesia; matheusjingga12@gmail.com
[4] Graduate School of Human Development and Environment, Kobe University, Kobe City 657-8501, Japan; yutanu4@gmail.com
[5] Graduate School of Environmental Studies, Nagoya University, Nagoya City 464-8601, Japan; quevedojaymar@gmail.com
* Correspondence: kohsaka@hotmail.com

Abstract: The nature and impacts of living in urban settings are gaining their saliences in developed and developing countries alike, particularly during the crisis of the COVID-19 pandemic. During the crisis, the wellbeing of urban society became intertwined with a so-called "new lifestyle", which involved quarantine and working in a home environment. Facing such challenges, urban gardening is deemed as an alternative intervention to enhance residents' wellbeing and the environmental sustainability of urban areas, including Indonesian cities. A preliminary study was conducted to monitor the wellbeing of urban gardening practitioners, as well as investigate the motivation and any association between gardening and wellbeing with the COVID-19 pandemic situation by analysing data from Indonesian metropolitan areas. The study utilized instruments of "satisfaction with life scale (SWLS)" and "scale of positive and negative experience (SPANE)" to investigate the subjective wellbeing of 67 respondents. Amongst others, we identified that urban gardening practitioners tend to be in positive moods and have better overall wellbeing; 52.24% of the respondents were highly satisfied with their life. Furthermore, we observed a variety of motivations to start gardening, with hobby and utilization of free space as prominent reasons, followed by other motivations such as environmental benefit and aesthetic. Integrating the environmental benefits of urban gardening and the implications for an individual's wellbeing can be reflected for sustainable urban development and policies during the COVID-19 pandemic.

Keywords: subjective wellbeing; community perceptions; urban gardening; Indonesia; COVID-19

Citation: Harding, D.; Lukman, K.M.; Jingga, M.; Uchiyama, Y.; Quevedo, J.M.D.; Kohsaka, R. Urban Gardening and Wellbeing in Pandemic Era: Preliminary Results from a Socio-Environmental Factors Approach. *Land* **2022**, *11*, 492. https://doi.org/10.3390/land11040492

Academic Editor: Alexandru-Ionuţ Petrişor

Received: 3 March 2022
Accepted: 24 March 2022
Published: 29 March 2022

Publisher's Note: MDPI stays neutral with regard to jurisdictional claims in published maps and institutional affiliations.

1. Introduction

Rapid urban expansions have generated a variety of issues such as poverty, socioeconomic gaps, and degradation of environmental quality through overconsumption of resources, water and air pollution, waste production, and the reduction in green spaces [1–5]. With half of the global human population living in urban areas, they are now increasingly disconnected from nature due to urbanization [6]. Furthermore, urban people often live in environments with low biodiversity, food insecurity, and social alienation [7], which can exacerbate the situation and cause a decline in wellbeing.

From the perspective of wellbeing, the rapid development in urban settings also poses a different set of challenges. For example, the population concentration in urban areas and changes in lifestyle have led to reduced opportunities for contact with nature in daily life, and the increased urbanization is associated with the stress of financial and health burdens, which have led to a demand for mental health enhancement strategies [8,9]. Reflecting on the current situation, there is the notion of creating cities that are more liveable and environmentally friendly, which is also reflected in the United Nation's development framework of Sustainable Development Goals (SDGs) through the interaction of provisioning, environment, and wellbeing [10]. Based on the SDGs, there are several interrelated goals and targets. The third goal, for example, is related to "Good Health and Wellbeing", which requires the cooperation of countries and societies [11]. In most cases, sustainable development emphasizes the reduction of environmental stress. For instance, in China, the effectiveness of urban development is measured by overall wellbeing, such as life satisfaction [12]. A critical view to alter urban lifestyles to decrease future energy demands and remove existing atmospheric carbon is also now being considered [13], with other approaches, such as the interaction between people and nature in urban areas, reported to have considerable impacts on environmental and life quality [14]. In another study, the importance of contact with nature in urban areas is highlighted in regard to mental wellbeing and health [15].

Furthermore, a study has shown that urban design and planning practices affect the psychological wellbeing of vulnerable groups [16]. For example, a study case from Scotland reported that living near a new motorway appeared to worsen residents' wellbeing and had negative impacts on health [17]. This highlights how urban planning is connected to the notion of individual wellbeing. As cities seek to become more liveable and environmentally friendly, there is a consideration from cities to invest in infrastructures to enhance the quality of life, such as household gardening [10]. Nonetheless, urban consolidation and expansion can threaten both private and public green space access for residents, therefore urban policies and planning need to carefully consider the benefit of green spaces [7]. In addition, the situation of the COVID-19 pandemic has only added to the challenges for urban settings and an emphasis on the mental wellbeing of urban societies is becoming more prominent.

The global COVID-19 pandemic, since late 2019 (and ongoing as of 2021), is one of the severest health crises, which is affecting human behaviour and wellbeing, given the unprecedented scope of COVID-19 stressors and challenges [18,19]. The COVID-19 pandemic has a more significant effect on vulnerable populations such as young children and elderly citizens, in particular to the health, wellbeing, and quality of education worldwide [20]. There are ample causes for concern regarding the impact of COVID-19 on the wellbeing of the general population. For instance, through the course of the pandemic, many governments implemented (some are still implementing as of this writing) lockdown systems, which severely affected the mental health of the people [21,22]. As there are still uncertainties surrounding COVID-19, further studies are needed to understand its impact on behavioural health, for example prolonged loneliness and its effects on the mental health and wellbeing of the public [23,24]. The challenge in this pandemic, as highlighted by Jakovljevic et al. [25], is the need to study the psychiatric and psychological aspects of COVID-19 from the perspectives of public and global mental health.

Promoting mental health and wellbeing during the pandemic are now emphasised in scientific literatures. For instance, the study from Stieger et al. [26] stated that in regard to emotional wellbeing during the COVID-19 pandemic, being outdoors was associated with higher emotional wellbeing, while greater loneliness was associated with poorer wellbeing. Though the COVID-19 pandemic restricted people's activities outdoors, with government-issued stay-at-home orders, outdoor interaction with green spaces such as gardening has been deemed important for mental health [27]. Gardening activities have been reported to be linked with improvements in human health and wellbeing [28]. In addition, urban greenery provides various ecosystem services, which play roles in the challenging context of urban deprivation and poverty [29], and networks of urban greenery can enhance urban biodiversity. Gardening activities in urban areas is also now becoming a trend towards more green areas in cities, which provide opportunities for regular contact with nature, physical activity, and allow people to consume homegrown fruit and vegetables, as well as reducing stress levels of gardeners, improving social cohesion and preventing health problems [15,30]. Considering the values of urban gardening spaces in the triple bottom lines, environment, society, and economy, they can be regarded as a catalyst to facilitate sustainable urban development.

The situation of the COVID-19 pandemic in Indonesia has led to the implementation of policies such as social distancing, work from home (WFH), and the closing down of facilities [31], where experts also advise people to limit their physical activities and contacts [32]. However, policies such as WFH and social restriction have been linked with negative mental wellbeing and internet addiction, respectively, which implies the need to mitigate the psychological risks from COVID-19-related policy actions [33,34]. As mentioned earlier, gardening is linked with mental health, with various reports indicating an increased interest in home gardening during the COVID-19 pandemic [27]. In this vein, there is an opportunity to observe the situation in Indonesia regarding urban gardening as an alternative approach to promote wellbeing during the COVID-19 pandemic era. Urban gardening and urban farming activities in Indonesia are growing in cities such as Jakarta, Surabaya, Palu, and Bandung, with various influencing backgrounds such as to promote green areas, food security, and environmental awareness [1,2,35,36]. Initially, these activities started after the economic crises in 1997–1998 with the utilization of open backyard areas and the common plantation of vegetables, lettuce, spinach, tomato, and onion [3,37]. Studies have shown that gardening activities in Indonesia provide positive benefits in promoting environmental awareness, strengthening community, and as a coping mechanism for anxiety and stress during the COVID-19 pandemic [38–40].

However, there is a knowledge gap on how gardening activities can influence wellbeing, particularly during the COVID-19 pandemic, and the overall benefits provided in urban settings. In this preliminary study, we aim to observe the wellbeing of urban gardening practitioners to understand what motivates them, and whether practitioners associate gardening with their wellbeing or not. We also examined the starting times of the gardening to identify whether those aspects influence wellbeing, in addition to providing discussions and insights for the improvement in wellbeing during the pandemic. Further understanding of the influence of urban gardening on the wellbeing of people and their motivations can serve as future policy implementation for urban areas in regard to initiating greenery and gardening policies. The outline of the paper included an introduction (Section 1) and methodology (Section 2), where we explained the concept of subjective wellbeing and the instruments used in this study, respectively. The results of the study are presented in Section 3, followed by Section 4 with discussions on ecosystem services and the benefits of gardening activities in urban areas and the links between the pandemic and people's wellbeing. In the last section, we present future studies and the general implications of this preliminary work on urban gardening and its influence on people's wellbeing.

2. Materials and Methods

2.1. Subjective Wellbeing (SWB) Using Satisfaction with Life Scale (SWLS)

The empirical science of subjective wellbeing (SWB) has grown significantly in the past decade [41], where SWB is defined as life satisfaction and depends on an individual's standards [42]. This study investigated SWB using the Satisfaction with Life Scale (SWLS) approach to explore overall life satisfaction [43,44]. SWLS is a widely used measure of life satisfaction and has been used across many socio-demographic groups [45]. The advantages of using SWLS are known due to the fast nature of the interview time, as a public domain, and reliability based on Cronbach alpha [46]. Environmental economists are interested in wellbeing or relational values beyond conventional monetary values [47]. Terms in psychology, such as happiness, are also analysed in the context of relationships with visitors and nature [48]. Thus, there are multi-disciplinary interests in SWB from psychology economics to environmental sciences.

The SWLS in this study consisted of five items, which were measured using a 7-point Likert scale ranging from 1 (completely disagree) to 7 (completely agree) based on the cognitive components [43]. Utilizing this structure, the total scores of SWLS can range from 5 to 35. According to a study from Pavot and Diener [49], the total scores of 5 to 9 indicate that the respondent is extremely dissatisfied with life, whereas scores ranging between 31 and 35 indicate that the respondents are extremely satisfied with life. In this study, the results from the SWLS interview were categorized as per definition from Pamungkas et al. [50] with (a) total scores of 5 to 15 as low category, (b) total scores of 16 to 25 as medium category, and (c) total scores of 26 to 35 as high category.

2.2. Supporting Instrument with Scale of Positive and Negative Experience (SPANE)

To support the results of SWLS, this study also used the Scale of Positive and Negative Experience (SPANE) on people's wellbeing. SPANE is a newly introduced instrument to measure positive–negative feelings and evaluate wellbeing [51], which consists of 12 items: 6 items for positive feeling (SPANE-P) and 6 items for negative feeling (SPANE-N). SPANE is stated as a valid and reliable scale to measure positive and negative experiences as a facet of wellbeing [52,53]. The results from the positive items were subtracted with the scores from negative items to obtain the balance score (SPANE-B), which was used to categorize the scale; (a) −24 to −8 scores of SPANE-B categorized as low, (b) −7 to 8 scores of SPANE-B categorized as medium, and (c) 9 to 24 scores of SPANE-B categorized as high, where if respondents are able to obtain a score of 24 (highest score), it can be interpreted that they never experience negative emotions.

2.3. Respondents and Questionnaire for Urban Gardening

The survey questionnaires, which were written in Bahasa, were initially broadcasted to gardening groups associated with the first-three authors using three platforms (Whatsapp, Line, and Instagram). The snowball sampling approach was then applied to spread the survey questionnaires to other related groups that were not necessarily associated with the authors. Initially, each of the respondents was asked for their agreement regarding the usage of the data and their confirmation of whether urban gardening influences their life satisfaction. All of the respondents agreed during this initial step. A total of 67 respondents were gathered from the 12 urban areas in Indonesia.

The survey questionnaire consisted of four sections, which included: (A) socio-demographic, (B) SWLS items, (C) SPANE items, and (D) wellbeing from gardening in the pandemic period. In section A, we asked respondents' age, current address, civil status, the period when they started gardening (before or after COVID-19 pandemic started), how long they have been gardening and their motivations for gardening. In section B, the SWLS consisted of five items: (1) in most ways my life is close to my ideal, (2) the conditions of my life are excellent, (3) I am satisfied with my life, (4) so far, I have gotten the important things I want in life, and (5) if I could live my life over, I would change almost nothing. In section C, there were 12 items, which were equally divided into 6 positive and 6 negative emotions. These included (1) positive, (2) negative, (3) good, (4) bad, (5) pleasant, (6) unpleasant, (7) happy, (8) sad, (9) joyful, (10) afraid, (11) contented, and (12) angry. SPANE is one of the most popular instruments used to measure experiences worldwide with 12 items consisting of six positive (SPANE-P) and six negative experiences (SPANE-N), and measuring general and specific emotions encompassing a wide range of human experiences [54]."We instructed the respondents to fill out the SPANE instrument by reporting their experiences during the last four weeks, using a 5-point scale: 1 = "almost never", 2 = "rarely", 3 = "sometimes", 4 = "often" and 5 = "almost every time". In the last section, we asked their perceptions of whether gardening activities contributed to their wellbeing during the pandemic period or not. We also inquired about the reasons for their respective answers.

2.4. Statistical Analysis

The results of the questionnaires were analysed and presented using descriptive statistics, which included Cronbach's alpha for the reliability test, mean scores, and standard deviations of SWLS and SPANE. In addition, Spearman's rank correlation analysis was performed to examine the relationship among the variables: (1) when the respondents start gardening, (2) what motivates them to do gardening, (3) SWLS, and (4) SPANE. Correlation analysis between SWLS and SPANE was carried out to examine the validity with other measures, since it is necessary to take SWLS and SPANE into both considerations of affective and cognitive components [55]. For the correlation analysis, we assigned dummy variables to define the times respondents started gardening and their motivations; for example, the respondents who started gardening after the COVID-19 pandemic were coded as 0, and respondents who started gardening before the pandemic as 1.

3. Results

3.1. Socio-Demographic Profile

The summary of the socio-demographic characteristics of the respondents is reflected in Table 1. A total of 67 respondents from 12 urban areas in Indonesia were collected in this study (mean age = 50.87 years, SD = 8.65). All of them satisfied the criteria of living in urban areas, which in this study was defined based on the classification from Indonesia Regulation Law No. 26/2007 that states "areas that consist of population at least one million and connected by integrated infrastructure network systems [56]". In terms of respondents' occupation, 24 were private employees, 8 were government employees, and 10 were entrepreneurs. Seventy-three percent of the respondents engaged in gardening activities before the pandemic, while 24% started after the pandemic.

Table 1. Respondents' socio-demographic profile.

Indicators	Number of Respondents
Age Group	
21–30	3 (4.5%)
31–40	3 (4.5%)
41–50	17 (25.4%)
51–60	39 (58.2%)
61–70	5 (7.5%)
Marital Status	
Married	64 (95.5%)
Single	3 (4.5%)
Occupation	
Government Employee	8 (11.9%)
Private Employee	24 (35.8%)
Housewife	22 (32.8%)
Entrepreneur	10 (14.9%)
Unemployed	3 (4.5%)
Start Doing Gardening	
Before Pandemic	49 (73.1%)
After Pandemic	18 (26.9%)

3.2. Motivations for Gardening

Respondents shared several reasons behind their motivation to engage in gardening activities. To present these motivations in a systematic way, we used codes, which were condensed into units from respondents' statements. We derived eight codes, namely "Happiness," "Environment," "Hobby," "Aesthetic," "Health," "Space," "Exercise," and "People." Examples of statements and the coding process are shown in Table 2. In cases where a respondent's statement generated multiple codes, we counted them separately (i.e., one respondent can include "Happiness," "Environment," and "Health").

Table 2. Coding frame for the statements of respondents' motivation on gardening.

Code	Example of Respondent's Statements (Bahasa)
Happiness	"Use leisure times joyfully" ("Menggunakan waktu kosong dengan sukacita")
	"Having fun" ("Bersenang-senang"); "Happy" ("Senang")
Environment	"To protect soil sustainability" ("Untuk menjaga kelestarian tanah")
	"For greenery" ("Penghijauan"); "For shade" ("Peneduh")
Hobby	"Stress reliever hobby" ("Hobi penghilang stres")
	"Activity for free time" ("Mengisi waktu luang")
Aesthetic	"To beautify the yard" ("Untuk memperindah pekarangan")
	"Love a beautiful yard" ("Menyukai pekarangan yang asri")
Health	"For medicinal plants" ("Untuk tanaman obat")
	"Creating fresh air" ("Menciptakan udara segar")
Space	"Utilize yard space" ("Memanfaatkan pekarangan")
Exercise	"Physical exercise" ("Olahraga")
People	"Inspiring neighbours" ("Menginspirasi tetangga")

Amongst the eight motivations, "Hobby" was the most frequented motivation, with 34.3% (23) of the respondents (Figure 1). They related gardening as a hobby to eliminate boredom, stress relief activity, or just as another activity in their free time. The second most frequented motivation was "Space," with 20 respondents motivated to utilize the empty space in their yard. This factor is related to the respondent's aim to plant productive vegetables and fruits in their empty yard as an alternative livelihood or just for self-consumption. "Environment (13)," "Aesthetic (13)," and "Happiness (12)" were also among the most mentioned motivations. For "Environment," the idea of gardening from several respondents was linked with various benefits, such as soil quality, water quality, air quality shading, temperature control, oxygen supply, and greenery. Meanwhile, statements pertaining to "Aesthetic" and "Happiness" did not differ much, which focused on the beautification of respondents' yard, personal satisfaction, and happiness they felt through gardening activities. The least stated factors for gardening were "Health," "Exercise," and "People." "Health" represented the idea that respondents felt that they can obtain positive health benefits from gardening (i.e., fresh air, planting medicinal plants). Meanwhile, "Exercise" is linked with gardening activities as a physical exercise, and "People" are derived from respondents' interaction with other people or becoming inspired (or inspiring) from others, such as family or neighbours.

3.3. SWLS and SPANE-B Categories

The results of SWLS and SPANE-B from 67 respondents are shown in Figure 2. For SWLS, the three categories, low, medium, and high, generated varying results among the respondents. The low category included a total of five respondents, with total SWLS scores ranging from 7 to 14. Meanwhile, 27 and 35 respondents classified themselves into medium and high categories, respectively. In the high category, which had total scores ranging from 26 to 35, respondents showed that they are satisfied with their current life. Scores between 21 and 25 represented slightly satisfied, while 31 and 35 indicated that respondents are extremely satisfied with their lives [49]. A similar study of wellbeing with SWLS showed that respondents with a total score of SWLS within the medium category can shift into the high category after psychotherapy sessions [50]. For the results of SPANE-B, 83.6% (56) of the respondents were classified under the medium category, while 16.4% (11) were under the high category. A respondent with a very high score SPANE-B of 24 indicates that she or he rarely or never experiences any of the negative feelings, and very often or always has all of the positive feelings [57].

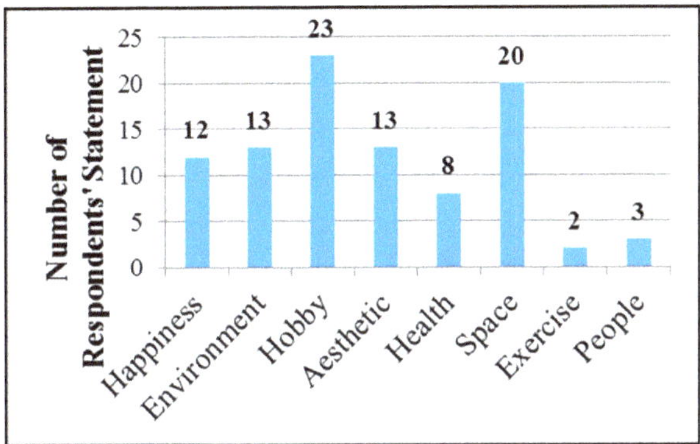

Figure 1. Frequency of the gardening motivation code across all 67 respondents.

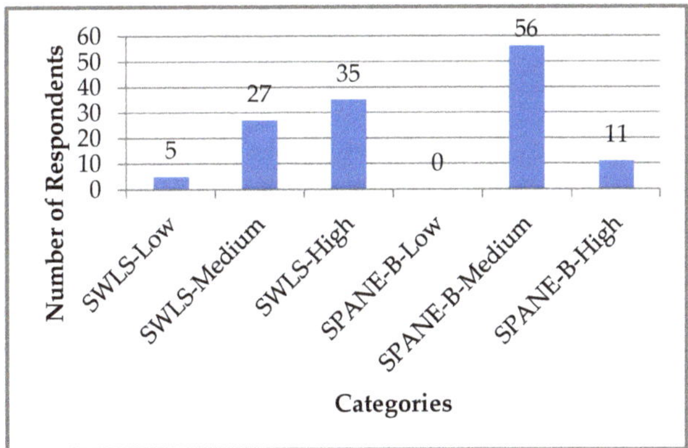

Figure 2. SWLS and SPANE-B results within three categories.

Overall, based on the SWLS, respondents are relatively satisfied with their lives (Table 3). The mean score of Items 1, 2, 3, and 4 is 5.104, 5.328, 5.418, and 5.448, respectively. For item 5, the average (4.0) was relatively lower than the first four items, suggesting that respondents are willing to change their lifestyle. The items listed under SWLS were fit and reliable based on Cronbach's alpha (0.886) generated in this study. According to Silva et al. [58], the reliability of SWLS ranged from 0.70 to 0.90. The average scores of SPANE-P were relatively high (above 4.0) for items (1) positive, (3) good, (5) pleasant, (7) happy and (9) joyful (Table 3). In contrast, the items associated with (11) contented has lower mean scores of 3.970. The opposite results can be seen for SPANE-N, where all items have low mean scores ranging from 2.0 to 2.7. The reliability results from SPANE were divided for positive and negative items. In this study, the Cronbach's alpha for positive affect is 0.58, and for negative affect is 0.71. We argue that the relatively low alpha scores can be due to the limited number of samples for this study. Despite the value of Cronbach's Alpha, which is lower than 0.7 is questioned by some authors; this consideration should not be taken as a "golden rule", in particular, an alpha that is too high could lead one to think that the items measure the same indicator of the construct [59].

Table 3. Item analysis for SWLS and SPANE.

SWLS Item	Mean	Std. Deviation	Cronbach's Alpha
Item 1	5.104	1.316	
Item 2	5.328	1.521	
Item 3	5.418	1.519	0.886
Item 4	5.448	1.329	
Item 5	4.000	1.977	
SPANE Item	**Mean**	**Std. Deviation**	**Cronbach's Alpha**
1-Positive	4.343	0.770	
2-Negative	2.269	0.750	
3-Good	4.448	0.610	
4-Bad	2.000	0.853	
5-Pleasant	4.164	0.687	
6-Unpleasant	2.358	0.847	SPANE-P = 0.580
7-Happy	4.224	0.623	SPANE-N = 0.713
8-Sad	2.343	0.827	
9-Joyful	4.119	0.729	
10-Afraid	2.612	1.058	
11-Contented	3.970	0.717	
12-Anger	2.463	0.785	

3.4. Correlation Analysis

The results of the correlation analysis between the starting times of gardening, motivations, SWLS and SPANE are shown in Table 4. There were no significant correlations observed between the starting times of gardening, either before or during the COVID-19 pandemic. However, for the correlation with motivation, Exercise was correlated significantly with SPANE-B ($r = 0.372$; $p \leq 0.001$). The other significant correlation we found is between SWLS and SPANE-B ($r = 0.496$; $p \leq 0.001$). According to Prado-Gascó et al. [55], this positive correlation is expected since in measuring subjective wellbeing it is necessary to take into consideration the affective component measured by SPANE and the cognitive component measured by SWLS.

Table 4. Spearman's rank correlation analysis between the starting times of gardening, motivations, SWLS, and SPANE.

Items	Correlation	
	SWLS	SPANE-B
Gardening before pandemic	0.037	0.037
Gardening after pandemic	0.035	−0.037
Happiness, Motivation	0.058	0.025
Environment, Motivation	−0.085	−0.068
Hobby, Motivation	−0.038	0.137
Aesthetic, Motivation	0.031	−0.220
Health, Motivation	0.187	0.067
Space, Motivation	−0.212	0.066
Exercise, Motivation	0.150	0.276 *
People, Motivation	0.099	−0.064
Number of Motivations	0.045	0.157
SWLS	1	0.476 **

*, ** Indicate significant correlation at p-value < 0.05 and <0.001, respectively.

3.5. Wellbeing from Gardening during the Pandemic Period

The majority (97.0%) of the respondents shared that gardening positively influenced their wellbeing. Based on their statements, the idea of gardening, which can influence respondents' wellbeing can be categorized into three reasons: (1) rewards received from gardening, (2) gardening as alternative activities during the pandemic period, and

(3) positive psychological impact. For the first reason, gardening was perceived to give a variety of rewards and benefits, for example, source of food, beautification of the yard, alternative for physical exercise, and environmental benefits (i.e., temperature control). Meanwhile, for the second reason, due to restricted movements outside coupled with government-issued laws, the isolation and WFH situation of the respondents gave them more free time to try different things, such as gardening. They shared that gardening motivated them to be more productive and reduced their boredom caused by staying longer indoors. The third reason was directly linked with their feelings. For example, gardening is linked with the psychological benefit of relieving stress and anxiety, sense of growth from monitoring plant, personal satisfaction and wellbeing, as well as a sense of comfort.

4. Discussion

4.1. Ecosystem Services of Gardening in Urban Setting

The activities of community gardens, private home gardens, and ordinary urban farms are known to be part of the urban agriculture concept to utilize the benefit of existing ecosystem services for non-farmer societies and contribute to the increase in wellbeing of urban society [60]. The results of this study also prove that urban gardening activities contribute to wellbeing with the variety of ecosystem services provided. First, the provisioning service, which is linked with the productive plants, influences the initiative to start the gardening activity. The condition of an empty backyard may motivate one to start gardening, either for self-consumption or as an alternative livelihood. This phenomenon should be noted, as the issue of food security in urban areas could become more apparent in the future. With the increasing population and demand for food, the role of urban areas to provide the food supply through urban gardening and agriculture will be more relevant [61]. A case study in the city of Yogyakarta showed that despite the limitation of space, the utilization of a yard in home gardening can provide productive and sustainable results such as the benefit of provisioning services and alternative livelihood [62]. In the context of Indonesia, the movement of urban farming has a different background in regard to the issue of food sovereignty [63]. Such initiatives now exist, for example in Surabaya city, as part of the local government program to achieve the goal of nutritious food fulfilment for the low-income society [64]. Despite the relatively smaller-scale activities from urban gardening, government and local stakeholders should take note of the momentum of various similar programs in Indonesia, particularly focusing on the idea of food provisioning as one of the main benefits of small-scale home gardening. Furthermore, looking at the pandemic situation, there is also another perspective on how provisioning from gardening can increase urban resilience. The issue of hunger during disasters can be mitigated through local food production, for example from urban agriculture, gardens, and community gardens, where understanding the amount of food that can be generated is important in a pandemic context [65]. That being said, collective gardens as part of common-pool resources initiatives were reported to be increasing, and the risk of overexploitation should be noted [66].

Aside from food provisioning benefits, there are other ecosystem services provided related to gardening, for example, environment-related and aesthetic services. For environmental benefits, the vegetation in garden and green roofs can serve as habitats for many organisms, controlling the temperature, managing runoff water, mitigation of urban heat island effect, carbon sequestration, and reducing air pollution from CO_2, NO_2, PM10, and SO_2 [67,68]. The observation from respondents' statements showed that there were certain environmental benefits that were commonly noticed, such as controlling temperature through the existence of a garden. As more environmental issues become more relevant in urban settings, such as the phenomenon of the urban heat island effect, the awareness and understanding of the society will be more prominent. For the sustainability of a region, a holistic view is necessary to understand the dynamics of cities in which collaborations between government sectors and various stakeholders can be implemented for municipal biodiversity management [69–71]. The involvement of the community, as seen in Bandung city with the Bandung Gardening (Bandung Berkebun) community, is reflected to contribute to character-building oriented for the environment [36]. The implementation of productive gardening places, as part of the urban farming concept, aside from the function of controlling pollution also provides a comfortable healthy environment and increases aesthetic value [72]. The findings of this study also complement the notion of aesthetic as a prominent motivation to start gardening activities compared to other benefits such as health. In a study by Chalmin-Pui et al. [28], the health benefits were seen to be an important component of gardening; however, it is not the main motivating factor for gardening, with joy, pleasure, and aesthetics presented as greater drivers.

Another perspective taken from the findings of this study is that for the most part, motivation to engage in gardening activities on its own does not have any influence on wellbeing, nor on positive emotions, as seen with Exercise as the only significant correlation. However, we argue that motivation is a driver to initiate the involvement of people in urban gardening activities, and through active participation in gardening activities, the respondents then associate their happiness and wellbeing. Gardeners are motivated by various reasons, for example connecting with nature, improved food access, or enhancing time spent with family [7]. Thus, exploring the aspect of motivation for gardening in Indonesia can be essential to better understand how local government, stakeholders, and communities, can effectively widen the concept of gardening and increase participation through the engagement of motivating factors and interests of the urban society. In the case of Chile and Switzerland, the strongest motivating factor is linked with the health aspect. However, motivation to gardening is also often seen not as a positive factor, such as the need to keep the garden tidy and peer pressure [28]. Understanding the motivating factors in regard to the ecosystem services from gardening should also be considered to understand the behaviour mechanism of society, as these insights can be used to enhance gardening from other perspectives, such as increasing aesthetic and offering a variety of environmental supporting services in the limitation of an urban setting.

4.2. Gardening in Pandemic Situation and Link to the Wellbeing

In the Indonesian context, the abrupt changes caused by the pandemic to social restriction, remote working, and home-based learning for extended periods will potentially affect mental wellbeing [73]. In addition, the implication of the COVID-19 pandemic is not only to physical health, but also to economic situations, social relationships, and wellbeing [32]. One of the insights of this study is that respondents' motivation to engage in gardening such as a hobby and the intention to try new experiences in their free time during the COVID-19 pandemic is associated with their wellbeing. Based on the results of SWLS and SPANE-B, the majority of the respondents who practiced urban gardening were within the high category, which indicated that they are satisfied with their lives and frequently experience positive emotions. Furthermore, the respondents also stated that practicing gardening has a positive influence on their wellbeing, with statements such as "gardening can be a stress-relieving activity", "focusing attention to productive activity", "eliminating boredom", and "sense of fulfilment by observing the plant's growth". A study from Ambrose et al. [10] reported that the psychological benefits of gardening are more significant on personal and social wellbeing levels through being more connected to nature and community. On the other hand, our study showed that there were no significant correlations in people's wellbeing between people who practiced gardening before and after the COVID-19 pandemic. The results might indicate that the pandemic itself is not an influencing factor for wellbeing, rather, the gardening activities themselves improve their wellbeing. One study highlighted that gardening on a frequent basis of at least 2–3 times a week corresponded with perceived benefits of health, which the pleasure of gardening drives as their motivation [28]. Insights gained from this study indicated policy implications; that urban people can contribute on the basis of ecosystem services benefit by gardening activities. Such an approach can provide multiple benefits to urban environment settings and the wellbeing of the people. Thus, local government should consider incorporating the contribution of urban people through individual gardening activities, as well as community gardening.

Comparing house gardening and community gardening, there is also the insight that community gardening is associated with higher health gains, resilience, and optimism, as well as an affordable and efficient way of promoting wellbeing in urban environments [5,9]. Visits to public green area facilities during the pandemic also gained salience; for instance, in Japan, visits to green areas during the COVID-19 pandemic are linked with socioeconomic attributes, where parks were mainly used by urban residents [74], and interaction with public green infrastructure in urbanized landscapes can enhance wellbeing and the urban liveability [75]. In addition, one study indicated that the high degree of participation of residents to visit parks or green spaces contributed to higher wellbeing, which showed that green spaces can have a very positive effect on people's welfare [76]. Nonetheless, the practice of community gardening might be difficult in the COVID-19 pandemic, and such practice should be carried out by following proper health protocol. Moreover, the practice of community gardens also has its own challenge, for example in South Africa with the theft of garden infrastructure or produce issues, which can hinder motivation and engagement for urban community gardening [77]. Some alternatives include the utilization of empty yards in housing environments. The situation of new social distancing mandates will require a breakthrough approach of public health policies that will effectively utilize outdoor spaces in order to bring benefits in terms of population wellbeing [26]. We argue that in this study, the potential to enhance wellbeing with urban gardening should be explored further, in particular through the effective use of outdoor empty spaces in private housing. The insights gained from this study complement past studies in regard to how gardening can support wellbeing [7,10,28], and builds a further foundation of expanding the discussion on how gardening can benefit the environmental aspects of the urban setting itself. The challenge of limited spaces in an urban setting, environmental awareness, and supporting policies for urban gardening during the COVID-19 pandemic should be addressed in future studies. In addition to considering the differences in the scale of urban settings, such as that observed in Japanese municipalities, where larger cities tended to have more discussions on conservation moves, such as with Geoparks [78,79]. The various characteristics of different cultures and socio-demographics in many Indonesian cities can be another discussion point to promote active participation in activities, as previous studies have shown that engagement from residents, for instance, in management activities, can be influenced by these attributes [80–82].

5. Conclusions

The wellbeing of 67 respondents, who were urban gardening practitioners in Indonesia during the COVID-19 pandemic, was observed in this preliminary study. Overall, respondents' wellbeing was categorized as satisfied with their current life and with positive emotions. Their motivations for engaging in urban gardening, as reflected in this work, included as a hobby, source of happiness, utilization of space, aesthetic, health benefits, environmental benefits, a form of exercise, and interacting with other people. Additionally, they considered urban gardening as positive and healthy for their wellbeing in the pandemic, particularly noting the benefits such as being productive, relieving stress, eliminating boredom, and interacting with family members during their free time and WFH policy.

In summary, this study is a preliminary in character; thus, the results cannot be used to generalize the entire urban population in Indonesia. However, despite the limited number of respondents, this study showed a general trend in understanding people's wellbeing and the environmental benefits in urban settings. In urban areas, people will most likely have similar perceptions (positive or negative) if they shared common activities [83], which in this study is urban gardening. We observed that the activity contributed to respondents' satisfaction with life, particularly in connection with the pandemic. Thus, future studies can consider addressing the topic of urban gardening for enhancing wellbeing during the pandemic in more specific contexts to better understand the influencing factors and to design a coping strategy for wellbeing, as well as expanding the scale and focusing on a certain region or site, for instance, in Jakarta and Indonesia. The COVID-19 crisis provided an unmatched opportunity of how collective behavioural change can help address it (i.e., travel restrictions, lockdowns), which can serve as an inspiration in fighting against climate change (the other global crisis) [84,85]. There is an opportunity for active participation from urban society members in gardening activities, where higher wellbeing can be achieved, and a variety of ecosystem services can be used as alternative intervention measures to various social and environmental issues in the urban setting. Moreover, the COVID-19 pandemic has restricted people's movements outdoors, which resulted in increased interest in home gardening [25]. Finally, combining urban gardening initiatives and other public policies is important for the sustainability of urban environments.

Author Contributions: Conceptualization, D.H., K.M.L. and M.J.; methodology, D.H., K.M.L. and M.J.; validation, D.H., K.M.L., M.J., Y.U., J.M.D.Q. and R.K.; data curation, M.J.; writing—original draft preparation, D.H. and K.M.L.; writing—review and editing, D.H., K.M.L., Y.U., J.M.D.Q. and R.K.; supervision, D.H., Y.U., J.M.D.Q. and R.K. All authors have read and agreed to the published version of the manuscript.

Funding: This study is supported and funded by JSPS KAKENHI Grant Numbers JP21K18456; JP20K12398; JP17K02105; JP16KK0053; JST RISTEX Grant Number JPMJRX20B3; JST Grant Number JP-MJPF2110; Heiwa Nakajima Foundation (2022); Asahi Group Foundation (2022); and Padjadjaran University.

Institutional Review Board Statement: Not applicable.

Informed Consent Statement: Informed consent was obtained from all subjects involved in the study.

Data Availability Statement: The data presented in this study are available on request from the first and second author. The data are not publicly available since this study contained personal information.

Conflicts of Interest: The authors declare no conflict of interest.

References

1. Anggrayni, F.M.; Andrias, D.R.; Adriani, M. Ketahanan Pangan dan Coping Strategy Rumah Tangga Urban Farming Pertanian dan Perikanan Kota Surabaya. *Media Gizi Indones.* **2015**, *10*, 173–178. [CrossRef]
2. Cahya, D.L. Analysis of urban agriculture sustainability in Metropolitan Jakarta (case study: Urban agriculture in Duri Kosambi). *Procedia Soc. Behav. Sci.* **2016**, *227*, 95–100. [CrossRef]
3. Fauzi, A.R.; Ichniarsyah, A.N.; Agustin, H. Urban Agriculture: Urgency, Role, and Best Practice. *J. Agrotenol.* **2016**, *10*, 49–62.
4. Indraprahasta, G.S. The potential of urban agriculture development in Jakarta. *Procedia Environ. Sci.* **2013**, *17*, 11–19. [CrossRef]
5. Lampert, T.; Costa, J.; Santos, O.; Sousa, J.; Ribeiro, T.; Freire, E. Evidence on the contribution of community gardens to promote physical and mental health and well-being of non-institutionalized individuals: A systematic review. *PLoS ONE* **2021**, *16*, e0255621. [CrossRef]
6. Davies, Z.G.; Fuller, R.A.; Loram, A.; Irvine, K.N.; Sims, V.; Gaston, K.J. A national scale inventory of resource provision for biodiversity within domestic gardens. *Biol. Conserv.* **2009**, *142*, 761–771. [CrossRef]
7. Egerer, M.H.; Philpott, S.M.; Bichier, P.; Jha, S.; Liere, H.; Lin, B.B. Gardener Well-Being along Social and Biophysical Landscape Gradients. *Sustainability* **2018**, *10*, 96. [CrossRef]
8. Imai, H.; Nakashizuka, T.; Kohsaka, R. An analysis of 15 years of trends in children's connection with nature and its relationship with residential environment. *Ecosyst. Health Sustain.* **2018**, *4*, 177–187. [CrossRef]
9. Koay, W.I.; Dillon, D. Community Gardening: Stress, Well-Being, and Resilience Potentials. *Int. J. Environ. Res. Public Health* **2020**, *17*, 6740. [CrossRef]
10. Ambrose, G.; Das, K.; Fan, Y.; Ramaswami, A. Is gardening associated with greater happiness of urban residents? A multi-activity, dynamic assessment in the Twin-Cities region, USA. *Landsc. Urban Plan.* **2020**, *198*, 103776. [CrossRef]
11. Guzel, A.E.; Arslan, U.; Acaravci, A. The impact of economic, social, and political globalization and democracy on life expectancy in low-income countries: Are sustainable development goals contradictory? *Environ. Dev. Sustain.* **2021**, *23*, 13508–13525. [CrossRef]
12. Long, L. Eco-efficiency and effectiveness evaluation toward sustainable urban development in China: A super efficiency SBM-DEA with undesirable outputs. *Environ. Dev. Sustain.* **2021**, *23*, 14982–14997. [CrossRef]
13. Okvat, H.A.; Zautra, A.J. Community Gardening: A Parsimonious Path to Individual, Community, and Environmental Resilience. *Am. J. Community Psychol.* **2011**, *47*, 374–387. [CrossRef]

14. Spano, G.; D'Este, M.; Giannico, V.; Carrus, G.; Elia, M.; Lafortezza, R.; Panno, A.; Sanesi, G. Are Community Gardening and Horticultural Interventions Beneficial for Psychosocial Well-Being? A Meta-Analysis. *Int. J. Environ. Res. Public Health* **2020**, *17*, 3584. [CrossRef]

15. Wood, C.J.; Pretty, J.; Griffin, M. A case-control study of the health and well-being benefits of allotment gardening. *J. Public Health* **2015**, *38*, e336–e344. [CrossRef]

16. Cassarino, M.; Shahab, S.; Biscaya, S. Envisioning Happy Places for All: A Systematic Review of the Impact of Transformations in the Urban Environment on the Wellbeing of Vulnerable Groups. *Sustainability* **2021**, *13*, 8086. [CrossRef]

17. Foley, L.; Prins, R.; Crawford, F.; Humphreys, D.; Mitchell, R.; Sahlqvist, S.; Thomson, H.; Ogilvie, D. Effects of living near an urban motorway on the wellbeing of local residents: Natural experimental study. *PLoS ONE* **2017**, *12*, e0174882. [CrossRef]

18. Plohl, N.; Musil, B. Modeling compliance with COVID-19 prevention guidelines: The critical role of trust in science. *Psychol. Health Med.* **2020**, *26*, 35–43. [CrossRef]

19. Volk, A.A.; Brazil, K.J.; Franklin-Luther, P.; Dane, A.V.; Vaillancourt, T. The influence of demographics and personality on COVID-19 coping in young adults. *Personal. Individ. Differ.* **2021**, *168*, 110398. [CrossRef]

20. Shammi, M.; Bodrud-Doza, M.; Islam, A.R.M.T.; Rahman, M.M. Strategic assessment of COVID-19 pandemic in Bangladesh: Comparative lockdown scenario analysis, public perception, and management for sustainability. *Environ. Dev. Sustain.* **2020**, *23*, 6148–6191. [CrossRef]

21. Brodeur, A.; Clark, A.E.; Flèche, S.; Powdthavee, N. COVID-19, Lockdowns and Well-Being: Evidence from Google Trends. *J. Public Econ.* **2020**, *193*, 104346. [CrossRef] [PubMed]

22. Prime, H.; Wade, M.; Browne, D.T. Risk and Resilience in Family Well-Being During the COVID-19 Pandemic. *Am. Psychol.* **2020**, *75*, 631–643. [CrossRef] [PubMed]

23. Saltzman, L.Y.; Hansel, T.C.; Bordnick, P.S. Loneliness, Isolation, and Social Support Factors in Post-COVID-19 Mental Health. *Psychol. Trauma Theory Res. Pract. Policy* **2020**, *12*, 55–57. [CrossRef] [PubMed]

24. White, R.G.; Van Der Boor, C. Impact of the COVID-19 pandemic and initial period of lockdown on the mental health and well-being of adults in the UK. *BJPsych Open* **2020**, *6*, e90. [CrossRef]

25. Jakovljevic, M.; Bjedov, S.; Jaksic, N.; Jakovljevic, I. COVID-19 Pandemia and Public and Global Mental Health from the Perspective of Global Health Security. *Psychiatr. Danub.* **2020**, *32*, 6–14. [CrossRef]

26. Stieger, S.; Lewetz, D.; Swami, V. Emotional Well-Being Under Conditions of Lockdown: An Experience Sampling Study in Austria During the COVID-19 Pandemic. *J. Happiness Stud.* **2021**, *22*, 2703–2720. [CrossRef]

27. Dzhambov, A.M.; Lercher, P.; Browning, M.H.E.M.; Stoyanov, D.; Petrova, N.; Novakov, S.; Dimitrova, D.D. Does greenery experienced indoors and outdoors provide an escape and support mental health during the COVID-19 quarantine? *Environ. Res.* **2021**, *196*, 110420. [CrossRef]

28. Chalmin-Pui, L.S.; Griffiths, A.; Roe, J.; Heaton, T.; Cameron, R. Why garden?—Attitudes and the perceived health benefits of home gardening. *Cities* **2021**, *112*, 103118. [CrossRef]

29. Gopal, D.; Nagendra, H. Vegetation in Bangalore's Slums: Boosting Livelihoods, Well-Being and Social Capital. *Sustainability* **2014**, *6*, 2459–2473. [CrossRef]

30. Schram-Bijkerk, D.; Otte, P.; Dirven, L.; Breure, A.M. Indicators to support healthy urban gardening in urban management. *Sci. Total Environ.* **2018**, *621*, 863–871. [CrossRef]

31. Wiguna, T.; Anindyajati, G.; Kaligis, F.; Ismail, R.I.; Minayati, K.; Hanafi, E.; Murtani, B.J.; Wigantara, N.A.; Putra, A.A.; Pradana, K. Brief Research Report on Adolescent Mental Well-Being and School Closures During the COVID-19 Pandemic in Indonesia. *Front. Psychiatry* **2020**, *11*, 598756. [CrossRef]

32. Saud, M.; Ashfaq, A.; Abbas, A.; Ariadi, S.; Mahmood, Q.K. Social support through religion and psychological well-being: COVID-19 and coping strategies in Indonesia. *J. Relig. Health* **2021**, *60*, 3309–3325. [CrossRef]

33. Siste, K.; Hanafi, E.; Sen, L.T.; Murtani, B.J.; Christian, H.; Limawan, A.P.; Siswidiani, L.P. Implications of COVID-19 and Lockdown on Internet Addiction Among Adolescents: Data From a Developing Country. *Front. Psychiatry* **2021**, *12*, 665675. [CrossRef]

34. Sutarto, A.P.; Wardaningsih, S.; Putri, W.H. Work from home: Indonesian employees' mental well-being and productivity during the COVID-19 pandemic. *Int. J. Workplace Health Manag.* **2021**, *14*, 386–408. [CrossRef]

35. Hamzens, W.P.S.; Moestopo, M.W. Pengembangan Potensi Pertanian Perkotaan di Kawasan Sungai Palu. *J. Pengemb. Kota* **2018**, *6*, 75–83. [CrossRef]

36. Prasetiyo, W.H.; Budimansyah, D. Warga Negara dan Ekologi: Studi Kasus Pengembangan Warga Negara Peduli Lingkungan Dalam Komunitas Bandung Berkebun. *J. Pendidik. Hum.* **2016**, *4*, 177–186.

37. Amir, H.M.; Saidin, S. Pengembangan Urban Farming Dalam Rangka Pemberdayaan Masyarakat di Kota Kendari. *J. Neo Soc.* **2020**, *5*, 227–237. [CrossRef]

38. Marietta, A.D.; Darmawani, E.; Noverina, R. Meningkatkan Karakter Peduli Lingkungan Melalui Kegiatan Berkebun Kelompok B Di RA Perwanida 4 Jakabaring Palembang. *PERNIK J. Pendidik. Anak Usia Dini* **2019**, *2*, 52–65. [CrossRef]

39. Prasetiyo, W.H.; Budimansyah, D.; Roslidah, N. Urban Farming as A Civic Virtue Development in The Environmental Field. *Int. J. Environ. Sci. Educ.* **2016**, *11*, 3139–3146. [CrossRef]

40. Vibriyanti, D. Society Mental Health: Managing Anxiety During Pandemic COVID-19. *J. Kependud. Indones.* **2020**, 69–74. [CrossRef]
41. Diener, E.; Oishi, S.; Tay, L. Advances in subjective well-being research. *Nat. Hum. Behav.* **2018**, 2, 253–260. [CrossRef]
42. Diener, E. Subjective well-being. *Psychol. Bull.* **1984**, 95, 542–575. [CrossRef] [PubMed]
43. Diener, E. *Assessing Well-Being*, 1st ed.; Springer: Dordrecht, The Netherlands, 2009.
44. Wicaksana, S.A.; Novasari, E.P.; Nunanisa, N.; Asrunputri, A.P. Gambaran Subjective Well Being Pada Tenaga Kerja Generasi Y. *JURISMA* **2019**, 9, 217–228. [CrossRef]
45. Emerson, S.D.; Guhn, M.; Gadermann, A.M. Measurement invariance of the Satisfaction with Life Scale: Reviewing three decades of research. *Qual. Life Res.* **2017**, 26, 2251–2264. [CrossRef] [PubMed]
46. Corrigan, J.D.; Kolakowsky-Hayner, S.; Wright, J.; Bellon, K.; Carufel, P. The Satisfaction With Life Scale. *J. Head Trauma Rehabil.* **2013**, 28, 489–491. [CrossRef] [PubMed]
47. IPBES. IPBES guide on the production of assessment. In *Secretariat of the Intergovernmental Science-Policy Platform on Biodiversity and Ecosystem Service*; IPBES: Bonn, Germany, 2018; p. 44.
48. Takahashi, T.; Uchida, Y.; Ishibashi, H.; Okuda, N. Subjective well-being as a potential policy indicator in the context of urbanization and forest restoration. *Sustainability* **2021**, 13, 3211. [CrossRef]
49. Pavot, W.; Diener, E. The Satisfaction with Life Scale and the emerging construct of life satisfaction. *J. Posit. Psychol.* **2008**, 3, 137–152. [CrossRef]
50. Pamungkas, C.; Wardhani, N.; Siswadi, A.G.P. The Effect of Positive Psychotherapy to Improve Subjective Well-Being of Early Adulthood Woman Who Hasn't Had a Partner. *J. Interv. Psikol.* **2017**, 9, 16–36. [CrossRef]
51. Senol-Durak, E.; Durak, M. Psychometric properties of the Turkish version of the Flourishing Scale and the Scale of Positive and Negative Experience. *Ment. Health Relig. Cult.* **2019**, 22, 1021–1032. [CrossRef]
52. du Plessis, G.A.; Guse, T. Validation of the Scale of Positive and Negative Experience in a South African student sample. *South Afr. J. Psychol.* **2017**, 47, 184–197. [CrossRef]
53. Telef, B.B. The Scale of Positive and Negative Experience: A validity and reliability study for adolescents. *Anatol. J. Psychiatry* **2013**, 14, 62–68. [CrossRef]
54. Espejo, B.; Checa, I.; Perales-Puchalt, J.; Lisón, J.F. Validation and Measurement Invariance of the Scale of Positive and Negative Experience (SPANE) in a Spanish General Sample. *Int. J. Environ. Res. Public Health* **2020**, 17, 8359. [CrossRef] [PubMed]
55. Prado-Gascó, V.; Romero-Reignier, V.; Mesa-Gresa, P.; Górriz, A.B. Subjective Well-Being in Spanish Adolescents: Psychometric Properties of the Scale of Positive and Negative Experiences. *Sustainability* **2021**, 12, 4011. [CrossRef]
56. Undang-Undang Republik Indonesia Nomor 26 Tahun 2007 Tentang Penataan Ruang. 2007. Available online: https://jdih.esdm.go.id/peraturan/uu-26-2007.pdf (accessed on 15 March 2022).
57. Diener, E.; Wirtz, D.; Tov, W.; Kim-Prieto, C.; Choi, D.; Oishi, S.; Biswas-Diener, R. New Well-being Measures: Short Scales to Assess Flourishing and Positive and Negative Feelings. *Soc. Indic. Res.* **2010**, 97, 143–156. [CrossRef]
58. Silva, A.D.; Taveira, M.C.; Marques, C.; Gouveia, V.V. Satisfaction with Life Scale Among Adolescents and Young Adults in Portugal: Extending Evidence of Construct Validity. *Soc. Indic. Res.* **2015**, 120, 309–318. [CrossRef]
59. Velasco, M.; Constanza, S.; Pérez, L.; Castañeda, I.A. Validation of questionnaire dispositional optimism using item response theory. *Divers. Perspect. Psicol.* **2014**, 10, 275–292.
60. Hashimoto, S.; Sato, Y.; Morimoto, H. Public-private collaboration in allotment garden operation has the potential to provide ecosystem services to urban dwellers more efficiently. *Paddy Water Environ.* **2019**, 17, 391–401. [CrossRef]
61. CoDyre, M.; Fraser, E.D.G.; Landman, K. How does your garden grow? An empirical evaluation of the costs and potential of urban gardening. *Urban For. Urban Green.* **2015**, 14, 72–79. [CrossRef]
62. Irwan, S.N.R.; Sarwadi, A. Productive Urban Landscape In Developing Home Garden In Yogyakarta City. *IOP Conf. Ser. Earth Environ. Sci.* **2017**, 91, 012006. [CrossRef]
63. Nasution, Z. Indonesian Urban Farming Communities and Food Sovereignty. Master's Dissertation, International Institute of Social Studies, The Hague, The Netherlands, 2015.
64. Junainah, W.; Kanto, S.; Soenyono, S. Program Urban Farming Sebagai Model Penanggulangan Kemiskinan Masyarakat Perkotaan (Studi Kasus di Kelompok Tani Kelurahan Keputih Kecamatan Sukolilo Kota Surabaya). *Wacana* **2016**, 19, 148–156.
65. Huff, A.G.; Beyeler, W.E.; Kelley, N.S.; McNitt, J.A. How resilient is the United States' food system to pandemics? *J. Environ. Stud. Sci.* **2015**, 5, 337–347. [CrossRef]
66. van Klingeren, F.; de Graaf, N.D. Heterogeneity, trust and common-pool resource management. *J. Environ. Stud. Sci.* **2021**, 11, 37–64. [CrossRef]
67. Li, W.C.; Yeung, K.K.A. A comprehensive study of green roof performance from environmental perspective. *Int. J. Sustain. Built Environ.* **2014**, 3, 127–134. [CrossRef]
68. Rahman, S.R.A.; Ahmad, H.; Mohammad, S.; Rosley, M.S.F. Perception of Green Roof as a Tool for Urban Regeneration in a Commercial Environment: The Secret Garden, Malaysia. *Procedia Soc. Behav. Sci.* **2015**, 170, 128–136. [CrossRef]
69. Uchiyama, Y.; Kohsaka, R. Application of the City Biodiversity Index to populated cities in Japan: Influence of the social and ecological characteristics on indicator-based management. *Ecol. Indic.* **2019**, 106, 105420. [CrossRef]
70. Kohsaka, R. Developing biodiversity indicators for cities: Applying the DPSIR model to Nagoya and integrating social and ecological aspects. *Ecol. Res.* **2010**, 25, 925–936. [CrossRef]

71. Quevedo, J.M.D.; Uchiyama, Y.; Kohsaka, R. A blue carbon ecosystems qualitative assessment applying the DPSIR framework: Local perspective of global benefits and contributions. *Mar. Policy* **2021**, *128*, 104462. [CrossRef]

72. Surya, B.; Syafri, S.; Hadijah, H.; Baharuddin, B.; Fitriyah, A.T.; Sakti, H.H. Management of Slum-Based Urban Farming and Economic Empowerment of the Community of Makassar City, South Sulawesi, Indonesia. *Sustainability* **2020**, *12*, 7324. [CrossRef]

73. Anindyajati, G.; Wiguna, T.; Murtani, B.J.; Christian, H.; Wigantara, N.A.; Putra, A.A.; Hanafi, E.; Minayati, K.; Ismail, R.I.; Kaligis, F.; et al. Anxiety and Its Associated Factors During the Initial Phase of the COVID-19 Pandemic in Indonesia. *Front. Psychiatry* **2021**, *12*, 634585. [CrossRef]

74. Uchiyama, Y.; Kohsaka, R. Access and Use of Green Areas during the COVID-19 Pandemic: Green Infrastructure Management in the "New Normal". *Sustainability* **2020**, *12*, 9842. [CrossRef]

75. Parker, J.; Simpson, G.D. Public Green Infrastructure Contributes to City Livability: A Systematic Quantitative Review. *Land* **2018**, *7*, 161. [CrossRef]

76. Ma, B.; Zhou, T.; Lei, S.; Wen, Y. Effects of urban green spaces on residents' well-being. *Environ. Dev. Sustain.* **2018**, *21*, 2793–2809. [CrossRef]

77. Roberts, S.; Shackleton, C. Temporal Dynamics and Motivaations for Urban Community Food Gardens in Medium-Sized Towns of the Eastern Cape, South Africa. *Land* **2018**, *7*, 146. [CrossRef]

78. Kohsaka, R.; Matsuoka, H. Analysis of Japanese Municipalities With Geopark, MAB, and GIAHS Certification: Quantitative Approach to Official Records With Text-Mining Methods. *SAGE Open* **2015**, *5*, 1–10. [CrossRef]

79. Kohsaka, R.; Matsuoka, H.; Uchiyama, Y.; Rogel, M. Regional management and biodiversity conservation in GIAHS: Text analysis of municipal strategy and tourism management. *Ecosyst. Health Sustain.* **2019**, *5*, 124–132. [CrossRef]

80. Quevedo, J.M.D.; Uchiyama, Y.; Lukman, K.M.; Kohsaka, R. How blue carbon ecosystems are perceived by local communities in the coral triangle: Comparative and empirical examinations in the Philippines and Indonesia. *Sustainability* **2020**, *13*, 127. [CrossRef]

81. Lukman, K.M.; Uchiyama, Y.; Kohsaka, R. Sustainable aquaculture to ensure coexistence: Perceptions of aquaculture farmers in East Kalimantan, Indonesia. *Ocean Coast. Manag.* **2021**, *213*, 105839. [CrossRef]

82. Lukma, K.M.; Uchiyama, Y.; Quevedo, J.M.D.; Kohsaka, R. Local awareness as an instrument for management and conservation of seagrass ecosystem: Case of Berau Regency, Indonesia. *Ocean Coast. Manag.* **2021**, *203*, 105451. [CrossRef]

83. Quevedo, J.M.D.; Uchiyama, Y.; Kohsaka, R. Linking blue carbon ecosystems with sustainable tourism: Dichotomy of urban–rural local perspectives from the Philippines. *Reg. Stud. Mar. Sci.* **2021**, *45*, 101820. [CrossRef]

84. Fang, C.C. The case for environmental advocacy. *J. Environ. Stud. Sci.* **2021**, *11*, 169–172. [CrossRef]

85. Schmidt, R.C. Are there similarities between the Corona and the climate crisis? *J. Environ. Stud. Sci.* **2021**, *11*, 159–163. [CrossRef]

 *land*

Article

Preliminary Experimental Trial of Effects of Lattice Fence Installation on Honey Bee Flight Height as Implications for Urban Beekeeping Regulations

Tomonori Matsuzawa [1,2] **and Ryo Kohsaka** [1,*]

[1] Graduate School of Environmental Studies, Nagoya University, Nagoya 464-8601, Japan; matsuzawa.tomonori@b.mbox.nagoya-u.ac.jp

[2] Biodiversity Research Center, IDEA Consultants, Inc., Yokohama 224-0025, Japan

* Correspondence: kohsaka@hotmail.com

Abstract: Urban beekeeping has gained salience because of its significance in biodiversity conservation and community building. Despite this, beekeeping practices in urban areas have received negative perceptions from residents, which stem from public safety concerns. There is, therefore, a need to enhance and/or work on appropriate rules for maximizing the profits while minimizing the risks. Amongst the present regulations, the installation of barriers and setbacks is the most common rule for public safety. However, only a limited number of empirical studies have reported on their effective location and height. Thus, in this study, an experimental apiary was set up with different types of barriers installed with varying distances to observe and measure flyway patterns of honey bees. We used a 3D laser scanner, which obtained 8529 points of highly accurate flight location data in about five hours. Results showed that the heights (1.8 and 0.9 m) of the barriers installed were effective in increasing the flight altitudes. The distance of the fence, which was installed as close as 1 m from the hives, was effective as well. These findings, which showed that barriers and setbacks are effective, can have regulatory implications in designing apiaries in urban spaces, where location is often restricted.

Keywords: urban beekeeping; regulations; barrier; setback; 3D laser scanner; remote sensing; fence location

Citation: Matsuzawa, T.; Kohsaka, R. Preliminary Experimental Trial of Effects of Lattice Fence Installation on Honey Bee Flight Height as Implications for Urban Beekeeping Regulations. *Land* **2022**, *11*, 19. https://doi.org/10.3390/land11010019

Academic Editors: Le Yu and Marta Debolini

Received: 9 December 2021
Accepted: 21 December 2021
Published: 23 December 2021

Publisher's Note: MDPI stays neutral with regard to jurisdictional claims in published maps and institutional affiliations.

1. Introduction

Beekeeping in the context of agroforestry systems is vital for supportinglocallivelihoods (e.g., production of honey and beeswax), particularly in forest villages [1,2]. As part of the biosphere, bees are considered major pollinators, influencing ecological relationships, genetic variation in the plant community, floral diversity, specialization and evolution, and ecosystem conservation and stability [2,3]. In agricultural settings, bees are essential for crop pollination, and pollination using bees is currently practiced globally (USA, Australia, New Zealand and Europe) [2,4]. For instance, in Canada, beekeepers are paid to provide pollination services for hybrid canola seed productions [5]. Despite their socio-ecological importance, there has been a decline in bee colonies over the last decade, which has renewed interest in honey bees, particularly in relation to colony collapse disorder [6]. This in turn has resulted in the global expansion of urban keeping [7], and scientists have argued that keeping bees in urban settings might be more beneficial for their survival due to the reduced exposure to agricultural pesticides and limited assortment of plants for foraging [8,9]. Furthermore, urban beekeeping has gained salience because of its significance in biodiversity conservation, food production, and community building in urban areas [10–12].

In recent years, increasing numbers of municipalities are actively adopting urban beekeeping as part of their environmental policies. For instance, in 2019, the German state of Bayer enacted Bavaria's nature protection law, which encourages the keeping of bees in urban areas [13]. In Los Angeles and New York, they changed their respective ordinances to allow urban beekeeping [14]. Moreover, pollinator-friendly cities, which promote the protection of pollinating insects including honey bees and the environment are on the rise [13]. However, as beekeeping practices in urban areas increase, concerns from local residents have also grown, stemming from property disputes (e.g., nuisance, trespass claims) by neighbors, negligence accusations against beekeepers, and challenges to the legal status of the honey bee by local communities [15]. Thus, there is a need to establish appropriate rules that maximize profits while minimizing the risks such as nuisance [16]. Regulations of urban beekeeping are usually motivated by concerns for public safety [7]. To reduce the probability of physical encounter between bees and the people, "Setbacks", which are defined as the distance of hives from the property boundary, adjacent dwelling, public facility, and/or street to raise the flight path upwards, "Flyaway Barriers", which refer to a solid wall or fence, or a dense hedge that helps increase the flight height of bees, as well as the number of hives that owners can keep on their property are common requirements to decrease the potential nuisance effect of beekeeping operations [16].

However, not many municipalities or other governments have set rules for urban beekeeping to date [17]. In the United States and Australia, where there are relatively more rules for urban beekeeping, there are approximately 8–10 regulatory items [16,18]. Among these, in "Lot Size and Colony Density", "Setbacks", "Flyway Barriers", and "Access to Water", we can observe cases where specific standards and criteria have been set [18]. It is unclear whether the present regulations are based on evidence. If not, it is important to understand what measures can be implemented based on science.

There are limited studies examining the effectiveness of regulatory items in urban beekeeping. A study by Garbuzov and Ratnieks [19] is one of the few cases that showed that barriers are effective for flyway control. The purpose of installing setbacks and flyway barriers is to control flyways and to lead the bees above human head height. However, there is a large range in the actual numbers specified. For example, the fence heights can range from 3 to 10 ft, and setbacks can range from 1 to 1000 ft [18]. In the regulations developed by local governments, there is no indication of scientific evidence regarding the quantities of regulatory items [18]. Garbuzov and Ratnieks [19] compared the flight height of honey bees with and without barriers and proved that barriers are effective at raising the mean honey bee flight height. However, their study did not provide data from more than 3.6 m above the ground because they measured the height by video recording honey bees passing across a 3.6 m high whiteboard. The "3.6 m" height appears to be quite insufficient when considering the flying height of honeybees in apiaries. In addition, major parallax error is included by using a single video image. These factors suggest that the observed heights could have contained considerable errors. To evaluate the effect of barrier location and height, multiple fence types and accurate measurement methods should be used in an experimental space with sufficient height.

To date, various methods have already been developed for the spatial localization of flying small animals such as insects in the wild [20,21]. In general, visual and camera measurements have weaknesses in terms of accuracy, labor, and observable range. Micro radio-transmitters, in addition to being laborious to install, can affect insect behavior. Radio-frequency identification tags are small and lightweight and can be mounted on insects as large as honey bees, but their relatively short measurable range (usually a few centimeters) limits their use [22]. Harmonic radar can be utilized over long distances, but requires large and expensive equipment, making it less practical. The retroreflector-based tracking system can track the behavior of honeybees within 35 m at low cost [21], but it is difficult to acquire numerous individuals in a short time. Tauc et al. [23] showed that their light detection and ranging system could accurately detect the spatial location of individual insects flying more than 100 m away. However, their light detection and ranging system was designed for long-range measurements, and thus has a limited measurement range for close-range apiary-scale measurements.

At present, there is a gap in the available observation methods for sufficiently measuring the flying altitude of honey bees at the apiary scale. This study aims to address this knowledge gap by examining the effects of fences and setbacks on honey bee flight height, as these are often set within the regulations of urban beekeeping. We also evaluated the methodological implications of using a 3D laser scanner to localize the bees. This method is non-destructive, does not attach observers to the insects, and can accurately acquire a large amount of data in a short time. The data gathered were then statistically analyzed to examine the effects of the fence location and height as well as the distance from the hives on flight height. These findings will provide valuable data for improving urban beekeeping management and implementing evidence-based regulations.

2. Materials and Methods

2.1. Study Area

We conducted a preliminary experiment and observations from 7:00 to 13:00 on 20 August 2021, in an experimental apiary site located in the center of Japan (Uchino, Oshino-mura village, Minamituru-gun, Yamanashi prefecture, Japan; 35°27′19.44″ N, 138°52′29.73″ E). A half-day experiment was preferred to minimize variables such as, weather, blooming conditions of nectar plants, and honey bee populations. The weather was sunny with occasional clouds and scattered rains at the end of the experiment. This experimental apiary was located on a perfectly flat agricultural field, with an area of 4 km in the east-west direction and 2 km in the north-south direction (Figure 1). There are some fields of cabbage and corn around this agricultural field. The elevation is 963 m and the site is surrounded by mountains with altitudes ranging from 1200 to 1500 m.

Figure 1. (**a**) Site of the experimental apiary in Japan; (**b**) Aerial image of the experimental apiary; (**c**) Hives and lattice fences installed in the experimental apiary, with a 3D laser scanner set next to the lattice fence (Photographed in the Yamanashi Prefecture, August 2021 by the first author).

2.2. Experimental Setup and Procedure

The experimental apiary site is a 32 × 42 m open cropland (Figure 2). The weeds were removed and reclaimed two weeks before conducting the study.

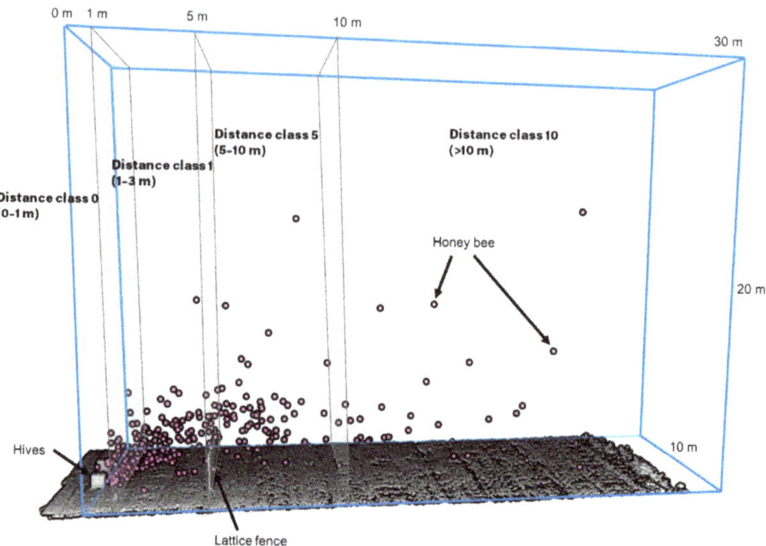

Figure 2. Layout of the experimental apiary. The lattice fence was placed 1 m or 5 m from the hive, and the 3D laser scanner was placed directly next to the lattice fence.

Three powerful hives of honey bees (*Apis mellifera*) were imported from another apiary, which was located 5 km away from the experimental site, at midnight two days prior to the start of the experiment. These hives were installed on concrete blocks facing west-northwest and kept at a 1 m distance from each other. The entrances of the hives were located 15 cm above the ground and opened during the daytime before the day of the study. The area for analyzing the location of honey bees was defined as 30 m in front of the hive, 10 m in the lateral direction, and 20 m towards the sky, using the entrance of the central hive as the origin point of the Euclidean space (Figure 2).

In a previous study [19], no significant difference was recorded between lattice fences and hedges as types of barriers; thus, in this experiment, we only used lattice fence, which was relatively easy to install. We constructed wooden skeleton lattice fences of two heights ("lattice fence 90 × 180 cm", Cain Co. Ltd., Honjo-shi, Saitama, Japan) as barriers, namely low and high. The low barriers had a height of 90 cm and a width of 540 cm while the high barriers had a height of 180 cm and a width of 540 cm. These heights were chosen based on the comprehensive review [18] conducted prior to this preliminary experiment. Based on the search of urban beekeeping in the United States, most cases have heights ranging from 90 (180 cases) to 180 cm (156 cases) [18].

These fences were installed within a few minutes and placed at 1 m or 5 m from the hive entrance (Figure 2). Honeybees memorize the height of the barrier, which can influence the flying height [19]. Thus, to minimize the effect of memory, five different fence types were installed, as reflected in Table 1: no fence, far low, far high, near low, and near high. Each of the lattice fences had gaps through which bees could pass. However, we rarely observed bees passing through the fences throughout the experiment. Although the lattice fence used in this experiment has gaps, the effect of raising the flight heights of bees can be enhanced by using a solid wall or dense vegetation.

Table 1. Fence type and experimental sequence.

Fence Type	Distance from Hives	Height of Barrier	Experimental Sequence and Time
A	None	None	1st (7:34–)
B	Far (5 m)	Low (0.9 m)	2nd (8:46–)
C	Far (5 m)	High (1.8 m)	3rd (9:52–)
D	Near (1 m)	Low (0.9 m)	4th (11:15–)
E	Near (1 m)	High (1.8 m)	5th (12:20–)

To determine any data error caused by animals other than honeybees, we conducted a flying animal capture test the day before the experiment. In this test, we installed an insect net with a diameter of 50 cm on a pole 250 cm in length and swept while walking 500 times at various heights to capture flying animals.

A total of eighteen individuals of six insect species were captured (Table 2). All insects except *Apis mellifera* and *Vespa simillima* were smaller than 2 mm, so they would not be detected by the laser beam profiler. During the experiment, butterflies (Nymphalidae gen. sp.), beetles (Scarabaeidae gen. sp.), and barn swallows (*Hirundo rustica*) passed through the experimental area less than five times. Hence, most of the plots detected by the laser scanner in this study were considered as honey bees.

Table 2. List of insects in the capture test. Sweeping was conducted 500 times in the experimental apiary.

Scientific Name	Body Size (mm)	Number of Individuals (%)
Apis mellifera	12–14	13 (72%)
Vespa simillima	22	1 (5.6%)
Psilopa polita	2	1 (5.6%)
Phoridae gen. sp.	1	1 (5.6%)
Drosophilidae gen. sp.	1	1 (5.6%)
Aphididae gen. sp.	1	1 (5.6%)

2.3. Measuring Flight Heights

The locations of honey bees were detected using a GLS-2200 laser scanner (Topcon Corporation, Tokyo, Japan), which was placed just beside the barriers outside the analysis space ($10 \times 30 \times 20$ m). This machine was developed to create high quality 3D images by irradiating more than 100,000 lasers per second in all directions. Due to the high density of lasers, flying objects in space can be detected, but such data are usually considered as noise. We used these noises as the location data of the honey bees.

Scanning was conducted using the high-speed mode, with 120,000 laser points/s (Class 3R) covering the entire experimental area. The laser was irradiated at a density of 12.5 mm and a distance of 10 m from the instrument. Since the size of a honey bee is approximately 12–14 mm, we considered that one individual would correspond to one point, but this has not been verified.

Though a single scan can capture the entire apiary, we completed 24 scans for each of the five fence types to increase the number of samples. Only 22 scans were completed for fence type 5 due to rain. Each scan took approximately 1 min and 30 s, and the data from the 24 or 22 scans were merged into one for each fence type in the application. The number of plots of honey bees captured by the 3D scanner is shown in Table 3. The obtained point cloud data were manually segmented into bees and background using the point cloud processing application QuickStitch (EIVA, Skanderborg, Denmark). The point cloud data were then converted to relative locations from the origin, and the distances from the hive were classified into four categories (distance0: 0–1 m, distance1: 1–5 m, distance5: 5–10 m, and distance10: >10 m) for statistical analysis.

Table 3. Number of honey bee plots observed by the 3D scanner.

Fence Type	Height and Distance	Number of Honey Bee Plots	Number of Honey Bee Plots in the Analysis Space (W 10 m × L 30 m × H 20 m)
A	No barrier	2007	845
B	Far–Low	3004	752
C	Far–High	1190	634
D	Near–Low	1099	633
E	Near–High	1329	671
Total	-	**8629**	**3535**

2.4. Statistical Analyses

The data collected were used to analyze the effects of the distance between the hive and the barrier and those of the height of the barrier on the flight height of the honey bees. All statistical analyses were performed with EZR [24], which is a modified version of R commander designed to provide statistical functions frequently used in biostatistics.

Friedman's test was used to determine the differences between the fence types (A–E) and distance classes (0, 1, 5 and 10). When statistically significant differences were detected among the groups, a Bonferroni correction was used for multiple comparisons (group to group). The significance level was set at 5%.

3. Results

Preliminary analysis was conducted to evaluate the null hypothesis of this study, which states that the population followed a normal distribution. Based on the Shapiro–Wilk normality test, all combinations excluding one case (fence type A × distance 10) were not normally distributed ($p < 0.05$), thus rejecting the null hypothesis. The same test was conducted for the flight height dataset and similar results were obtained, rejecting the null hypothesis. Based on the normality test results, we used non-parametric methods to further analyze the data.

The Friedman test performed for fence type and distance class showed significant differences between all types and classes ($p < 0.001$). Bonferroni's multiple comparisons were then conducted as a post-hoc test. For fence type, fence type C had a lower flying height than that of all the other fence types ($p < 0.001$, Figure 3a). For the distance class, all combinations were significantly different ($p < 0.001$, Figure 3b), and the flying height increased with the distance from the hives.

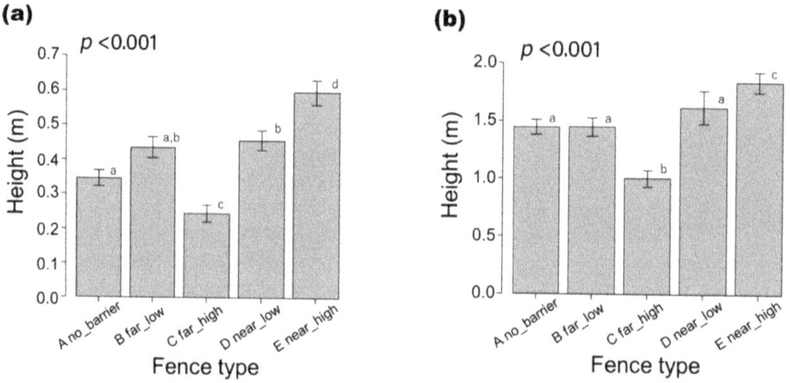

Figure 3. Difference in flying height by (**a**) fence type and (**b**) distance class. Letters above the error bars represent the results of Bonferroni's post-hoc pairwise comparison test. Error bar heights are means ± standard error.

Therefore, to examine the effect of fence type, the data were divided into distance classes, and the Kruskal–Wallis test was conducted within the same distance class. When significant differences were observed, the Bonferroni multiple comparison test was conducted.

We conducted the Kruskal–Wallis test based on the conditions set, and the results showed a significant difference for distance class 0 ($p < 0.001$) (Figure 4a), distance class 1 ($p < 0.001$) (Figure 4b), and distance class 5 ($p < 0.001$) (Figure 4c). However, for distance class 10, the results did not show any significant difference ($p = 0.69$) (Figure 4d). Subsequently, Bonferroni's post-hoc pairwise comparison tests were performed on combinations for which significant differences were detected ($p < 0.05$). For distance class 0 (0–1 m from the hives), the results showed a significant difference in fence types D (D > A, $p = 0.025$), E (E > A, $p < 0.001$), and C (C < A, <0.001) (Figure 4a). For distance class 1 (1–5 m from the hives), the results showed a significant difference in fence types E (E > A, $p < 0.001$) and C (C < A, $p < 0.001$) (Figure 4c). For distance class 5 (5–10 m from the hives), the results showed a significant difference in fence types E (E > A, $p < 0.022$) and C (C > A, $p < 0.001$ and C > B, $p < 0.001$) (Figure 4c).

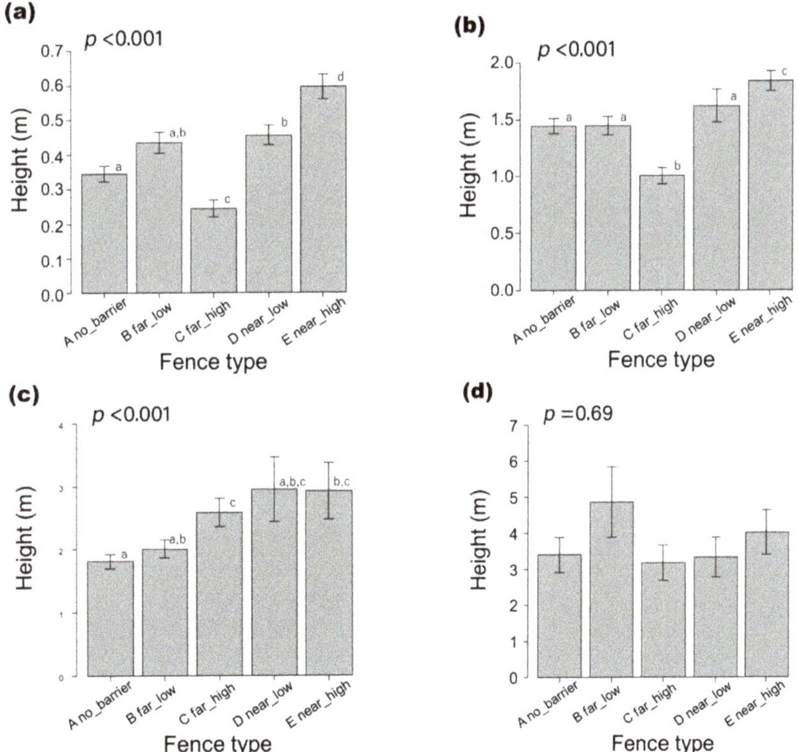

Figure 4. Effect of fence type on flight altitude. Average flying height for (**a**) distance class 0 (0–1 m), (**b**) distance class 1 (1–5 m), (**c**) distance class 5 (5–10 m), and (**d**) distance class 10 (>10 m). (**a**–**c**) showed significant differences among fence types, while (**d**) showed no significant differences. Letters above the error bars represent the results of Bonferroni's post-hoc pairwise comparison test. Bar heights are the means ± standard error.

In summary, the 1.8 m high barrier was effective at leading the honey bees to fly higher. The 0.9 m high barrier had a similar effect, but a less effective one. The flight height increased with distance from the hives, and the effect of the barriers became less significant when the honey bees were more than 10 m away from the hives.

4. Discussion

4.1. Effectiveness of the Barriers (Location and Height of the Lattice Fence)

The purpose of this study is to reveal the effect of barrier installation on the flight height of honey bees. The three evaluated parameters, which included the fence position, height, and distance from the hives, were found to affect the flight height of honey bees. The barriers were effective at increasing the flight height both at 1 and 5 m from the hives. However, the honey bees will most likely increase their flight height with distance from the hives regardless of the presence of the fences [19]. Therefore, barriers are more likely to be effective if placed closer to the hives.

In this study, 1.8 m (6 ft) and 0.9 m (3 ft) high barriers were used, and both showed positive effects on the flight height, although the effect of the 1.8 m barrier was greater. Fence type E (1.8 m barrier placed 1 m away from the front of the hives) showed an average flight height of 1.84 m ($\pm$0.09, standard error) at distance class 1 (1–5 m from the hives), and fence type C (1.8 m barrier placed 5 m away from the front of the hives) showed an average flight height of 2.59 m ($\pm$0.22, standard error) at distance class 5 (5–10 m from the hives). Based on these results, it can be expected that this system will be effective at preventing nuisance regardless of its location. These further suggest that fences installed as close as 1 m from the hives are sufficiently effective. These findings can have regulatory implications for designing apiaries in urban spaces, where the location of fences is often restricted.

For all five fence types, the flying height tended to increase with distance. Even in the case of no barrier (fence type A), the average height in distance classes 5 (5–10 m) and 10 (>10 m) was 1.81 and 3.41 m, respectively. This indicates that even without a fence, if there is enough distance to the property boundary, nuisance to people is unlikely to occur, illustrating the effectiveness of the setback.

If the setback is too large, it could be a disincentive for urban beekeeping. The City of Ontario, Canada, has a 30-m setback, while the Osaka Prefecture, Japan, has a 20-m setback requirement, which essentially prohibits urban beekeeping [9,18]. It may be worthwhile for these cities to re-examine whether they can make their setback provisions shorter.

In conclusion, this study confirmed that using 3D scanners represents an effective method for measuring small flying insects. The flyway was raised by installing a lattice fence. Higher fences can increase the flying height of honey bees. A fence 1 m from the hive was adequately effective at raising the flying height. Moreover, the flight heights tended to increase as the bees moved further away from the hive, and the effect of the barrier could not be confirmed when the distance was >10 m.

4.2. Methodological and Management Implications

Various methods have been used to record the location of flying insects. For example, attaching radio transmitters and radio-frequency identification tags have weaknesses such as the influences on the insects, time required for attaching, limited number of samples, and short observable distance [19]. Visual and video camera analyses are also disadvantageous in terms of the accuracy and time required [21]. The present study used a 3D laser scanner, which has rarely been used for determining the location of flying insects. Using this device, we were able to obtain 8529 points of highly accurate flight location data in about 5 h, without attaching any sensing devices to the bees. Furthermore, the data can be analyzed and processed in just a few hours, which is much faster than analyzing video images or using radio transmitters. With a more multifunctional laser device, it may be possible to identify species and analyze their migration speed and direction.

Urban beekeeping brings a variety of benefits, but also risks, so it is crucial that appropriate regulations exist [10,18]. Existing regulations on urban beekeeping often include regulatory items, such as the number and density of hives, water supply, as well as barriers and setbacks. The effects of barriers, however, have rarely been tested [19]. The approach of this study will be helpful to examine and provide evidence for the effects of these regulatory items.

4.3. Limitations and Future Studies

Though this study is a preliminary experiment, significant results were obtained. Quite simply, both the barrier and the setback had the effect of increasing the flying height. Nevertheless, future studies can increase the number of experiments, days, and sites, as Garbuzov and Ratnieks [19] observed. Long-term effects of fencing relative to different seasons can also be investigated.

Another limitation of this work is the accuracy of detecting flying objects other than honey bees flying more than 3 m above the ground. To improve this methodological flaw, it is recommended to use a more multifunctional laser device such as wing-beat modulation LiDAR. These devices have been successfully used in [23,25].

Moreover, this study only focused on the average flying height of honey bees. There might be a number of honey bees flying more than the average height, which could bother people. Thus, increasing the number of flight variations (e.g., lower altitudes) in future studies is recommended. Garbuzov and Ratnieks [19] argued that honey bees memorize the height of the barriers, so the raising effect is unlikely to appear immediately after the installation of the barriers. Contrastingly, the results of this study revealed that the barrier raised the flight height even immediately after installation. In the future, we plan to conduct more long-term observations, taking into account the memory effects, not only immediately after the installation of the fence.

The regulations on urban beekeeping may not be based on scientific evidence, although there are various provisions such as the number of hives, density, setbacks, and barrier height. It is not recommended that over-regulation reduces the various benefits that urban beekeeping could provide. Governments need to develop rules to enable urban beekeeping while ensuring safety [16], but more work needs to be done, e.g., the application of environmental DNA analysis to honey bees' behavior [26] to provide a scientific basis for the regulations, as was done in this preliminary study.

Future studies could look into socio-ecological contexts of urban beekeeping, e.g., [17,27]. Fostering collaborations among different stakeholders (e.g., citizens, research institutions) is important to secure commitments to urban beekeeping in the context of pollinator conservation [15,28]. However, at the local scale, collaborations among different sectors are common challenges in biodiversity monitoring and management practices [29,30]. Nevertheless, from methodological perspectives, stakeholders' perceptions in urban–rural settings are found to record appropriate information regarding the environment (e.g., [31–33]) and effective in providing feedback to management policies (e.g., [34–36]) for future development.

Author Contributions: Conceptualization, R.K.; methodology, T.M.; software, T.M.; validation, T.M.; investigation, T.M.; writing—original draft reparation, T.M.; writing—review and editing, T.M. and R.K.; visualization, T.M.; supervision, R.K.; project administration R.K.; funding acquisition, R.K. and T.M. All authors have read and agreed to the published version of the manuscript.

Funding: This work was supported by JSPS KAKENHI Grant Numbers JP21K18456; JP20K12398; JP16KK0053; JP17K02105; JST RISTEX Grant Number JPMJRX20B3. This work was also supported by IDEA Consultants, Inc.

Institutional Review Board Statement: Not applicable.

Informed Consent Statement: Not applicable.

Data Availability Statement: Not applicable.

Acknowledgments: The authors would like to thank Jay Mar D. Quevedo, Yuta Uchiyama, Yoshitaka Miyake, Kenichiro Nishibayashi, Naoki Koyanagi, Hiroshi Asanuma, Ryosuke Komatsu, Satoshi Kashiwabara, and Keita Kosoba for their continuous support.

Conflicts of Interest: The authors declare no conflict of interest.

References

1. Ayan, S.; Ayan, Ö.; Altunel, T.; Yer, E.N. Honey forests as an example of agroforestry practices in Turkey. *For. Ideas* **2014**, *20*, 141–150.
2. Bradbear, N. *Bees and Their Role in Forest Livelihoods: A Guide to the Services Provided by Bees and the Sustainable Harvesting, Processing and Marketing of Their Products*; Food and Agriculture Organization of the United Nations: Rome, Italy, 2009.
3. IPBES. Summary for Policymakers of the Assessment Report of the Intergovernmental Science-Policy Platform on Biodiversity and Ecosystem Services (IPBES) on Pollinators, Pollination and Food Production. Available online: https://www.researchgate.net/publication/329503936_Summary_for_policymakers_of_the_assessment_report_of_the_Intergovernmental_Science-Policy_Platform_on_Biodiversity_and_Ecosystem_Services_IPBES_on_pollinators_pollination_and_food_production (accessed on 31 December 2019).
4. Toni, C.; Djossa, B.; Yedomonhan, H.; Zannou, E.; Mensah, G. Western honey bee management for crop pollination. *Afr. Crop. Sci. J.* **2018**, *26*, 1–17. [CrossRef]
5. Hoover, S.E.; Ovinge, L.P. Pollen Collection, Honey Production, and Pollination Services: Managing Honey Bees in an Agricultural Setting. *J. Econ. Entomol.* **2018**, *111*, 1509–1516. [CrossRef]
6. Watson, K.; Stallins, J.A. Honey Bees and Colony Collapse Disorder: A Pluralistic Reframing. *Geogr. Compass* **2016**, *10*, 222–236. [CrossRef]
7. Moore, L.J.; Kosut, M. *Buzz: Urban Beekeeping and the Power of the Bee*; New York University Press: New York, NY, USA, 2013.
8. Henry, M.; Béguin, M.; Requier, F.; Rollin, O.; Odoux, J.-F.; Aupinel, P.; Aptel, J.; Tchamitchian, S.; Decourtye, A. A Common Pesticide Decreases Foraging Success and Survival in Honey Bees. *Science* **2012**, *336*, 348–350. [CrossRef]
9. Askham, B. Urban buzz. *Sanctuary Mod. Gr. Homes* **2013**, *25*, 76–79.
10. Egerer, M.; Kowarik, I. Confronting the Modern Gordian Knot of Urban Beekeeping. *Trends Ecol. Evol.* **2020**, *35*, 956–959. [CrossRef] [PubMed]
11. Baldock, K. Opportunities and threats for pollinator conservation in global towns and cities. *Curr. Opin. Insect Sci.* **2020**, *38*, 63–71. [CrossRef]
12. Skelton, J. Adventures in Urban Beekeeping. *Briarpatch* **2006**, *35*, 14–17.
13. Wilk, B.; Rebollo, V.; Hanania, S. A Guide for Pollinator-Friendly Cities: How Can Spatial Planners and Land-Use Managers Create Favourable Urban Environments for Pollinators? Available online: https://www.iucn.org/sites/dev/files/local_authorities_guidance_document_en_compressed.pdf (accessed on 30 October 2021).
14. City News Service Backyard Beekeeping OK'd by City of Los Angeles—Daily News. Available online: https://www.dailynews.com/2015/10/14/backyard-beekeeping-okd-by-city-of-los-angeles/ (accessed on 13 January 2020).
15. Gallay, C. Beeware of the Consequences: The Importance of Urban Apiaries and Environmental ADR. *Cardozo J. Confl. Resol.* **2018**, *20*, 417–442.
16. Salkin, P.E. Honey, It's All the Buzz: Regulating Neighborhood Beehives. *BC Envtl. Aff. L. Rev.* **2012**, *39*, 55–71.
17. Larson, K.L.; Andrade, R.; Nelson, K.C.; Wheeler, M.M.; Engebreston, J.M.; Hall, S.J.; Avolio, M.L.; Groffman, P.M.; Grove, M.; Heffernan, J.B.; et al. Municipal regulation of residential landscapes across US cities: Patterns and implications for landscape sustainability. *J. Environ. Manag.* **2020**, *275*, 111132. [CrossRef] [PubMed]
18. Matsuzawa, T.; Kohsaka, R. Status and Trends of Urban Beekeeping Regulations: A Global Review. *Earth* **2021**, *2*, 933–942. [CrossRef]
19. Garbuzov, M.; Ratnieks, F.L.W. Lattice fence and hedge barriers around an apiary increase honey bee flight height and decrease stings to people nearby. *J. Apic. Res.* **2014**, *53*, 67–74. [CrossRef]
20. Reynolds, D.; Riley, J. Remote-sensing, telemetric and computer-based technologies for investigating insect movement: A survey of existing and potential techniques. *Comput. Electron. Agric.* **2002**, *35*, 271–307. [CrossRef]
21. Smith, M.T.; Livingstone, M.; Comont, R. A method for low-cost, low-impact insect tracking using retroreflective tags. *Methods Ecol. Evol.* **2021**, *12*, 2184–2195. [CrossRef]
22. Nunes-Silva, P.; Hrncir, M.; Guimarães, J.T.; Arruda, H.; Costa, L.; Pessin, G.; Siqueira, J.O.; de Souza, P.; Imperatriz-Fonseca, V.L. Applications of RFID technology on the study of bees. *Insectes Sociaux* **2019**, *66*, 15–24. [CrossRef]
23. Tauc, M.J.; Fristrup, K.M.; Repasky, K.S.; Shaw, J.A. Field demonstration of a wing-beat modulation lidar for the 3D mapping of flying insects. *OSA Contin.* **2019**, *2*, 332–348. [CrossRef]
24. Kanda, Y. Investigation of the freely available easy-to-use software 'EZR' for medical statistics. *Bone Marrow Transplant.* **2013**, *48*, 452–458. [CrossRef]
25. Chen, Y.; Why, A.; Batista, G.; Mafra-Neto, A.; Keogh, E. Flying Insect Classification with Inexpensive Sensors. *J. Insect Behav.* **2014**, *27*, 657–677. [CrossRef]
26. Matsuzawa, T.; Kohsaka, R.; Uchiyama, Y. Application of Environmental DNA: Honey Bee behavior and Ecosystems for Sustainable Beekeeping. In *Modern Beekeeping-Bases for Sustainable Production*; In Tech Open: Rijeka, Croatia, 2020. [CrossRef]
27. Sponsler, D.B.; Bratman, E.Z. Beekeeping in, of or for the city? A socioecological perspective on urban apiculture. *People Nat.* **2021**, *3*, 550–559. [CrossRef]
28. Nicholls, A.A.; Epstein, G.B.; Colla, S.R. Understanding public and stakeholder attitudes in pollinator conservation policy development. *Environ. Sci. Policy* **2020**, *111*, 27–34. [CrossRef]

29. Uchiyama, Y.; Kohsaka, R. Application of the City Biodiversity Index to populated cities in Japan: Influence of the social and ecological characteristics on indicator-based management. *Ecol. Indic.* **2019**, *106*, 105420. [CrossRef]
30. Shih, W.-Y.; Mabon, L.; de Oliveira, J.A.P. Assessing governance challenges of local biodiversity and ecosystem services: Barriers identified by the expert community. *L. Use Policy* **2020**, *91*, 104291. [CrossRef]
31. Uchiyama, Y.; Kohsaka, R. Strategies of Destination Management Organizations in Urban and Rural Areas: Using Text Analysis Method for SWOT Descriptions at Meta-level. *Int. J. Hosp. Tour. Adm.* **2021**, *1*, 1–19. [CrossRef]
32. Quevedo, J.M.D.; Uchiyama, Y.; Kohsaka, R. Linking blue carbon ecosystems with sustainable tourism: Dichotomy of urban–rural local perspectives from the Philippines. *Reg. Stud. Mar. Sci.* **2021**, *45*, 101820. [CrossRef]
33. Imai, H.; Nakashizuka, T.; Kohsaka, R. An analysis of 15 years of trends in children's connection with nature and its relationship with residential environment. *Ecosyst. Health Sustain.* **2018**, *4*, 177–187. [CrossRef]
34. Kohsaka, R.; Park, M.S.; Uchiyama, Y. Beekeeping and honey production in Japan and South Korea: Past and present. *J. Ethn. Foods* **2017**, *4*, 72–79. [CrossRef]
35. Quevedo, J.M.D.; Uchiyama, Y.; Kohsaka, R. A blue carbon ecosystems qualitative assessment applying the DPSIR framework: Local perspective of global benefits and contributions. *Mar. Policy* **2021**, *128*, 104462. [CrossRef]
36. Kohsaka, R. Developing biodiversity indicators for cities: Applying the DPSIR model to Nagoya and integrating social and ecological aspects. *Ecol. Res.* **2010**, *25*, 925–936. [CrossRef]

 *land*

Article

Ecosystem and Driving Force Evaluation of Northeast Forest Belt

Zhihong Liao [1,†], Kai Su [1,*,†], Xuebing Jiang [2], Xiangbei Zhou [1], Zhu Yu [3], Zhongchao Chen [4], Changwen Wei [1], Yiming Zhang [1] and Luying Wang [1]

[1] College of Forestry, Guangxi University, Nanning 530004, China
[2] School of Mechanical Engineering, Guangxi University, Nanning 530004, China
[3] Guangxi Forest Inventory & Planning Institute, Nanning 530011, China
[4] Guizhou Linfa Survey and Design Co., Ltd., Guiyang 550001, China
* Correspondence: sukai_lxy@gxu.edu.cn
† These authors contributed equally to this work.

Citation: Liao, Z.; Su, K.; Jiang, X.; Zhou, X.; Yu, Z.; Chen, Z.; Wei, C.; Zhang, Y.; Wang, L. Ecosystem and Driving Force Evaluation of Northeast Forest Belt. *Land* **2022**, *11*, 1306. https://doi.org/10.3390/land11081306

Academic Editors: Jay Mar D. Quevedo, Norie Tamura, Yuta Uchiyama and Ryo Kohsaka

Received: 18 July 2022
Accepted: 11 August 2022
Published: 13 August 2022

Publisher's Note: MDPI stays neutral with regard to jurisdictional claims in published maps and institutional affiliations.

Abstract: The ecosystem in the Northeast Forest Belt (NFB) can provide various ecosystem services, such as soil conservation, habitat provision, water conservation, and so on. It is essential for maintaining the ecological environment in Northeast China and the entire country. In the face of increasingly severe environmental problems, the comprehensive and accurate evaluation of ecosystem conditions and their changes is significant for scientific and reasonable recovery and protection measures. In this study, the NFB was taken as the research area. The spatio-temporal changes in ecological quality from 2005 to 2015 and the main driving factors behind them were analyzed by constructing the comprehensive ecosystem evaluation index. The results showed that: The landscape types of the NFB were mainly forest, cropland, and grassland. And the better ecological environment of the NFB was mainly distributed in the south of Changbai Mountains (CBM), the middle of Lesser Khingan Mountains (LKM), and the northwest of Greater Khingan Mountains (GKM). In contrast, the northeast of CBM, the southwest of LKM, and the edge of southern GKM were relatively poor. During 2005–2015, the ecosystem in the NFB was in a relatively good state as a whole, showing a steady-to-good development trend. However, more attention needed to be paid to some areas where degradation still existed. Land use/cover, climate (annual average rainfall, etc.), and human disturbance were potential factors affecting ecosystem evolution in the NFB. This study aims to provide an effective scientific basis and policy reference for the environmental protection and construction of the NFB.

Keywords: land use; land cover; temporal and spatial change; ecosystem assessment; GIS; northeast forest belt

1. Introduction

As the largest ecosystem on land, the forest ecosystem can provide a variety of ecosystem services (ES) such as soil conservation service (SCS), carbon sequestration service (C), habitat provision service (HP), sand-stabilization service (SSS), water conservation service (WCS) and so on. The Northeast Forest Belt (NFB) contains China's typical and essential forest ecosystem [1]. Due to the rich forest and biological resources, it is an important forest resource and biodiversity protection base in China. At the same time, because of the fertile soil and rich mineral resources, the NFB is an important food base and an old industrial base in China. To further protect the environment and promote regional sustainable development, China proposed the strategic pattern of building "two ecological barriers and three shelters" ecological stability in 2006 [2].

However, under the dual influence of global climate change and human disturbance, a series of ecological environment problems have been triggered, affecting the ecological environment stability in Northeast China. Understanding ecosystem evolution and driving factors is significant for ecosystem protection and sustainable development [3–10]. In the face of increasingly serious ecological problems, it is urgent to scientifically evaluate the ecosystem status, changing trends, and driving factors of the NFB. With the construction of the pattern of "two ecological barriers and three shelters" in China, many studies of the ecosystem of the NFB have also been carried out. Sun [1] assessed the ecosystem pattern, quality, and service function of the NFB by remote sensing data combined with monitoring statistical data through model simulation and sample surveys, and analyzed the changing trend between 2000 and 2010. It was found that there was no significant change in ecosystem quality (EQ) and service function of the NFB during the period. Still, the interference of social activity factors was increasing continuously. Based on this study, the calculation model of net ecosystem production was constructed to evaluate the carbon sequestration in the forest ecosystem of NFB [2]. It was concluded that the overall performance of carbon sequestration was carbon sinks, and human disturbances such as urban expansion, poor vegetation growth, and high temperature were significant factors affecting carbon sequestration. It provided an effective scientific basis for formulating ecological and environmental protection measures and policies in the NFB. After that, Su et al. [3] analyzed the changes in landscape pattern in the NFB from 2000 to 2015, and pointed out that the ecosystem in the NFB remained stable as a whole for 15 years, and simulated the changing trend in 2020. Qi et al. [11] explored the spatial differentiation characteristics and mechanism of trade-offs and synergies among six different ES in the NFB, indicating that the six ES showed a significant aggregation distribution, and the trade-offs and synergies had apparent spatial differentiation. Zhu et al. [12] evaluated the spatial-temporal distribution and changes in ecological vulnerability in the NFB in two different periods. They concluded that the overall ecological vulnerability was at a good level. Net primary productivity and land use types were the main driving factors affecting the spatial differentiation pattern of ecological vulnerability. This series of studies have provided a certain theoretical and scientific basis for the ecological system protection and management and its sustainable development in the NFB. However, previous studies have not constructed a comprehensive evaluation index for the NFB, so as to conduct a more integrated and comprehensive evaluation of its ecosystem.

Due to the limitations of single-factor evaluation results, more and more studies have begun to construct remote sensing ecological index (RSEI) to comprehensively evaluate and analyze the ecosystem changes in different regions [13–16]. In recent years, analyzing ecosystem evolution based on RSEI is also a hot topic worldwide. Mohammad et al. [17] used RSEI and impervious surface percentage feature space to map the urban surface ecological poorness zone (USEPZ) by combining land surface temperature (LST), humidity, normalized difference vegetation index (NDVI), and normalized difference soil index (NDSI), and thus proposed a new method for quantifying USEPZ. The results showed that the significant differences between surface ecological status (SES) and USEPZI in different cities were mainly caused by the physical characteristics of surface organisms. Based on this study, Mohammad et al. [18] continued to develop a new surface ecological condition index–land surface ecological status composition index (LSESCI) and compared it with RSEI. It was concluded that LSESCI was superior to RSEI in simulating the spatial and temporal changes of SES. At the same time, Karbalaei et al. [19] also synthesized RSEI under the PSR (pressure-state-response) framework by using principal component analysis combined with NDVI, LST, land surface moisture (LST) and normalized differential built-up, and bare soil index (NDBSI). The results showed that the disturbance of human activities and climate change disturbed the ecological balance, resulting in the reduction of urban EQ. Then based on RSEI, Maity et al. [20] used Landsat image data to evaluate the temporal and spatial variation of ecological environment quality in Kolkata urban agglomeration. The results accurately described the ecological environment quality of the study area, which could help the policy makers to scientifically guide ecological management decisions. In China, compared with other relevant studies, Xu et al. [21] early proposed a new remote sensing ecological index by coupling four evaluation indexes: vegetation index, surface temperature, humidity component, and soil index, and integrating each index with principal component transformation. Based on previous studies, the proposed remote sensing ecological index was improved, and the time series of ecological status images were generated by using the improved index and surface temperature images to realize the detection of ecological changes at different scales [22]. Then Yuan et al. [23] constructed remote sensing ecological index by integrating NDVI, humidity, LST, and NDBSI, and analyzed the temporal and spatial changes of EQ in Dongting Lake Basin from 2001 to 2019. Wang et al. [24] selected four indicators of humidity, greenness, dryness, and heat, and used the principal component analysis method to construct the ecological environment quality evaluation index. Combined with the land use data of Xinjiang and the central urban area of Urumqi, they explained the ecological environment quality and land use status in each period, thus revealing its dynamic development trend. Although the RSEI has high effectiveness and reliability and is widely used in different regions and ecosystem assessments, it ignores the impact of land use/cover change on the structure and function of the ecosystem [23,25,26]. Single-factor (or specific-factor) assessments are oriented and can be used to assess ecosystem conditions from a specific perspective. Although multi-factor (pattern-quality-function) multi-level assessments can systematically and comprehensively assess ecosystems, the results of different levels of assessments have opposite conclusions among certain regions [27–29].

To sum up, using a comprehensive evaluation index to evaluate the changes in the ecosystem in each region effectively reduces the uncertainty based on single-factor evaluation results. In addition, combined with the analysis of various influencing factors, we can more comprehensively and accurately understand the impact of natural and human disturbance on the ecosystem. Therefore, this study used remote sensing (RS) and GIS technology, based on the "pattern-quality-service" evaluation framework system, using the analytic hierarchy process (AHP) to determine the weight of each index, and constructed a comprehensive index to evaluate the ecosystem of the NFB. The purpose of this study was: (1) To assess the long-term dynamic changes of EQ in the NFB from 2005 to 2015; (2) To explore the driving forces of these changes. This study can provide decision support and a scientific basis for the ecosystem stability of the NFB.

2. Materials and Methods

2.1. Study Area

The study area is located in the NFB. According to its geographical characteristics, it can be divided into three regions: Greater Khingan Mountains (GKM) area, Lesser Khingan Mountains (LKM) area, and Changbai Mountains (CBM) area. The NFB is located in 118°48′~134°22′ East longitude and 40°52′~53°34′ North latitude, with a total area of about 670,000 km^2, of which the forest area accounts for about 400,000 km^2. The terrain is diverse, mainly mountainous and hilly, the landscape is undulating, and the soil is fertile. The climate belongs to the temperate monsoon climate, hot and rainy in summer, dry and less rain in winter, and relatively cold; four seasons are distinct; annual precipitation is about 300~1000 mm. As the ecological stability barrier in Northeast China, the NFB is a forest region with a high conservation value in Northeast China [3,12]. The location of the study area is shown in Figure 1.

which was divided into three levels, including goal layer A, labeling layer B, and indicator layer C. Secondly, seven experts in this study scale the factors at each level and constructed judgment matrix, that is, the number 1~5 and its reciprocal were used as the scale to objectively judge the factors of each level. Subsequently, the eigenvector and the maximum eigenvalue of the matrix were obtained, and the consistency test of the matrix was carried out to obtain the weight coefficients. In this study, $CR < 0.1$ passed the consistency test. The judgment matrix of different levels is shown in Table 2. The consistency test index formulas are [33,34]:

$$CR = CI/RI \tag{1}$$

$$CI = \frac{\lambda_{\max} - n}{n - 1} \tag{2}$$

where CI is the deviation consistency index; RI is the average consistency index; $\lambda_{\max}$ is the maximum eigenvalue; n is the order of the judgment matrix; ensuring that the ratio of CI to RI is less than 0.1 [34].

Table 2. Judgment matrix of different levels.

Hierarchical Model	Judgment Matrix							Consistency Test
	A	B1	B2	B3	Wi			
A–B	B1	1	1/3	2	0.2969			$CR = 0.052$
	B2	3	1	3	0.5401			$\lambda_{\max} = 3.054$
	B3	1/2	1/3	1	0.1630			
	B1	C1	C2	C3	Wi			
B1–C	C1	1	3	2	0.5466			$CR = 0.009$
	C2	1/3	1	1/2	0.1810			$\lambda_{\max} = 3.009$
	C3	1/2	2	1	0.2724			
	B2	C4	C5	C6	C7	C8	Wi	
	C4	1	3	4	1/2	2	0.2620	
B2–C	C5	1/3	1	2	1/4	1/4	0.0984	$CR = 0.015$
	C6	1/4	1/2	1	1/5	1/3	0.0623	$\lambda_{\max} = 5.068$
	C7	2	4	5	1	3	0.4167	
	C8	1/2	2	3	1/3	1	0.1606	

Note: A is the goal layer; B is the labeling layer, and B1–B3 are ecosystem quality, ecosystem service, and ecosystem pattern; C is the indicator layer, C1–C8 are FVC, NPP, LAI, WCS, SCS, SSS, C, HP; Wi is the weight value.

Then the obtained weight coefficients are used to calculate the comprehensive index. The calculation method is [35]:

$$V_{cei} = \sum_{i=1}^{n} V_i \times \varepsilon_i \tag{3}$$

where, V_{cei} is the composite index; V_i is the evaluation result of indicator i; ε_i is the weight of the first index.

2.3.2. Ecosystem Service

One of the important purposes of the construction of the NFB is to improve regional ES and promote regional ecological stability. The rich ES assessment in the NFB is added to the assessment framework system, which can better realize the protection, improvement, and sustainable development of the ecosystem. Therefore, this study selected SSS, SCS, WCS, C, and HP five ES for analysis and evaluation.

SSS is usually estimated by the amount of sand-stabilization of vegetation. By calculating the difference between the potential and the actual wind erosion, the sand stabilization of vegetation can be estimated. The potential wind erosion refers to the wind erosion when there is no vegetation, and the actual wind erosion refers to the wind erosion when there is vegetation. The calculation formulas are [36]:

$$SR = S_{L1} - S_{L2} \tag{4}$$

$$S_{L1} = \frac{2 \cdot x}{S_1^2} Q_{max1} \cdot e^{-\left(\frac{x}{S_1}\right)^2} \tag{5}$$

$$S_{L2} = \frac{2 \cdot x}{S_2^2} Q_{max2} \cdot e^{-\left(\frac{x}{S_2}\right)^2} \tag{6}$$

$$Q_{max1} = 109.8\left[WF \times EF \times SCF \times K'\right] \tag{7}$$

$$S_1 = 150.71\left(WF \times EF \times SCF \times K'\right)^{-0.3711} \tag{8}$$

$$Q_{max2} = 109.8\left[WF \times EF \times SCF \times K' \times COG\right] \tag{9}$$

$$S_2 = 150.71\left(WF \times EF \times SCF \times K' \times COG\right)^{-0.3711} \tag{10}$$

where SR is the sand-stabilization, kg/m²; S_{L1} and S_{L2} are potential wind erosion and actual wind erosion, kg/m², respectively; Q_{max1} and Q_{max2} are the maximum migration capacities of potential and actual soil erosion, kg/m; S_1 and S_2 are the critical field lengths of potential and actual soil erosion, respectively, m; x is the maximum wind erosion distance, m; WF is the meteorological factor; EF is the soil erodibility factor; SCF is a soil crust factor; K' is the surface roughness factor; COG is the vegetation cover factor.

SCS is usually estimated by the soil conservation of vegetation. The soil conservation of vegetation can be estimated by calculating the difference between the potential and the actual soil conservation. The potential soil conservation refers to the amount of soil erosion generated without vegetation, and the actual soil conservation refers to the amount of soil erosion generated with vegetation. This study selected USLE to evaluate SCS in the NFB ecosystems. The calculation formula is as follows [37]:

$$SC = SE_p - SE_a = R \cdot K \cdot LS \cdot (1 - COG) \tag{11}$$

where, SC is soil conservation, [t/(hm² a)]; SE_p and SE_a are potential and actual soil conservation, respectively, [t/(hm² a)]; R is rainfall erosivity factor, MJ·mm/(hm²·ha); K is soil erodibility factor, t·hm²·h/(hm²·MJ·mm); LS and COG are terrain factor, vegetation cover factor, no dimension.

WCS is usually estimated by the water balance equation. WCS can be estimated using precipitation minus evaporation, evapotranspiration, and rainstorm runoff. The calculation formula is as follows [38]:

$$WR = PRE - ET - QF \tag{12}$$

where WR is water conservation, mm; PRE is annual precipitation, mm; QF is rainstorm runoff, mm. ET is the actual evaporation and emission, mm.

C is usually calculated using above-ground biomass multiplied by the biomass-carbon conversion coefficient. The main calculation formula is as follows [39]:

$$COS = \sum_{i=1}^{j} AGB_i \times C_i \tag{13}$$

where COS is the ecosystem carbon storage; i is the type i ecosystem; j is the total number of ecosystem types; AGB_i is the above-ground biomass of ecosystem type i; C_i is the biomass-carbon conversion coefficient of i ecosystem types.

HP is usually estimated by the distribution of indicator species in county units. The calculation formula is [11]:

$$E_a = \sum_{i=1}^{n} E_i \tag{14}$$

where E_a indicates the total number of species (number); E_i is the number of species (number).

2.3.3. Ecosystem Quality

Vegetation is an important part of the ecosystem, which connects the natural processes such as the atmosphere, water, and soil. Its change will directly affect the regional climate, hydrology, and soil conditions, and has an important impact on the regional energy cycle and the biochemical cycle of substances. It is an important indicator of regional EQ change. It is also significant to add the in-depth evaluation of vegetation growth status and growth vigor into the evaluation framework system. Therefore, this study selected fractional vegetation coverage (FVC), net primary productivity (NPP), leaf area index (LAI), three EQ analyses, and evaluations.

FVC refers to the ratio of the vertical projection area of plants in a region to that of the region, which is one of the most essential indicators to measure EQ and surface vegetation status. Normally, the FVC is usually calculated by NDVI. The calculation formulas are [40]:

$$\text{NDVI} = (\rho_{\text{NIP}} - \rho_{\text{R}})/(\rho_{\text{NIP}} + \rho_{\text{R}}) \tag{15}$$

where ρ_{NIP} and ρ_{R} are the reflectivity of near-infrared band and red band, respectively.

$$\text{FVC} = (\text{NDVI} - \text{NDVI}_{\text{soil}})/(\text{NDVI}_{\text{veg}} - \text{NDVI}_{\text{soil}}) \tag{16}$$

where FVC is fractional vegetation coverage, NDVI_{veg} is the NDVI value of pure vegetation pixel, and $\text{NDVI}_{\text{soil}}$ is the NDVI value of no vegetation pixel.

NPP reflects the efficiency of fixing and converting light energy into compounds in plants. NPP can well reflect the vegetation productivity and EQ. At present, there are many models to calculate NPP. In this study, the CASA model estimates the NPP of the NFB. In the CASA model, NPP is mainly determined by the two variables of APAR and light energy conversion rate (ε) absorbed by vegetation. The calculation formulas are [41]:

$$\text{NPP} = \text{APAR}(t) \times \varepsilon(t) \tag{17}$$

$$\text{APAR} = \text{FPAR} \times \text{PAR} \tag{18}$$

$$\varepsilon(t) = \varepsilon \times T1(t) \times T2(t) \times W(t) \tag{19}$$

where $\text{APAR}(t)$ is the photosynthetically active radiation absorbed by plants; $\varepsilon(t)$ is the conversion efficiency of APAR into organic carbon; FPAR is the effective absorptivity of APAR by plants; PAR is the driving energy of plant photosynthesis. ε is the maximum light energy conversion rate under ideal conditions; $T1(t)$ and $T2(t)$ are the effects of temperature and photosynthesis in varying degrees, respectively; $W(t)$ is the influence coefficient of water stress.

LAI refers to the ratio of the total leaf area of plants to the land area per unit of land area. LAI can reflect the vegetation growth in vertical structure, which is also one of the important indicators for evaluating EQ. There are many methods to calculate LAI, among which remote sensing technology can realize dynamic real-time and large-scale monitoring of LAI. The vegetation canopy radiative transfer model is one of the common models to calculate LAI by using remote sensing technology. The corresponding radiative transfer process is described as follows [42]:

$$-\mu\frac{\partial L(z,\Omega)}{\partial\tau} + G(\tau,\Omega)L(z,\Omega) = \frac{\omega}{4\pi}\int_{4\pi} p(\Omega' \to \Omega)G(\Omega')L(z,\Omega')d\Omega' \tag{20}$$

where L is brightness; $\mu = cos\theta$ is the chord value of transmission direction zenith angle; G is the leaf angle distribution function; ω is the leaf albedo; $p(\Omega' \to \Omega)$ is the vegetation canopy phase function.

2.3.4. Evolution of Ecosystem Pattern

The land use/cover change has an important function in maintaining ecosystem service function. Land use/cover change will affect the function and structure of the ecosystem, and the evolution of land use/cover mode will directly affect the types and intensity of ES. It is significant to study the evolution of ecosystem service value driven by land use/cover change, which is also an essential quantitative index of environmental effects of land use/cover change. Therefore, this study selected the ecosystem transfer matrix and PNTIL model to analyze and evaluate the land use/cover change in the NFB.

The ecosystem transfer matrix can effectively list the conversion relationship between different ecosystems in different periods in the form of a matrix to reflect the change characteristics of ecosystems and the flow direction between systems in detail, and quantify the conversion status between systems [43–46]. Using the following formula to calculate the proportion of different types of landscape areas [47]:

$$P_{ij} = S_{ij}/TS(i = 1, 2, \ldots, n; j = 1, 2, \ldots, n) \tag{21}$$

where P_{ij} is the area ratio of type i ecosystem based on classification at all levels in the first year of the ecosystem classification system; S_{ij} is the area of category i ecosystems in the ecosystem classification system at all levels in the year j; TS is the total area of the evaluation area; n is the number of ecosystem types.

PNTIL model can analyze the evolution trend of ecosystem patterns [47–49]. Based on the stability and biodiversity of each ecosystem, the ecosystem transformation was divided into positive change and reverse change, and the positive/reverse transformation rules of the ecosystem were established (Table 3). The ecosystem transfer matrix of the NFB from 2005 to 2015 was used to quantitatively evaluate the ecosystem PNTIL of the NFB. The calculation formula of the ecosystem conversion index is [47]:

$$LCTR_{i,k} = (\Delta S_{i,k}/S_i) \times 1/t \times 100 \tag{22}$$

where $LCTR_{i,k}$ is the conversion rate of ecosystem type to k in the i region; S_i is the total area of all ecosystems in region i; $\Delta S_{i,k}$ is the total area converted into k type; $k = 1$ represents the positive conversion rate of ecosystems in the region; $k = 2$ represents the inverse conversion rate of the ecosystem, t being a time variable.

Table 3. Rules of forward/reverse ecosystem transformation.

Ecosystem Type	Types after Conversion	Conversion Direction
Forest (I)	III	+
	II, IV, V, VI, VII	-
Grassland (II)	I, III	+
	IV, V, VI, VII	-
Wetland (III)	-	+
	I, II, IV, V, VI, VII	-
Cropland (IV)	I, II, III	+
	V, VI, VII	-
Built-up land (V)	I, II, III, IV	+
	VI, VII	-
Desert (VI)	I, II, III, IV, V, VII	+
	-	-
Others (VII)	I, II, III, IV, V	+
	VI	-

2.3.5. Quantitative Spatio-Temporal Analysis of EQ

A quantitative spatio-temporal analysis of EQ in the NFB was carried out. The results of EQ were divided into five grades, and the transfer matrix of five grades of EQ in three periods was counted to reflect the changes in EQ in time and space. Using the following formula to calculate the proportion of different levels of EQ area [47]:

$$P_{ij} = S_{ij}/TS(i = 1, 2, \ldots, n; j = 1, 2, \ldots, n) \tag{23}$$

where P_{ij} is the area ratio of type i EQ in year j based on classification at all levels in the EQ classification system; S_{ij} is the area of category i EQ in year j based on classification at all levels in the EQ classification system; *TS* is the total area of the evaluation area; *n* is the number of types of EQ at different levels.

2.3.6. Ecosystem Driving Force Analysis

The geo-detector model is mainly a new statistical method to reveal the driving factors behind spatial differentiation, which can detect whether a certain factor is a reason for the evolution of the ecosystem. This model mainly uses the q-value measurement factor to detect the results. The larger the q-value, the stronger the driving effect of independent variables for the spatial differentiation of dependent variables and vice versa [50–55]. Based on the availability of relevant research [12,35] and regional data, 18 potential socio-ecological driving factors were selected as independent variables X. The comprehensive evaluation indexes of the ecosystem in the three periods of 2005, 2010, and 2015 were selected as the dependent variable Y, including terrain-related factors (elevation, slope, aspect), climate (annual average precipitation, annual average temperature), social economics (population density, GDP, road), and land cover (cropland, forest, grassland, wetland, built-up land, desert, other types), soil texture (silt, clay, sand). Next, the natural breakpoint method was used to change the independent variable X from numerical value to category quantity. The independent variable X was matched with the dependent variable Y. Then, the geographical detector model was used for factor detection analysis. Finally, the influence value and driving effect of each variable X for the dependent variable Y were obtained.

3. Results and Analysis

3.1. Evolution of Ecosystem Pattern

From 2005 to 2015, the classification result of ecosystem patterns in the NFB is shown in Figure 3 and Table 4. It can be seen that forest, grassland, and cropland were the main ecosystems. Statistics showed that the area of these three ecosystems was about 580,000 km², accounting for more than 94%. Wetland, built-up land, desert, and other ecosystems were scattered in the NFB. During 2005–2015, the structure of ecosystems in the NFB was relatively stable, but conversions between different types were frequent due to climate change, human activities, and other interference factors. Ecosystem transfer in the NFB during 2005–2015 is shown in Figure 4.

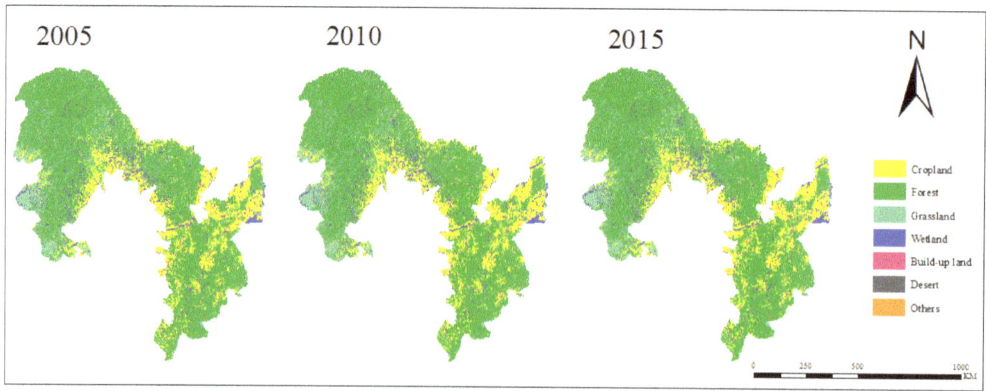

Figure 3. Spatial distribution of different land use types in the Northeast Forest Belt (NFB) from 2005–2015.

Table 4. Statistics of different land use types of ecosystem patterns of the NFB from 2005–2015.

Ecosystem Pattern	2005		2010		2015	
	Area (km²)	Percent (%)	Area (km²)	Percent (%)	Area (km²)	Percent (%)
Forest	110,260	17.92	110,164	17.90	110,838	18.01
Grassland	391,951	63.68	391,810	63.66	391,497	63.61
Wetland	80,262	13.04	80,377	13.06	80,080	13.01
Cropland	26,738	4.34	26,799	4.35	26,492	4.30
Built-up land	5607	0.91	5665	0.92	5900	0.96
Desert	172	0.03	172	0.03	173	0.03
Others	468	0.08	471	0.08	478	0.08
Total amount	615,458	100	615,458	100	615,458	100

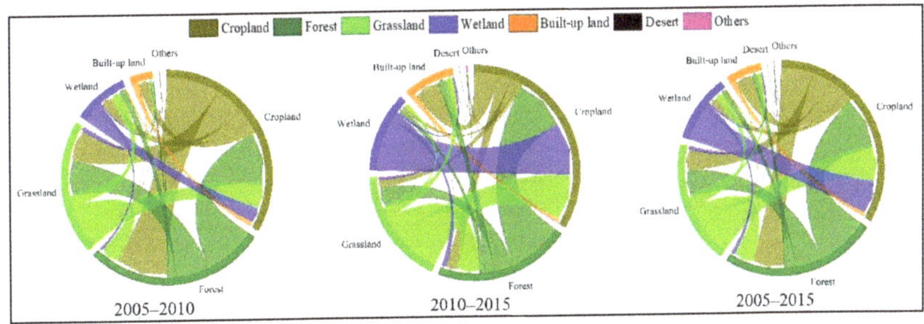

Figure 4. Area transfer matrix chord of landscape pattern types in the NFB from 2005–2015.

From 2005 to 2010, the areas of cropland and forest decreased, and the areas of grassland, wetland, and built-up land increased. Cropland was mainly converted into forest, about 378 km², followed by grassland, about 219 km². However, some grassland and forest were still transformed into cropland, with a conversion area of about 537 km². This showed that the project of returning cropland to forest and grassland in the important ecological function areas of the NFB had achieved good results. However, there were still some cases of deforestation and increasing human disturbance.

From 2010 to 2015, large areas of grassland, forest, and wetland were transformed into cropland and built-up land. During this period, cropland conversion was the main driver of the increase in built-up land areas, with about 153 km² of cropland converted into built-up land. However, there were still 951 km² of grassland, forest, and wetland transformed into cropland. During 2010–2015, the area of cropland and built-up land increased steadily, and the interference of social factors such as human activities was significantly higher than that between 2005–2010.

Between 2005 and 2015, there were more mutual transformations between cropland and forest, forest and grassland, cropland and grassland, built-up land and cropland, cropland and wetland. Among them, the forest type was transferred out of 1214 km², mainly to cropland and grassland. The area transferred to other land was relatively small, indicating that the protection of forests had a certain effect in 10 years. A total of 854 km² area of grassland changed, mainly converted to forest and cropland. The outflow of cropland mainly flowed to forest, grassland, built-up land, and wetland, with areas of 449 km², 260 km², 230 km², and 112 km², respectively. The built-up land area increased yearly, and the conversion speed was gradually accelerating. The increasing area mainly came from cropland, forest, and grassland. The wetland area mainly flowed to cropland, and other landscape types had almost no change. Overall, the project of returning cropland to forest and grassland had achieved a good result. Still, the areas of cropland and the built-up land areas had gradually increased in the past 10 years, indicating that the impact of human interference, such as the blind expansion of cultivated land and construction land on the NFB, was gradually increasing, which required us to pay more attention to and the protection of its ecosystem needed to be continued.

3.2.2. Dynamic Characteristics of LAI

It can be seen from Figure 5 that the distribution of high LAI areas was similar to that of FVC, which was mainly distributed in the middle of the three mountains in the NFB, and the edge areas of northeast CBM, southwest LKM and southern GKM were relatively low. In 2015, the area with an LAI of greater than 48 was mainly distributed in the central part of LKM and the central and southern part of CBM, with an area of 206,334.50 km^2, accounting for 33.50%. The area with an LAI of less than 32 was mainly located northeast of CBM and on the edge of the southern GKM, with an area of 146,782.5 km^2, accounting for 23.82%.

It can be seen from Figure 7 and Table 7 that in addition to the small area of poor quality, the other four grades were almost similar, and the area of excellent grade had always been the highest. During 2005–2010, areas at the general and middle levels showed a decreasing trend, while areas at the poor, good and excellent levels showed an increasing trend. The largest transformation area was from fair grade to middle grade, and the transfer area was 44,022.50 km^2. The period generally showed a trend of transformation in a good direction. During 2010–2015, the area of some middle and general grades increased and decreased, but the area of excellent and good grades continued to increase. The largest transformation area was from middle to fair grade, and the transfer area was 127,125 km^2. The overall trend was relatively stable. It showed that the LAI of the NFB had been continuously improved and enhanced during 2005–2015, and the policy of returning cropland to forest and grassland had protected and improved its ecosystem.

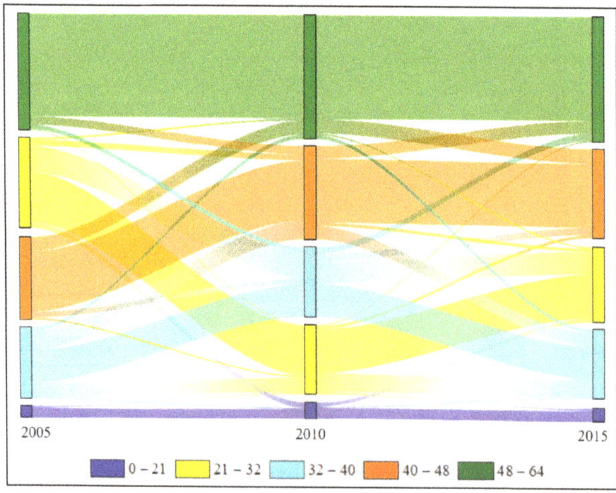

Figure 7. Sankey diagram of the area transfer matrix of different LAI levels in the NFB from 2005–2015.

Table 7. Area transfer matrix of different LAI levels in the NFB from 2005–2015.

Year	LAI Level	Poor (km^2)	Fair (km^2)	Middle (km^2)	Good (km^2)	Excellent (km^2)
	Poor		3037.25	534.50	72.00	383.25
	Fair	9115.00		44,022.50	10,289.75	3443.00
2005–2010	Middle	2052.00	23,814.50		30,719.25	7123.00
	Good	159.00	3325.25	10,524.75		23,651.75
	Excellent	113.50	1179.75	7669.50	13,580.00	
	Poor		6478.25	2084.50	381.00	387.75
	Fair	3873.50		30,428.50	5348.50	2170.00
2010–2015	Middle	644.50	37,092.00		12,495.75	11,521.75
	Good	77.50	5725.00	22,008.25		20,833.25
	Excellent	338.25	2606.00	6227.00	23,994.50	
	Poor		3700.25	877.75	193.50	465.75
	Fair	6822.00		39,639.50	11,397.50	4577.00
2005–2015	Middle	1096.50	27,113.00		23,052.25	11,783.50
	Good	226.50	4274.25	13,668.00		21,271.25
	Excellent	107.00	1915.75	6896.75	15,372.75	

3.2.3. Dynamic Characteristics of FVC

It can be seen from Figure 5 that the high-level regions were mainly distributed in most parts of the NFB, and most of them were concentrated in the middle of the three mountains of the NFB. The areas with low FVC were mainly distributed in the edge area of the southern GKM and the northeast of CBM. The cropland area in this area was large, and the FVC value was low. During 2005–2015, except for the southern edge of the GKM, FVC decreased, and the other regions showed an upward trend.

It can be seen from Figure 8 and Table 8 that the area with good and excellent grades quality FVC accounted for a large proportion, accounting for about 81% of the total area. During 2005–2010, the area of excellence level had increased significantly, and its proportion exceeded half of the total area. The poor grade mainly turned to the general grade; the general grade was mainly transferred to the middle grade; the middle grade mainly turned to the good grade; the good grade mainly turned to the excellent grade, and the excellent grade mainly turned to a good grade. During 2010–2015, the area at the excellent and middle levels continued to increase, while the area at the poor and general levels decreased; the poor grade mainly turned to the general grade, the general grade primarily turned to the middle grade, the middle grade mainly turned to the good grade, the good grade mainly turned to the excellent grade, the excellent grade mainly turned to good grade. During 2005–2015, it generally showed a trend of good transformation, indicating that FVC in the NFB had been in a better situation during this period and had continued to improve and enhance.

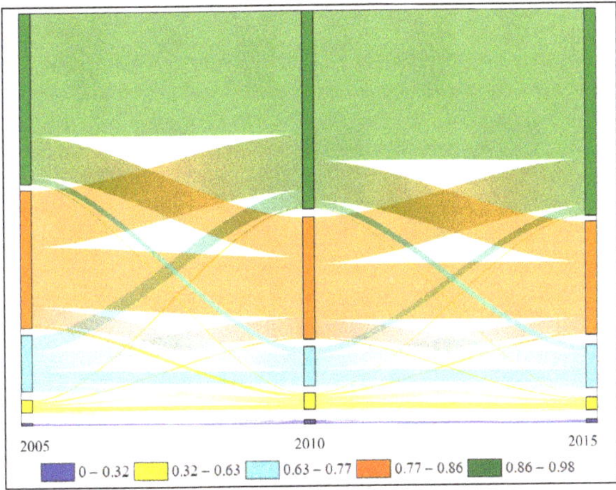

Figure 8. Sankey diagram of area transfer matrix of different FVC levels in the NFB from 2005–2015.

Table 8. Area transfer matrix of different FVC levels in the NFB from 2005–2015.

Year	FVC Level	Poor (km²)	Fair (km²)	Middle (km²)	Good (km²)	Excellent (km²)
	Poor		811.75	230.75	126.75	157.25
	Fair	2063.50		4124.00	3458.75	3274.50
2005–2010	Middle	761.50	11,163.50		31,493.50	26,591.25
	Good	438.25	5585.00	27,511.75		91,299.25
	Excellent	126.00	1643.50	10,801.25	64,285.25	
	Poor		1970.75	545.00	318.50	149.25
	Fair	1698.75		9140.25	3963.25	1783.00
2010–2015	Middle	273.75	5699.75		22,891.00	13,337.00
	Good	73.75	2252.25	27,844.25		76,032.50
	Excellent	47.00	1069.25	11,636.00	64,940.00	
	Poor		937.00	256.75	152.50	178.25
	Fair	1929.00		4743.50	3284.75	3353.25
2005–2015	Middle	534.00	8397.75		29,159.75	27,175.25
	Good	184.75	3720.50	31,297.00		93,027.75
	Excellent	49.25	945.00	8591.50	56,073.00	

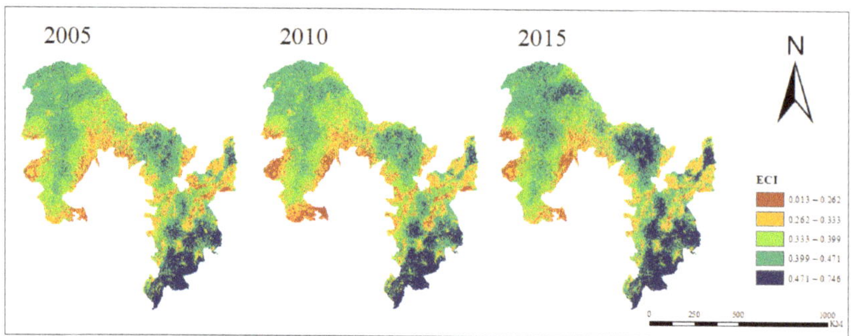

Figure 10. Spatial distribution of different levels of the comprehensive index in the NFB from 2005–2015.

Table 9. Statistics of different levels of comprehensive index data of the NFB from 2005–2015.

Comprehensive Index Level	2005		2010		2015	
	Area (km²)	Percent (%)	Area (km²)	Percent (%)	Area (km²)	Percent (%)
Poor (0.013–0.262)	33,456	5.21	45,326	7.06	27,521	4.29
Fair (0.262–0.333)	131,000	20.40	132,402	20.62	118,302	18.42
Middle (0.333–0.399)	177,052	27.57	174,635	27.19	147,437	22.96
Good (0.399–0.471)	229,065	35.67	223,391	34.78	230,204	35.85
Excellent (0.471–0.746)	71,641	11.16	66,460	10.35	118,750	18.49
Total amount	642,214	100	642,214	100	642,214	100

The spatial and temporal distributions of the ecosystem comprehensive evaluation index in the NFB in 2005, 2010, and 2015 are shown in Figure 10. The areas with a high spatial distribution value of the ecosystem in the NFB were mainly distributed in the southern part of CBM, the central part of LKM, and the northwestern part of GKM, while the northeastern part of CBM, the southwestern part of LKM and the marginal area of southern GKM were relatively low. During 2005–2010, the value of the southern edge of GKM showed a significant decline, and the eastern and central CBM and the central LKM also showed a slight decline. During 2010–2015, the change in the ecosystem comprehensive evaluation index showed an almost opposite trend; that is, the values of southern GKM and central LKM increased significantly, and other regions also increased to varying degrees. During the whole study period from 2005 to 2015, except for the decrease in the value around the marginal area in the southern GKM, the value in other regions showed an upward trend, which was closely related to a series of protection and improvement measures such as the "two ecological barriers and three shelters" national ecological stability pattern strategy, grain for green project, and comprehensive management of soil and water loss carried out in the NFB.

According to the change level of the ecosystem comprehensive evaluation index (0.013–0.746, Table 9), the change of the ecosystem comprehensive evaluation index was divided into five grades: excellent, good, middle, general, and poor. The area of the five grades of the ecosystem comprehensive evaluation index in 2005, 2010, and 2015 is shown in Table 4. Specifically, in 2005, the area with good and middle grades accounted for a large proportion (total 63.24%), and the proportion of poor grade was very small (5.21%). This indicated that the overall quality of ecosystems in 2005 was good. The grade area ratio in 2010 was consistent with 2005, and the overall area changed little. However, on the whole, some areas fell from the middle, good and excellent grades to the poor and general grades, indicating that during this period, the NFB faced some problems such as ecological destruction and environmental degradation. In 2015, the good grade area was the largest (230,204 km²), and the poor and middle-grade areas significantly decreased. Good and excellent areas had been significantly increased, mainly from poor, general, and middle regions, indicating the effectiveness of the "two ecological barriers and three shelters" national ecological stability pattern strategy implemented in the NFB.

3.5. Driving Force Analysis

In this study, a 10 km × 10 km grid was used to generate 7119 points in the study area in ArcGIS. After analysis, the influence value and driving effect of each index factor on the ecosystem evolution in the NFB were obtained (as shown in Figure 11).

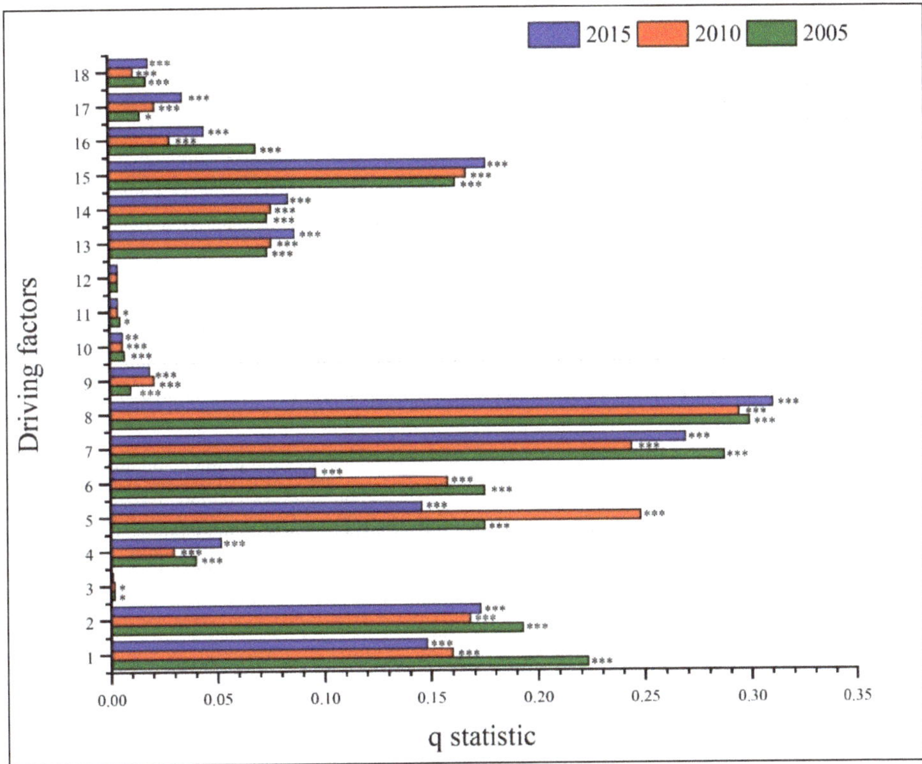

Figure 11. Geo-detector model results of the original index system in the NFB. Note: 1 is elevation, 2 is slope, 3 is slope direction, 4 is temperature, 5 is precipitation, 6 is NPP, 7 is cropland, 8 is forest, 9 is wetland, 10 is grassland, 11 is built-up land, 12 is clay, 13 is silty sand, 14 is sand, 15 is soil type, 16 is population density, 17 is GDP, and 18 is road. *** indicates that the correlation is significant at the 0.001 level; ** expresses relevance at the 0.01 level; * indicates that the correlation is significant at the 0.05 level.

It can be seen from Figure 11 that land use/cover, climate (annual average rainfall, etc.), and human disturbance (GDP, etc.) were related potential factors affecting the ecosystem evolution of the NFB. Most of the driving factors in the three periods of 2005, 2010, and 2015 were sufficient for the evolution of the NFB ecosystem ($p < 0.05$), indicating that the evaluation system constructed in this study was reasonable. In 2005, the three most influential factors in the evolution of the NFB ecosystem were forest, cropland, and elevation, and their influence values were 0.299, 0.287, and 0.223, respectively. The influence of driving factors from high to low were terrain factors, soil factors, meteorological factors, and human social factors. The influence of some driving factors in 2010 and 2015 differed from 2005, but the impact of the overall driving factors showed little change. Land use types had the greatest influence on ecosystem evolution in the NFB, and cropland and forest had a more significant influence than wetland, grassland, built-up land, and other types. The influence value of NPP remained at a middle-to-high level during the study period, and its influence value was more obvious than other factors. Regarding topographic factors, the influence value of elevation and slope was relatively significant, while the influence of slope direction was not big. In terms of soil factors, the influence of sand was larger than the other three factors. The influence of temperature on climate factors had been small, while the influence of precipitation is more obvious; even in 2010, it was only lower than that of the forest. In terms of social factors, the influences of human activity factors such as GDP and population density were small. Still, their visibility was extremely significant ($p < 0.001$, except for the GDP factor in 2005), indicating that human disturbance factors such as GDP and population density were also potential factors affecting the evolution of the NFB ecosystem.

4. Discussion

4.1. Advantages of the Integrated Ecosystem Assessment Index

In this study, the AHP was used to obtain the comprehensive evaluation index of the ecosystem, and a variety of methods were used to evaluate the ecosystem of the NFB comprehensively. The results showed that the landscape types of the NFB were mainly forest, cropland, and grassland, and the study showed a stable state as a whole. However, the areas of cropland and built-up land were still expanding, which needed more attention. This result was consistent with the research of others [3]. Compared with the principal component analysis method [56–60] that has been studied more before, the results showed that the ecosystem comprehensive evaluation index created by AHP in this study can also carry out a comprehensive ecosystem assessment. The results showed that the areas with a good ecosystem environment in the NFB were mainly distributed in the south of CBM, the middle of LKM, and the northwest of GKM and had a significant improvement and improvement trend with time. In contrast, the areas around the southeast and southwest of GKM were relatively poor, similar to some previous research results [2,12]. The reason may be that the good region was mainly located in the main part of the three mountains and was dominated by forests, with lush vegetation and less human disturbance, so the comprehensive index was high. The poor area may be due to the northeast plain, arable land to the NFB continued to expand, and more serious interference from human activities, resulting in the poor composite index. Throughout 2005–2015, the ecosystem in the NFB was in a relatively good state as a whole, and maintained a trend of improvement, which was closely related to a series of protection and improvement measures such as the "two ecological barriers and three shelters" national ecological stability pattern strategy [61], the project of returning cropland to forest and grassland [62–64], and the comprehensive control of soil and water loss. By comparing the results of landscape pattern, ES, and EQ in the NFB in this study, it could be seen that the evaluation results based on a single factor were uncertain and limited. And there were conflicts and contradictions between the evaluation results of different factors. Thus the ecosystem cannot be accurately evaluated. Based on these shortcomings and deficiencies, some researchers chose the method of constructing RSEI to analyze the ecosystem evolution [23,65,66] to obtain more accurate and comprehensive results. However, the construction of the remote sensing ecological index will ignore the impact of the land use/cover change process on the structure and function of the ecosystem. Therefore, in this study, combined with multiple indicators such as land use/cover, ES, and EQ, it was more accurate, systematic and comprehensive to evaluate the evolution of the NFB ecosystem by constructing the ecosystem comprehensive evaluation index.

4.2. Uncertainty of Driving Force Analysis

The driving force analysis showed that the natural factors had a great influence on the ecosystem of the NFB. Among them, land use type was an important factor affecting the ecosystem evolution of the NFB, which was consistent with the research results of Su and Zhu et al. [3,12]. However, the influence value of NPP in the study of Zhu et al. was more significant than in this study, which may be related to the difference in data sources and the number of grid points. The influence of elevation and slope was larger, probably because the higher and steep areas were covered by vegetation or less affected by human activities; the ranking of precipitation had been ahead, and its influence value rose to second in 2010, which was consistent with some studies [67–69]. Precipitation was one of the main factors affecting vegetation growth, so it also affected the evolution of ecosystems. The impact of human activity factors such as population density and GDP on the evolution of the ecosystem in the NFB was small, consistent with the results of Zhu et al. [12]. This may be because the overall social and economic activities in the NFB were less, and the population density and GDP were relatively sparse. In addition, this study only selected population density, GDP, and road as the three human activity factors for analysis. It did not comprehensively and specifically analyze what human social factors will affect the evolution of the NFB ecosystem, resulting in uncertainty in the final analysis of the results. Currently, there is no consistent method to evaluate the driving force that affects the evolution of the ecosystem. The methods for analyzing the driving force of the ecosystem mainly included correlation analysis, residual or redundancy analysis, and different methods may produce different results. Zhu et al. [12] found that the influence of social factors such as GDP and population density on the ecosystem of the NFB was not obvious; Sun et al. [3] indicated that human disturbance factors such as urban expansion were the main reason for the decrease of carbon sequestration in the NFB. As for the drought index, Wang et al. [70] considered that it was the main factor driving the growth of forest vegetation and the change of carbon sink function in Northeast China. The results of driving force analysis in this study were similar to those in Zhu et al. [12], although there were still differences in some factors and effects. The selection of driving force indicators and methods may have an important impact on the final results and the subsequent management of the ecosystem. Therefore, the accurate analysis of driving force still needs further continuous research.

4.3. Limitations of Current Research

This study still had some deficiencies and needed to be further studied. Although the landscape pattern of the NFB ecosystem analyzed in this study could roughly reflect the changes during the entire study period. It cannot accurately reflect the evolution of the NFB ecosystem pattern and the relationship between horizontal and vertical pattern structure and quality function. This still needs more detailed research. And the remote sensing data used in this study, due to sources and other limitations, only selected 2005, 2010, and 2015 data, and remote sensing data resolution was relatively low, which may affect the accuracy of the results. In addition, this study only conducted quantitative spatio-temporal assessments of ES and EQ in the NFB. In the future, the collaboration and trade-off between ES and quality will be explored. In addition, in the driving force analysis, due to the limitation of data sources, only three human activity factors, namely GDP, population density, and road, were selected for analysis. However, these three factors were relatively macroscopic, which cannot reflect the specific human disturbance activities that affect the evolution of the ecosystem in the NFB, which was not conducive to the application in a specific practice. In future research, more comprehensive driving factors can be added to analyze, and the evaluation system framework can be further optimized to more comprehensively and accurately evaluate the changes in the ecosystem environment and the impact of driving factors.

5. Conclusions

In this study, the multi-source remote sensing data and multiple indicators were used to construct the comprehensive evaluation ecosystem index of the NFB. The changes and driving forces of the NFB ecosystem from 2005 to 2015 were evaluated based on the PNTIL model, transfer matrix, geographical detector, and other methods. Using an ecosystem comprehensive evaluation index can avoid the shortcomings of uncertainty and limitation based on single factor evaluation results to evaluate the evolution of the NFB ecosystem more systematically, comprehensively, and accurately. The main conclusions are as follows:

The landscape pattern types of the NFB were mainly forest, cropland, and grassland, and the overall landscape pattern of the NFB was stable during 2005–2015. The comprehensive assessment of the ecosystem index showed that the areas with a good ecosystem environment in the NFB were mainly distributed in the southern CBM, the central LKM, and the northwestern GKM, while the northeastern CBM, the southwestern LKM, and the marginal areas of southern GKM were relatively poor. During 2005–2015, the overall ecosystem of the NFB was in a relatively good state and maintained a trend of improvement in a good direction. However, during 2005–2010, there were still some anthropogenic disturbances that led to a decline in the comprehensive index of some regions. The driving force analysis showed that natural factors had a great influence on the ecosystem of the NFB. Land use/cover was an important driving factor affecting the ecosystem evolution of the NFB. Climate (annual average rainfall, etc.) and human disturbance (GDP, etc.) were potential factors affecting the ecosystem evolution of the NFB. The results showed that in the past 10 years, the EQ of the NFB was generally in a good state and continued to improve, but the expansion of cropland and built-up land in some areas continued to increase, which needed to be paid more attention.

In the future, relevant research will continue to analyze the synergy and trade-off relationship of various ES and their spatial differentiation and further regionalize ES to explore the formation mechanism of the synergy and trade-off relationship of different ES. The correlation between synergy and trade-off relationship with the main driving forces obtained in the driving force analysis of this study can also be further studied. At the same time, the optimization of the evaluation system framework and the reasonable and comprehensive selection of impact factors still need to be further improved, including the prospect of the change characteristics of the ecosystem comprehensive index with the above factors. It is hoped that a series of studies will provide a more scientific and comprehensive reference for the ecosystem stability and improvement of the NFB.

Author Contributions: Conceptualization, Z.L. and K.S.; methodology, Z.L. and K.S.; software, Z.L.; validation, X.Z., Z.Y. and Z.C.; formal analysis, Z.L. and K.S.; resources, K.S.; data curation, L.W., C.W., Z.L. and Y.Z.; writing-original draft preparation, Z.L. and K.S.; writing-review and editing, K.S. and X.J.; visualization, L.W., C.W., Z.L. and Y.Z.; supervision; project administration, K.S.; funding acquisition, K.S. All authors have read and agreed to the published version of the manuscript.

Funding: This research was funded by [Youth Science Foundation of Natural Science Foundation of Guangxi] grant number [2022GXNSFBA035570] and [Talent Introduction Program of Guangxi University] grant number [A3360051018].

Institutional Review Board Statement: Not applicable.

Informed Consent Statement: Not applicable.

Data Availability Statement: The DEM data, meteorological data, soil data, land cover data are available in this research, upon any reasonable request, by emailing the authors.

Conflicts of Interest: The authors declare no conflict of interest.

References

1. Sun, B. *Decade Change of Ecosystem Pattern, Quality, Service Function and Stress in the Northeastern Forest Regions (2000–2010)*; University of Chinese Academy of Sciences: Beijing, China, 2015.
2. Sun, B.; Zhao, H.; Lu, F.; Wang, X. Spatial and temporal patterns of carbon sequestration in the Northeastern Forest Regions and its impact factors analysis. *Acta Ecol. Sin.* **2018**, *38*, 4975–4983.
3. Su, K.; Wang, Y.; Sun, X.; Yue, D. Landscape pattern change and prediction of Northeast Forest Belt based on GIS and RS. *Trans. Chin. Soc. Agric. Mach.* **2019**, *50*, 195–204.
4. Kong, L.; Zhang, L.; Zheng, H.; Xu, W.; Xiao, Y.; Ouyang, Z. Driving forces behind ecosystem spatial changes in the Yangtze River Basin. *Acta Ecol. Sin.* **2018**, *38*, 741–749.

5. Borzì, L.; Anfuso, G.; Manno, G.; Distefano, S.; Urso, S.; Chiarella, D.; Di Stefano, A. Shoreline evolution and environmental changes at the NW Area of the Gulf of Gela (Sicily, Italy). *Land* **2021**, *10*, 1034. [CrossRef]

6. Aho, K.; Weaver, T. Alpine nodal ecology and ecosystem evolution in the North-Central Rockies (Mount Washburn; Yellowstone National Park, Wyoming). *Arct. Antarct. Alp. Res.* **2010**, *42*, 139–151. [CrossRef]

7. Huang, J.; Sun, L.; Wang, X.; Wang, Y.; Huang, T. Ecosystem evolution of seal colony and the influencing factors in the 20th century on Fildes Peninsula, West Antarctica. *J. Environ. Sci.* **2011**, *23*, 1431–1436. [CrossRef]

8. Li, C.; Zhao, J.; Zhuang, Z.; Gu, S. Spatiotemporal dynamics and influencing factors of ecosystem service trade-offs in the Yangtze River Delta urban agglomeration. *Acta Ecol. Sin.* **2022**, *14*, 1–13.

9. Apostolopoulos, D.N.; Avramidis, P.; Nikolakopoulos, K.G. Estimating quantitative morphometric parameters and spatiotemporal evolution of the Prokopos Lagoon using remote sensing techniques. *J. Mar. Sci. Eng.* **2022**, *10*, 931. [CrossRef]

10. Weng, H.; Gao, Y.; Su, X.; Yang, X.; Cheng, F.; Ma, R.; Liu, Y.; Zhang, W.; Zheng, L. Spatial-temporal changes and driving force analysis of green space in coastal cities of Southeast China over the past 20 years. *Land* **2021**, *10*, 537. [CrossRef]

11. Qi, L.; Zhang, Y.; Xu, D.; Zhu, Q.; Zhou, W.; Zhou, L.; Wang, Q.; Yu, D. Trade-offs and synergies of ecosystem services in forest barrier belt of Northeast China. *Chin. J. Ecol.* **2021**, *40*, 3401–3411. [CrossRef]

12. Zhu, Q.; Wang, Y.; Zhou, W.; Zhou, L.; Yu, D.; Qi, L. Spatiotemporal changes and driving factors of ecological vulnerability in Northeast China forest belt. *Chin. J. Ecol.* **2021**, *40*, 3474–3482. [CrossRef]

13. Liang, B.; Shi, P.; Wang, W.; Tang, X.; Zhou, W.; Jing, Y. Integrated assessment of ecosystem quality of arid inland river basin based on RS and GIS: A case study on Shiyang River Basin, Northwest China. *Chin. J. Appl. Ecol.* **2017**, *28*, 199–209. [CrossRef]

14. Zewdie, W.; Csaplovics, E. Identifying categorical land use transition and land degradation in northwestern drylands of Ethiopia. *Remote Sens.* **2016**, *8*, 408. [CrossRef]

15. Foroumandi, E.; Nourani, V.; Dąbrowska, D.; Kantoush, S.A. Linking spatial-temporal changes of vegetation cover with hydroclimatological variables in terrestrial environments with a focus on the Lake Urmia Basin. *Land* **2022**, *11*, 115. [CrossRef]

16. Gao, Z.; Bai, Y.; Zhou, L.; Qiao, F.; Song, L.; Chen, X. Characteristics and driving forces of wetland landscape pattern evolution of the city belt along the Yellow River in Ningxia, China. *Chin. J. Appl. Ecol.* **2020**, *31*, 3499–3508. [CrossRef]

17. Karimi, F.M.; Majid, K.; Mehdi, H.; Jokar, A.J.; Kazem, A.S. A novel method to quantify urban surface ecological poorness zone: A case study of several European cities. *Sci. Total Environ.* **2021**, *757*, 143755.

18. Mohammad, K.F.; Solmaz, F.; Majid, K.; Asim, B.; Mehdi, H.; Kazem, A.S. Land surface ecological status composition index (LSESCI): A novel remote sensing-based technique for modeling land surface ecological status. *Ecol. Indic.* **2021**, *123*, 107375.

19. Sajjad, K.S.; Solmaz, A.; Mostafa, G. Spatiotemporal ecological quality assessment of metropolitan cities: A case study of central Iran. *Environ. Monit. Assess.* **2021**, *193*, 305.

20. Maity, S.; Das, S.; Pattanayak, J.M.; Bera, B.; Shit, K.P. Assessment of ecological environment quality in Kolkata urban agglomeration, India. *Urban Ecosyst.* **2022**, *25*, 3. [CrossRef]

21. Xu, H. A remote sensing index for assessment of regional ecological changes. *China Environ. Sci.* **2013**, *33*, 889–897.

22. Xu, H.; Wang, Y.; Guan, H.; Shi, T.; Hu, X. Detecting ecological changes with a remote sensing based ecological Index (RSEI) produced time series and change vector analysis. *Remote Sens.* **2019**, *11*, 2345. [CrossRef]

23. Yuan, B.; Fu, L.; Zou, Y.; Zhang, S.; Chen, X.; Li, F.; Deng, Z.; Xie, Y. Spatiotemporal change detection of ecological quality and the associated affecting factors in Dongting Lake Basin, based on RSEI. *J. Clean. Prod.* **2021**, *302*, 126995. [CrossRef]

24. Wang, J. *Study on the Spatiotemporal Changes of Ecological Environment and Land Use in the Central Area of Urumqi*; Xinjiang University: Urumqi, China, 2021.

25. Hu, X.; Xu, H. A new remote sensing index for assessing the spatial heterogeneity in urban ecological quality: A case from Fuzhou City, China. *Ecol. Indic.* **2018**, *89*, 11–21. [CrossRef]

26. Boori, M.S.; Choudhary, K.; Paringer, R.; Kupriyanov, A. Spatiotemporal ecological vulnerability analysis with statistical correlation based on satellite remote sensing in Samara, Russia. *J. Environ. Manag.* **2021**, *285*, 112138. [CrossRef]

27. Thiesen, T.; Bhat, M.G.; Liu, H.; Rovira, R. An ecosystem service approach to assessing agro-ecosystems in urban landscapes. *Land* **2022**, *11*, 469. [CrossRef]

28. Chen, Q.; Chen, Y.; Wang, M.; Jiang, W.; Hou, Y.; Li, Y. Ecosystem quality comprehensive evaluation and change analysis of Dongting Lake in 2001—2010 based on remote sensing. *Acta Ecol. Sin.* **2015**, *35*, 4347–4356.

29. Chen, H. Establishment and application of wetlands ecosystem services and sustainable ecological evaluation indicators. *Water* **2017**, *9*, 197. [CrossRef]

30. Yilmaz, F.C.; Zengin, M.; Cure, C.T. Determination of ecologically sensitive areas in Denizli province using geographic information systems (GIS) and analytical hierarchy process (AHP). *Environ. Monit. Assess.* **2020**, *192*, 589. [CrossRef]

31. Li, W.; Sun, B.; Wang, R.; Cui, L.; Ma, Q.; Zhang, M.; Lei, Y.; Zhu, Y. Various beneficiaries-based analysis on service weight of ecosystem of Momoge Wetland. *Water Resour. Hydropower Eng.* **2017**, *48*, 10–15. [CrossRef]

32. Morandi, D.T.; França, L.C.d.J.; Menezes, E.S.; Machado, E.L.M.; Silva, M.D.d.; Mucida, D.P. Delimitation of ecological corridors between conservation units in the Brazilian Cerrado using a GIS and AHP approach. *Ecol. Indic.* **2020**, *115*, 106440. [CrossRef]

33. Zhang, R.; Zhang, X.; Yang, J.; Yuan, H. Wetland ecosystem stability evaluation by using analytical hierarchy process (AHP) approach in Yinchuan Plain, China. *Math. Comput. Model.* **2013**, *57*, 366–374. [CrossRef]

34. Shafaghat, A.; Ying, O.J.; Keyvanfar, A.; Jamshidnezhad, A.; Ferwati, M.S.; Ahmad, H.; Mohamad, S.; Khorami, M. A treatment wetland park assessment model for evaluating urban ecosystem stability using analytical hierarchy process (AHP). *J. Environ. Treat. Tech.* **2019**, *7*, 81–91.

35. Guo, Z.; Wei, W.; Pang, S.; Li, Z.; Zhou, J.; Jie, B. Spatio-temporal evolution and motivation analysis of ecological vulnerability in Arid Inland River Basin based on SPCA and remote sensing index: A case study on the Shiyang River Basin. *Acta Ecol. Sin.* **2019**, *39*, 2558–2572.

36. Chi, W.; Bai, W.; Liu, Z.; Dang, X.; Kuang, W. Wind Erosion in Inner Mongolia Plateau Using the Revised Wind Erosion Equation. *Ecol. Environ. Sci.* **2018**, *27*, 1024–1033. [CrossRef]

37. Kong, L.; Zheng, H.; Rao, E.; Xiao, Y.; Ouyang, Z.; Li, C. Evaluating indirect and direct effects of eco-restoration policy on soil conservation service in Yangtze River Basin. *Sci. Total Environ.* **2018**, *631–632*, 887–894. [CrossRef]

38. Fang, Q.; Wang, G.; Zhang, S.; Peng, Y.; Xue, B.; Cao, Y.; Shrestha, S. A novel ecohydrological model by capturing variations in climate change and vegetation coverage in a semi-arid region of China. *Environ. Res.* **2022**, *211*, 113085. [CrossRef] [PubMed]

39. Deng, Y.; Yao, S.; Hou, M.; Zhang, T.; Lu, Y.; Gong, Z.; Wang, Y. Assessing the effects of the Green for Grain Program on ecosystem carbon storage service by linking the InVEST and FLUS models: A case study of Zichang county in hilly and gully region of Loess Plateau. *J. Nat. Resour.* **2020**, *35*, 826–844.

40. Nie, Q.; Xu, J.; Ji, M.; Cao, L.; Yang, Y.; Hong, Y. The vegetation coverage dynamic coupling with climatic factors in Northeast China Transect. *Environ. Manag.* **2012**, *50*, 405–417. [CrossRef]

41. Potter, C.S.; Randerson, J.T.; Field, C.B.; Matson, P.A.; Vitousek, P.M.; Mooney, H.A.; Klooster, S.A. Terrestrial ecosystem production: A process model based on global satellite and surface data. *Glob. Biogeochem. Cycles* **1993**, *7*, 811–841. [CrossRef]

42. Berdjour, A.; Dugje, I.Y.; Rahman, N.A.; Odoom, D.A.; Kamara, A.Y.; Ajala, S. Direct estimation of maize leaf area index as influenced by organic and inorganic fertilizer rates in Guinea Savanna. *J. Agric. Sci.* **2020**, *12*, 66. [CrossRef]

43. Zhuo, J.; Zhu, Y.; He, H.; Wang, J.; Dong, J.; Quan, W. Impacts of ecological restoration projects on the ecosystem in the Loess Plateau. *Acta Ecol. Sin.* **2020**, *40*, 8627–8637.

44. Rotich, B.; Kindu, M.; Kipkulei, H.; Kibet, S.; Ojwang, D. Impact of land use/land cover changes on ecosystem service values in the cherangany hills water tower, Kenya. *Environ. Chall.* **2022**, *8*, 100576. [CrossRef]

45. Nuppenau, E.-A. Soil fertility management by transition matrices and crop rotation: On spatial and dynamic aspects in programming of ecosystem services. *Sustainability* **2018**, *10*, 2213. [CrossRef]

46. Müller, F.; Bicking, S.; Ahrendt, K.; Kinh Bac, D.; Blindow, I.; Fürst, C.; Haase, P.; Kruse, M.; Kruse, T.; Ma, L.; et al. Assessing ecosystem service potentials to evaluate terrestrial, coastal and marine ecosystem types in Northern Germany–An expert-based matrix approach. *Ecol. Indic.* **2020**, *112*, 106116. [CrossRef]

47. Su, K.; Sun, X.; Wang, Y.; Chen, L.; Yue, D. Spatial and temporal variations of ecosystem patterns in Northern Sand Control Barrier Belt based on GIS and RS. *Trans. Chin. Soc. Agric. Mach.* **2020**, *51*, 226–236.

48. Su, K.; Liu, H.; Wang, H. Spatial–temporal changes and driving force analysis of ecosystems in the Loess Plateau Ecological Screen. *Forests* **2022**, *13*, 54. [CrossRef]

49. Ouyang, Z.; Zheng, H.; Xie, G.; Yang, W.; Liu, G.; Shi, Y.; Yang, D. Accounting theories and technologies for ecological assets, ecological compensation and scientific and technological contribution to ecological civilization. *Acta Ecol. Sin.* **2016**, *36*, 7136–7139.

50. Wang, J.; Xu, C. Geodetector: Principle and prospective. *Acta Geogr. Sin.* **2017**, *72*, 116–134.

51. Zhang, Z.; Song, Y.; Wu, P. Robust geographical detector. *Int. J. Appl. Earth Obs. Geoinf.* **2022**, *109*, 102782. [CrossRef]

52. Bai, L.; Jiang, L.; Yang, D.; Liu, Y. Quantifying the spatial heterogeneity influences of natural and socioeconomic factors and their interactions on air pollution using the geographical detector method: A case study of the Yangtze River Economic Belt, China. *J. Clean. Prod.* **2019**, *232*, 692–704. [CrossRef]

53. Ji, J.; Wang, S.; Zhou, Y.; Liu, W.; Wang, L. Studying the eco-environmental quality variations of Jing-Jin-Ji urban agglomeration and its driving factors in different ecosystem service regions from 2001 to 2015. *IEEE Access* **2020**, *8*, 154940–154952. [CrossRef]

54. Li, H.; Wang, J.; Zhang, J.; Qin, F.; Hu, J.; Zhou, Z. Analysis of characteristics and driving factors of wetland landscape pattern change in Henan Province from 1980 to 2015. *Land* **2021**, *10*, 564. [CrossRef]

55. Polykretis, C.; Grillakis, M.G.; Argyriou, A.V.; Papadopoulos, N.; Alexakis, D.D. Integrating multivariate (GeoDetector) and bivariate (IV) statistics for hybrid landslide susceptibility modeling: A Case of the vicinity of Pinios Artificial Lake, Ilia, Greece. *Land* **2021**, *10*, 973. [CrossRef]

56. Mavhura, E.; Manyangadze, T. A comprehensive spatial analysis of social vulnerability to natural hazards in Zimbabwe: Driving factors and policy implications. *Int. J. Disaster Risk Reduct.* **2021**, *56*, 102139. [CrossRef]

57. Uddin, M.N.; Saiful Islam, A.K.M.; Bala, S.K.; Islam, G.M.T.; Adhikary, S.; Saha, D.; Haque, S.; Fahad, M.G.R.; Akter, R. Mapping of climate vulnerability of the coastal region of Bangladesh using principal component analysis. *Appl. Geogr.* **2019**, *102*, 47–57. [CrossRef]

58. Zhang, H.; Yu, D.; Yang, H. Quantitative evaluation of ecological environment improvement in Saihanba based on principal component analysis. *Int. Core J. Eng.* **2022**, *8*, 608–615.

59. Liu, Y.; Dang, C.; Yue, H.; Lv, C.; Qian, J.; Zhu, R. Comparison between modified remote sensing ecological index and RSEI. *Natl. Remote Sens. Bull.* **2022**, *26*, 683–697.

60. Li, Y.; Zhang, G.; Lin, T.; Ye, H.; Liu, Y.; Chen, T.; Liu, W. The spatiotemporal changes of remote sensing ecological index in towns and the influencing factors: A case study of Jizhou District, Tianjin. *Acta Ecol. Sin.* **2022**, *42*, 474–486.

61. Wang, X.; Lesi, M.C.; Zhang, M. Ecosystem pattern change and its influencing factors of "two barriers and three belts". *Chin. J. Ecol.* **2019**, *38*, 2138–2148. [CrossRef]

62. Zhang, H.; Fan, J.; Shao, Q.; Zhang, Y. Ecosystem dynamics in the 'Returning Rangeland to Grassland' programs, China. *Acta Prataculturae Sin.* **2016**, *25*, 1–15.

63. Wang, J.; Peng, J.; Zhao, M.; Liu, Y.; Chen, Y. Significant trade-off for the impact of Grain-for-Green Programme on ecosystem services in North-western Yunnan, China. *Sci. Total Environ.* **2017**, *574*, 57–64. [CrossRef]

64. Jia, X.; Fu, B.; Feng, X.; Hou, G.; Liu, Y.; Wang, X. The tradeoff and synergy between ecosystem services in the Grain-for-Green areas in Northern Shaanxi, China. *Ecol. Indic.* **2014**, *43*, 103–113. [CrossRef]

65. Cheng, Z.; He, Q. Remote sensing evaluation of the ecological environment of Su-Xi-Chang city group based on remote sensing ecological Index (RSEI). *Remote Sens. Technol. Appl.* **2019**, *34*, 531–539.

66. Singh, B.M.; Komal, C.; Rustam, P.; Alexander, K. Eco-environmental quality assessment based on pressure-state-response framework by remote sensing and GIS. *Remote Sens. Appl. Soc. Environ.* **2021**, *23*, 100530.

67. Siles, A.J.; Tomas, C.; Stefano, M.; Rosa, M. Effect of altitude and season on microbial activity, abundance and community structure in Alpine forest soils. *FEMS Microbiol. Ecol.* **2016**, *92*, fiw008. [CrossRef]

68. Silvia, A.C.; Mack, M.C. Influence of precipitation on soil and foliar nutrients across nine Costa Rican forests: Influence of precipitation eegimes on nutrients. *Biotropica* **2011**, *43*, 433–441.

69. Li, J.; Zhang, C.; Zhu, S. Relative contributions of climate and land-use change to ecosystem services in arid inland basins. *J. Clean. Prod.* **2021**, *298*, 126844. [CrossRef]

70. Wang, X.; Liu, Z.; Jiao, K. Spatiotemporal dynamics of normalized difference vegetation index (NDVI) and its drivers in forested region of Northeast China during 2000–2017. *Chin. J. Ecol.* **2020**, *39*, 2878–2886. [CrossRef]

Review

State of the Art Review on Land-Use Policy: Changes in Forests, Agricultural Lands and Renewable Energy of Japan

Ryo Kohsaka [1,*] **and Satomi Kohyama** [2]

[1] Graduate School of Environmental Studies, Nagoya University, Nagoya 464-8601, Japan
[2] Graduate School of Sustainability Studies, University of Toyama, Toyama 930-8555, Japan; kohyama@eco.u-toyama.ac.jp
* Correspondence: kohsaka@hotmail.com

Abstract: Policies in Japan are shifting focus on sustainable land-use management-related policies through consensus building, given the complex options for the community and the landowners. For instance, conversion of agricultural lands to renewable energy sites, which is an example of "land-use conversion for a newly found objective", is rapidly progressing, and actions on "managing of croplands in a minimal (low labor demand) way" has been embodied in certain policies. Currently, there are political and scientific efforts to balance environmental conservation with production activities in agriculture and forestry sectors based on science and evidence. With policies catching up, it is possible to confirm what has been moved from the planning to the implementation stage of the proposed consensus-building system by summarizing and discussing the current progress of the project. More specifically, we highlighted the trends in reusing agricultural lands under the current national-level policies and management options for croplands, such as the "less maintenance way." We also discussed and presented the preliminary results, insights, and prospects from the ongoing project, which then led to the discussion of future considerations in sustainable land-use management in Japan.

Keywords: evidence-based policymaking; agricultural land; forests; consensus building; Japan

Citation: Kohsaka, R.; Kohyama, S. State of the Art Review on Land-Use Policy: Changes in Forests, Agricultural Lands and Renewable Energy of Japan. *Land* **2022**, *11*, 624. https://doi.org/10.3390/land11050624

Academic Editor: Benedetto Rugani

Received: 25 March 2022
Accepted: 22 April 2022
Published: 23 April 2022

Publisher's Note: MDPI stays neutral with regard to jurisdictional claims in published maps and institutional affiliations.

1. Introduction

There is increased attention to evidence-based policy makings (EBPMs) in the field of agriculture and forestry, for instance, towards agricultural sustainability and food security [1] and sustainable management of forests [2]. Supporting decision-making with evidence-based policy analysis tends to do better in the long run [3] and provides confidence to policymakers in developing policies [4]. In Japan, government officials, academic researchers, and the general public recognize the necessity of EBPM, yet, the evaluation of the actual implementation of EBPM is not as frequent as initially expected [5]. Sugitani [6] noted that if EBPM relied on fewer experts in the formulation of policies, it would not produce a positive contribution; however, if EBPM adopted a participatory policy-creation process, it would provide solid evidence, because the responses of stakeholders can be compiled, reflected upon, and consequently, directions of policy-makings would be re-directed with accountability, which can be translated into practice and ordinary words [6].

With the decreasing population and labor forces in Japan, strategic decisions, for instance, on environmental conservation, animal damage, and disaster prevention are required, and downsizing the labor cost, demand, and management areas should be taken into consideration for EBPMs [7]. Shrinkage or strategic downsizing has been a buzzword for policy making and the sciences alike. For instance, there are ongoing discussions in the field of urban planning and biodiversity conservation [8–10], and increasingly so in the field of forests and agricultural lands [11–13]. In existing literatures, rural areas with unoccupied houses, abandoned farmlands, unmanaged forest lands, or unknown owners, are increasingly becoming challenges in contemporary Japan [14]. The decreasing number of residents can have multiple negative effects such as: (i) there are fewer workers in the field, resulting in the deterioration of the productivity of agriculture and forests lands, (ii) damage from wild animals accompanied by land abandonment, (iii) dangers from unoccupied houses increases the risk of disasters and social safety, and (iv) increased difficulties to maintain the cost of infrastructures including water supply, bridges, roads, and general social services. There is an increased awareness that rural land areas require policies and scientific attention. It is forecasted that the declining and aging population in Japan is expected to accelerate, albeit with regional differences, which in turn, will result in a decrease in the usage of land areas (Figure 1). For instance, in mountainous agricultural areas, the population will be halved in the next 30 years, and the majority will be 65 years old or older (Figure 1) [15].

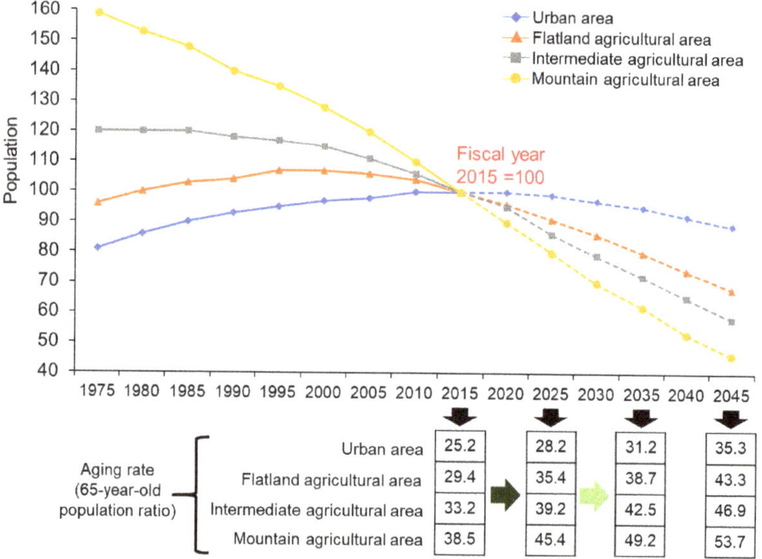

Figure 1. Population trends and future forecasts by agricultural area type (modified from [15], MAFF 2019, [Ministry of Agriculture, Forestry and Fisheries]).

Currently, there are political and scientific efforts in Japan to balance environmental conservation with production activities in the agriculture and forestry sectors by applying a evidence-based approach (referred here as EBPM), which encompasses social, economic, and environmental aspects (e.g., land-use, land abandonment, community group discussions and designs, wildlife management, landowner preferences, and trans-generational knowledge transmission) at the local and regional levels [7]. We conducted preliminary reviews of the newly evolving topic with a focus on academic literatures, relevant policy documents, newspaper articles, and other non-peer-reviewed documents. We reviewed different documents since the scope ranged from both practical and scientific areas, and peer-reviewed literatures were still limited, given that changes in the policies are fairly recent. Local contexts were highlighted because landowners have different perceptions on land-use characteristics, which can either be positive (environment, tourism, identity) or negative (damage by wildlife, deterioration of ecosystems, and landscapes) attributes, or both. In other words, it is increasingly recognized that "landowners have (public) responsibilities," in addition to rights, that have been conventionally recognized in legal terms.

In this work, we focused on the broader consensus building beyond individual landowners since we observed during the initial phase of the project that decisions at local levels were frequently formed by a series of groups, instead of aggregations of individual landowners. First, we reviewed the rapidly changing state-of-the-art policy changes at the national level. Second, we provided project-level results to share the experiences and unique insights obtained from the implementation. The project-level insights were obtained from a project entitled "Development and Implementation of Consensus Building Method for Policies on Balanced Conservation, Agriculture and Forestry," which is based in the Iidaka area, Matsusaka City, Mie, Japan [7]. This 3.5-year project (October 2020 to March 2024) is supported by the Research Institute of Science and Technology for Society, Japan Science and Technology Agency (JST RISTEX) [7].

The review presented here is conducted in two layers (both national and local—project site level) since the changes at the national level can affect decisions at the local level (with possible feedback mechanisms). Moreover, increased policy attention is given to the consensus of the community at national levels, particularly for agricultural lands (i.e., the *Hito Nouchi* Plan or the Agricultural Land Management Plan), and the scientific communities are responding by designing science-based tools, attuned to local policy settings and processes, that support the decisions of the community.

Currently, policies at the national level are still progressing, with a gradual shift on sustainable land-use management-related policies through consensus building, given the complexity of communities and landowners. As shown in Figure 2, there are several options that can be considered by landowners in sustainably managing their lands. For example, "the conversion of agricultural and forest lands to renewable energy sites," is the usual way of repurposing abandoned lands [16]. Additionally, management activities requiring less maintenance and low labor costs are being embedded in policies, such as the "conversion of croplands to grasslands" [17].

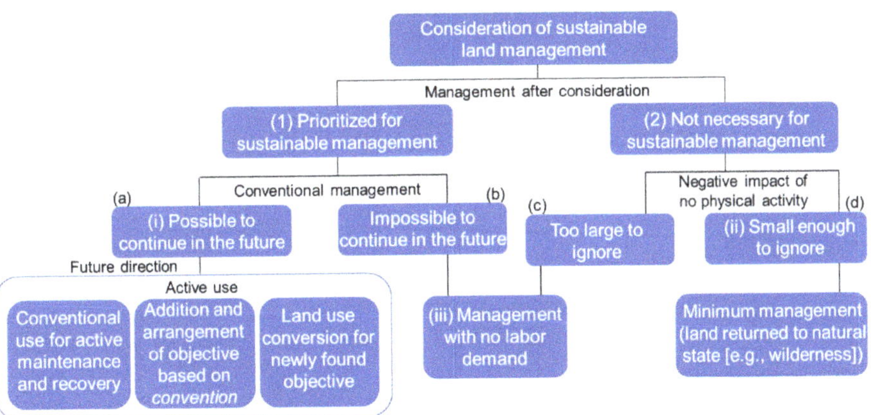

Figure 2. Development and planned implementation of the consensus-building mapping system to promote policies that balance agriculture and forestry production with environmental conservation. To achieve sustainable land management goals, croplands that are: (a) possible to manage in the future, (b) cannot be manage, (c) too large to ignore, and (d) small enough to ignore are identified (modified from [7]).

With policies still progressing, we aim to capture the trends, particularly in the planning and the implementation phase of the proposed consensus-building system. More specifically, we highlighted the trends in reusing agricultural lands under the current national-level policies (Section 2) and management options for croplands such as "low labor costs" (Section 3). In addition, we discussed and presented the preliminary results, insights, and prospects from the project site (Section 4). Finally, we elaborated and discussed the general trends observed and the future considerations that can be tackled as the project progresses (Section 5).

2. Diversifying Roles of Agricultural Lands under National-Level Policy Framework

2.1. Promotion of Active Utilization of Abandoned Agricultural and Degraded Cropland

Article 2 (1) of the Cropland Act (Act No. 249 of 1952) defines "cropland (*Nochi*)" as "land used for cultivation" [18]. Based on this law, there are measures to manage abandoned croplands (Figure 3). In Article 32 (1), there are two types of "abandoned cropland (*Yukyu-Nochi*)" based on the progress of abandoning cultivation. First, Article 32 (1) (i) states that if the abandonment progresses further, the cropland will be described as "not used for cultivation and is not expected to be used in the future (*Ichigo-Yukyu-Nochi*)." Second, Article 32 (1) (ii) describes croplands as "croplands where agricultural use is found to be significantly inferior compared to other croplands in the surrounding areas (*Nigo-Yukyu-Nochi*)". The first type can also be described as a "degraded agricultural land (*Kohai-Nochi*)." As the abandonment of cultivation practices progresses and the degraded agricultural lands continue not to be used, they will be recognized as "difficult to recycle" by the Agricultural Commission (*Nogyo-Iinkai*) under the Cropland Act [18].

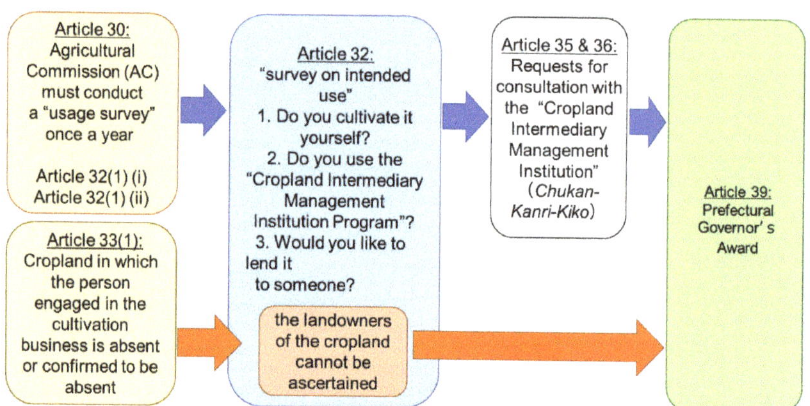

Figure 3. Measures in managing abandoned croplands [18].

The Ministry of Agriculture, Forestry and Fisheries (MAFF) categorizes agricultural lands that are not cultivated into four types: (1) abandoned cultivated land (*Kosa-ku-Hokichi*), (2) abandoned land (*Fu-Sakuzukechi*), (3) degraded agricultural lands (*Kohai-Nochi*) that can be reused, and (4) degraded agricultural lands that cannot be reused. In the first category, abandoned cultivated land is defined as "formerly cultivated land without production in the past one year and the owner does not intend to produce crops [in the next] few years" [19]. The first category is problematized in policy debates that aggregate areas that are equal to the size of Saitama or Shiga prefectures [20]. Although, it is noted that the definition of the first category is related to "subjective elements" of the intent of landowners, and not simply the "objective elements" of agricultural lands, which are defined in categories 3 and 4. The second category describes abandoned land as the "land that has not been planted in the past year but might be (re-)cultivated with the landowner's willingness [19]. Moreover, the third category is "cropland that is not actually used for cultivation and is not expected to be used in the future (Article 32 (1) (i) [18])" and the fourth category notes the cropland that is "currently not used for cultivation and cannot be cultivated again due to . . . abandonment and [is] objectively impossible to cultivate crops with [using] conventional measures" [21] (Table 1).

Table 1. Classification of degraded cropland (determined annually by field surveys of the MAFF [21]).

Type of Degraded Cropland	Definition
Category 3—degraded cropland that can be reused (approximately 90,000 ha of farmland; 55,000 ha of this total is accounted for agricultural areas).	This category of cropland is not actually used for cultivation and is not expected to be used for cultivation in the future (Article 32 (1) (i) of Cropland Act [18]).
Category 4—degraded cropland that is difficult to reuse or cultivate again (approximately 192,000 ha of farmland; 81,000 ha of this total is accounted for agricultural areas).	This category can also refer to land that will not be continuously utilized even if it is restored as cropland due to surroundings or physical conditions, such as forested areas, where restoring or revitalizing the cropland is extremely difficult due to long abandonment.

Currently, 6% of the total agricultural land area (4,654,000 ha) in Japan falls under categories 3 and 4, which were identified as "degraded agricultural lands" by the Agricultural Commission and staff of municipalities (Figure 4) [22]. According to the MAFF [22], a certain portion of these degraded croplands is expected to be converted to renewable energy lands; however, conversions of agricultural lands are challenging because of the presence of the Cropland Act, which protects and regulates the conversion of "cropland" into other land-use types [18]. Although, recent developments concerning conversion to other land-use types are gaining traction in Japan because of the increasing number of abandoned arable lands resulting from a declining labor force [23]. For example, certain portions of cropland were converted to an installation site of solar power generation facilities.

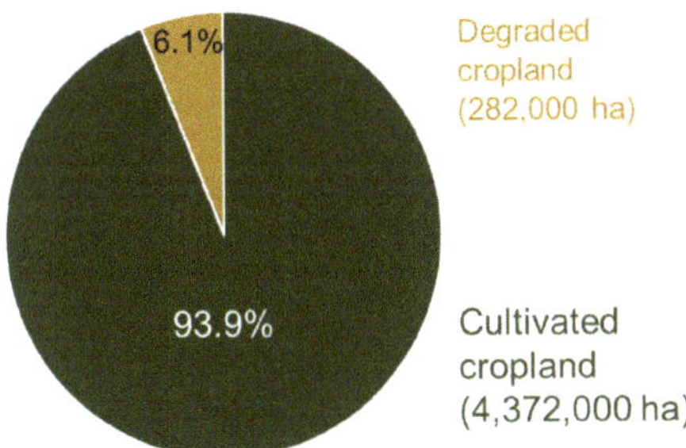

Figure 4. Percentage of cultivated and degraded croplands in Japan as of the end of 2020 [22].

2.2. Conversion of Agricultural Lands to Renewable Energy Sites

There are two ways to introduce solar power generation facilities, either the whole croplands are converted, or shared-use systems are implemented, where land continues to be used for agriculture with additional renewable energy purposes. To date, the latter approach of "solar sharing" is adopted in the majority of croplands, in which farming activities are being continued while the agricultural power plant is generating electricity [23]. First, it is frequently not realistic to immediately convert the whole cropland to have another use. There are financial reasons related to tax; the fixed asset tax of the land will increase if the land is converted from cropland or farmland [23]. This shared land-use system (both farming and solar power) is allowed when a "permit to convert" is granted and built panels of the solar power are specified [23]. In this framework, solar power plants built on cropland were referred to as agricultural power (*Einogata-Hatsuden*) facilities or solar sharing.

There are complications when the installer of solar power generation facilities differs from those who practice farming on site. If the operator wishes to install superficies for underground or overhead structures (*Chijyo-ken*) including solar panels (as set forth in

Article 269-2 (1) of the Civil Code Act [24]), they are requested to obtain the permission set forth in Article 3 (1) of the Cropland Act [18].

The move to consider agricultural power generation (solar sharing) is promoted under the framework of the "Rural Renewable Energy Act (Act No. 81 of 2013)" or the "Act promoting the sound development of agriculture, forestry and fisheries and power generation of renewable energy" [25]. The law was enforced in 2014 to introduce renewable energy in rural areas and improve regional income through renewable energy regeneration [25]. The enforcement of the law encouraged municipalities in developing a system that establishes and approves a renewable facility generation plan without interfering with the flow of food production or national land preservation [26]. There is a strict monitoring of the solar sharing system in Japan, which mandates that the average production volume of the farm should not be decreased by more than 20% (at least maintaining 80% of the production level before the introduction of the solar system) to continue farming activities efficiently with solar panels, and to prevent utilizing the farmland solely as a power generation site [26].

2.3. Deregulation of Cropland Use

On the ground, the monitoring and maintenance of the production level are not well maintained. The MAFF, for example, admitted in March 2022 that 80% of the cropland yield was not met due to interference of solar power plants in farming activities; 308 cases out of 2591 farming cases did not meet the requirement as of the end of 2019 [27]. Amongst the 308 cases, 247 have had inadequate cultivation management [27]. The MAFF documented that 60% of these cases had solar power generators installed by people who lack farming knowledge; thus, the cropland is geared towards electricity sales revenue rather than farm production yields [27]. If the required production yield is not met, the prefectural government or the AC will order farmers to improve their yields or convert cultivated crops. In cases of non-compliance, where farmers do not follow the rules, solar panels are de jure removed. Although, as of February 2021, there were no cases ordering the removal of solar panels [27].

Due to this accelerating trend, these regulations were deregulated recently in a drastic manner, and the mandate on maintaining a cropland yield of 80% has been retracted in Japan [28]. Instead of requiring landowners to produce a yield of 80% in the converted cropland, the MAFF decided to simplify the requirement by examining if the cropland along the solar power plant is utilized properly and efficiently [28]. Thus, when the permit to convert cropland to other land-uses expires (after 10 years), the operators of the solar power plants can renew their permits without considering agriculture production yields. In addition, with the renewal of permits becoming a less tedious process, financial support during cropland conversion is more attainable, and a management system through "project finance" schemes have begun to be sought after in Japan [29].

Recently, the MAFF submitted a draft amendment of the "Agriculture, Forestry and Fisheries Vitalization Act (Act No. 48 of 2007)" or the "Act on settlement for the revitalization of rural areas and promotion of inter-regional exchange" that will enable a collective transfer of cropland rights as a measure to counter degraded croplands [30]. When the AC determines that the cropland is categorized as degraded, and is difficult to reuse or cultivate again (Category 4 in Table 1), they will notify the owners, municipalities, and other relevant parties or organizations [28]. In return, the recipient of the notice is requested to send a notification, ex officio, to the Legal Affairs Bureau to change the land category, for example, as "non-cropland" or "degraded cropland" [28]. With regard to degraded cropland that is exempt from conversion under the Rural Renewable Energy Act [25], when production conditions continue to be unsuitable and non-cultivation for a considerable period has been observed, the law will be relaxed allowing conversion to other land-uses [28].

Another reason for the deregulation of cropland law is due to a general surge of demands to address climate change and increasing global warming countermeasures. There has been an increase in the number of municipalities and populations declaring that they will become carbon-neutral by 2050 or "zero-carbon cities by 2050" in Japan. As of

February 2022, there were a total of 598 municipalities, which is composed of 40 prefectures, 365 cities, 20 special wards, 144 towns, and 29 villages, that showed a strong commitment to counter global warming (Figure 5) [31]. In terms of population density, 115.2 million or 92.0% of the total population of Japan expressed a strong social demand and motivation for climate change countermeasures (Figure 5) [31].

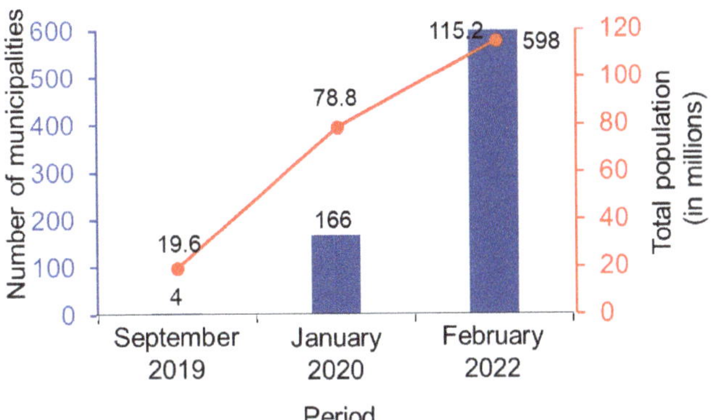

Figure 5. Number of municipalities and population (in millions) with a strong social demand and motivation to increase global warming countermeasures [31].

In addition, the revision of the "Act on Promotion of Global Warming Countermeasures (Act No. 117 of 1998)" [32] that will take effect on April 2022, stated that prefectural governments and government-designated cities are required to set and disclose targets for the introduction of renewable energy, and that the local government is required to designate areas ("promotion area" or *Sokushin-Kuiki*) to the promotion of renewable energy (Figure 6) [33]. These can become zones where solar or other renewable energy installations will be encouraged.

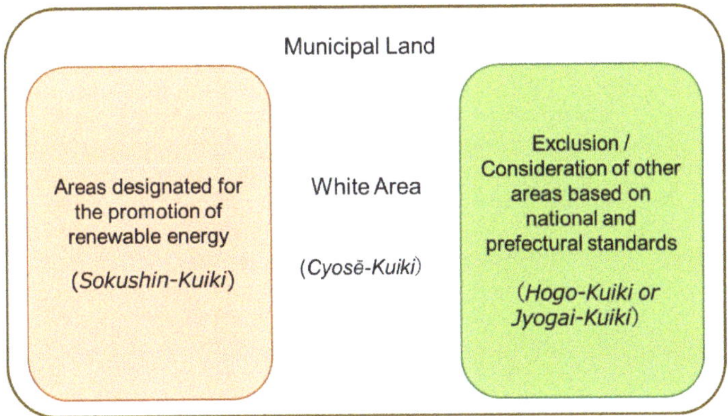

Figure 6. Local governments designate "promotion areas" or *Sokushin-Kuiki* for renewable energy sites [33].

3. Management Options for Croplands at National Level

The MAFF has provided support for businesses in districts that engage community members in the maintenance of croplands to prevent the increase of "degraded cropland", brought about by shortages within the labor force [34]. In relation to this support, "optimal land-use measures" (*Saiteki Tochi Riyo Taisaku*) is one of the policy measures that the MAFF initiated in the fiscal year 2021, concerning the willingness to utilize croplands [35]. In the framework, areas with 10 ha of cropland are covered by the measure [35]. This initiative, which plans the use of croplands, is a collaborative work among different stakeholders including the Agricultural Commission, regional agricultural cooperatives (e.g., Japan Agricultural Cooperative), the Cropland Intermediate Management Organization (Cropland Bank), land improvement district offices, municipalities, farmers, and local residents. The MAFF aims to achieve the maintenance and strengthening of the communities through this project in 100 areas nationwide by the fiscal year 2026. The regional development division of the MAFF said that "If you are having trouble maintaining a village due to aging and lack of successors, I would like you to use this project to discuss sustainable land-use measures" [34].

During the planning process, agricultural districts can divide croplands into "croplands that can be cultivated and concentrated with farmers" and "degraded croplands that are difficult to manage or cultivate." In the latter land type, landowners can decide the management method to use, either (1) grazing, (2) cultivating labor-saving crops like honey-source plants, or (3) afforestation with wildlife buffer zone functions. In addition to these three management methods, financial support and infrastructure improvements such as the leveling of ground and the necessary installation of electric fences will be carried out [36]. In the first management method, the "recommendation of grazing on abandoned cultivated lands" published by the National Livestock Improvement Center was introduced on the website of the MAFF [37], and financial support was provided for conditioning the land for electrical pasture fences. For example, during the fiscal year 2021, there were five districts nationwide that applied low-cost land-use projects for grazing and planting of local crops such as those located in the Hokkaido and Oita prefectures [34,37].

In the second management method, which considers cultivating labor-saving crops, financial assistance was provided for the procurement of seeds and seedlings of honey-source crops and necessary equipment [34]. Moreover, in the third management method, financial support for afforestation methods included, for instance, subsidies for project meetings (up to 5000 Japanese yen per 0.1 ha) and hardcore projects such as land development (approximately 36,000 Japanese yen per 0.1 ha) [34].

The addition of the third management option (afforestation) is one of the unique features added to the project in the fiscal year 2022. This new addition seemed to show a certain gap between the agriculture and forestry sectors, and the decision to start "afforestation" in the agricultural sector raised concerns in terms of how far the sector will be involved in the disaster prevention (ecosystem-based disaster risk re-duction (Eco-DRR)) and habitat provision (wildlife buffer zone) services that come with the afforested area (green infrastructure [GI]). It is important that the agricultural sector clearly delineates these services, especially since they are investing in public subsidies to implement Eco-DRR/GI using planting trees methods.

As an example of how important it is to clearly set boundaries in policies, a similar afforestation method was adopted in the 1997 revision of the River Act (Act No. 167 of 1964) [38], where "forest belt zones" (*Jyurin-Tai*) were added to the "river management facility" (*Kasen-Kanri-Shisetsu*) in Article 3 (2) of the aforementioned law. In this law, the forest zone is controlled by the River Bureau of the Ministry of Land, Infrastructure, Transport and Tourism (MLIT), indicating that they shall become the defendant in cases where there is a defect in the control of the forest belt zone. This is regarded as a different scene from the vertically divided administration of Japan. This is because, in the case of river management, if there are any defects in the management of the afforestation area, and there are serious damages caused by the wildlife, the agricultural authority of the MAFF will be in charge to deal and implement preventive measures, instead of the forestry authority.

4. Insights and Prospects from Matsusaka City Project Site

To date, the project is conducting social experiments in Matsusaka City, Mie Prefecture, where agriculture and forestry are being practiced [7]. This site was selected because the area is active in both agriculture and forestry. There were two approaches involved in the investigation. First, the labor schedule was obtained by interviewing the actual workers. This was done to estimate the labor required in maintaining such land-use types (e.g., cropland). Based on the 2020 agriculture and forestry census in the Iida area (Miyamae, Kabata Mori, Haze districts), there were 96 and 50 management entities in agriculture and forestry, respectively. Amongst this group, we were able to conduct preliminary interviews with 18 workers from forestry, tea, rice, and other agricultural industries from July to October 2021. Then, the information gathered is used in comparing and simulating future scenarios with decreasing populations and possible changes to products. Second, we organized group discussions with the local communities about their present and future preferences for society from a general perspective. In that discussion, we also presented the changes to the legal system (as summarized in this paper), and shared that these changes do not force members of the community to do anything, but rather, they expand their options. As an example, subsidies will be provided to landowners who prefer a minimal management–demand system (e.g., grazing). There are also options to convert agricultural lands to renewable energy sites, which is expected to have an economic spillover effect of up to about 180 million yen per year for local residents and businesses [39]; however, this option is challenging because in order to create a ripple effect in the region, "an increase of 188 migrants for measures against vacant houses and 18,880 for tourism promotion is needed" [39].

The trade-offs between the presentation of future scenarios (selective only), and the advantages and disadvantages of each scenario were discussed as clearly as possible. At that time, we focused on (1) the grand model scaled for the entire region, (2) the decision-making of the Agricultural Commission, which is the representative organization of the farmers, and (3) the decision-making of the individual residents (Figure 7). We tried to give advice, where possible, on each of the steps (1–3 in Figure 7), and suggested specific models and trade-off factors for each.

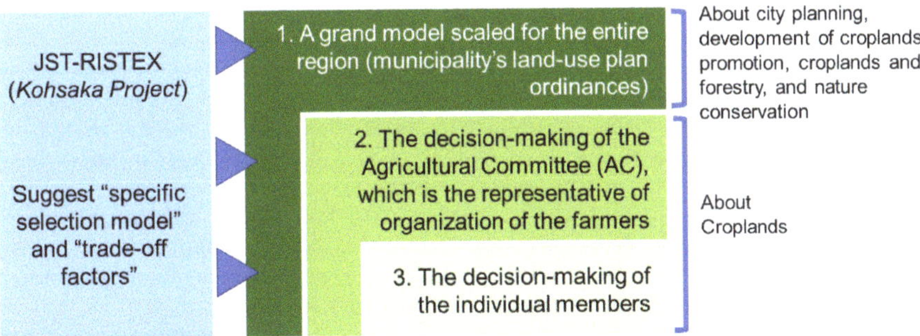

Figure 7. Multi-layered decision-making of community members about croplands.

We documented that the most important one is the "grand model" scale of the entire region. If there is a higher-level unified view, it will integrate normativity. Each piece of land belongs to an individual, but if it is used differently between neighboring lands, it will be inefficient, and the unity of the area will be lacking. Moreover, it does not quite establish regional brand products. After the controvertible problem arose, community members tended to think of the problem as their own, which made it difficult to develop a unified and objective view; therefore, we deduced that it is important to prescribe a unified image of the area when there are no specific problems. In other words, it is important to establish a higher-level norm of the region and a unified image of the future that can play a normative role when deciding things.

The preliminary results of the project showed that broader topics allowed the community members to discuss and share their general perspectives on planning issues, including, for instance, transportation, education, and employment issues. We drew the possible implications for land-use in their areas from these general discussions. For instance, from interviews and discussions, we noted that there are land use-related implications such as the critical points in selecting potential sites for downsizing, which included (i) areas that can still be managed in the future, (ii) areas that are "returning to nature" with minimal management, and (iii) areas that can be managed with a minimal labor force (e.g., strategic zoning) (cf. Figure 2).

From the group discussions, we gained further insight into community members' perspectives. The discussions intentionally avoided focusing on land-use, and instead focused on broader topics, since the community members tended to avoid directly expressing their views on sensitive topics such as land use. Initiating discussions with less direct matters was one of the insights and lessons learned.

The project is still in progress. Based on our understanding thus far, the project aims to propose a consensus-building mapping system in the long-term, aiming to produce maps with fundamental information on agriculture, forestry, and environmental conservation to supporting local land-use policies and decision making. The planned mapping system intends to classify areas for management plans, in the case of forestry, the forests, and other areas, for the introduction of coniferous and broad-mixed forests. Furthermore, based on the stage of development of the project, the proposed system will consider specific businesses and management methods that can simultaneously improve productivity and environmental conservation in the areas.

5. Discussion and Future Considerations

The recent progress of the project (i) presents ongoing discussions on the active promotion of reusing degraded cropland and (ii) provides valuable insights to be considered moving forward. Here, we raised three major points for consideration that are paramount in supporting the central government (e.g., MAFF of Japan) in its strategic zoning of cropland, particularly those categorized as "degraded", to make a balanced system between agriculture, forestry sectors, and environmental conservation.

The first point is the addition of the new management method—"afforestation"—in the agricultural sector of the MAFF. This method is, of course, not new to the forestry sector, but for the agricultural sector, the new addition to the system would be the first policy change since the end of World War II. The MAFF of Japan covers three jurisdictions, including agriculture, forestry, and fisheries. In the current system ("agriculture > forestry > fisheries"), the use of a "forestry tool (afforestation)" and investment of the subsidies (e.g., public subsidies) that come with it from the agricultural sector, must make a major change to the ministry. The current system mainly refers to orders in budget allocation, staff sizes (14,199 for MAFF headquarter (mainly in agriculture), 4705 for the Forestry Agency, and 987 for Fishery Agency as of 2022), and hierarchical orders on human resources, which are critical for the consciousness of insider bureaucrats. The differences in roles are conventionally clear, and afforestation-related measures were under the control of the Agency of Forestry in the post-war period; thus, the newly introduced measures indicated that there is a shift in such authoritative boundaries (e.g., agricultural-related departments will handle afforestation measures). This point is very crucial moving forward, since the ministry is the primary actor in the "optimal land-use measures" (or *Saiteki-Tochi-Riyou* in Japanese) of the government; thus, the change in the policy should be further evaluated to ensure that there is no overlapping of projects with other sectors (e.g., forestry), and to implement it efficiently and effectively.

The second point we raised is the difficulty in balancing carbon reduction measures with cropland conservation. Though recent trends showed an improvement, as discussed in Chapter 2, there are still many challenges in achieving a balanced system between conservation and economic goals. For instance, the MoE of Japan said that "the person who controls carbon reduction measures shall be adopted by the next generation"; however, Japan's agriculture is the "home industry" of the country, so, even if the next generation is governed by energy-related policies, food security will always be one of the foundations of national security.

The MoE has shown that the introduction of solar power generation (5000 kW or 1000 households at 5 kW per household) will benefit the local economies (e.g., migration and tourism), and will have an economic ripple effect of up to approximately JPY 180 million per year for local residents and enterprises. To create such an economic spillover effect in the region, there should be an increase of 188 migrants to occupy vacant houses, and 18,800 tourists for tourism promotion [39,40]; thus, effective promotion of migration and tourism is needed to entice people. Alternatively, it is also possible to evaluate the attitude of promoting the introduction of renewable energy that is cost-effective. For instance, the MoE suggested that "it is important to make renewable energy projects that benefit the region such as revitalizing regional economies and building disaster-resilient regions", since there is a problem with "regional consensus-building" [39].

It is necessary for the region to decide what is best for them in terms of revitalizing the regional economy. In the past, opportunities were limited in terms of utilizing degraded cropland based on the Cropland Act [18]; however, to date, there are now other possibilities such as conversion to renewable energy sites or coexistence with renewable energy facilities (e.g., solar sharing). With a series of cropland policies concerned with deregulation, and abandoned cropland marketed as "degraded cropland," the freedom to use alternative management has increased, and croplands have been flexibly converted and operated. This rapid increase has, in turn, raised a question from the local government and residents: "what kind of region should be created?" Thus, local actors play an important role in regional consensus building. We suggested to the local communities to think about ideal conditions in 30 years, which covered temporal and spatial scales. This was suggested because the owners frequently think about the past, particularly what their ancestors did or what the current difficulties are. Thus, the issues are locked down in individual ownership-related topics; however, thinking towards the long-term promotes communities to think holistically beyond land types and ownerships. In a similar vein, changing the viewpoints on different scales can promote discussions from different angles and perspectives.

Finally, regarding the third point, coordination among various plans such as landscape (target area), renewable energy (promotion area), and land-use (utilization area) plans is extremely critical. In areas where there is no interest in conserving cropland in the future, we documented that there were individuals with a "desire to install solar panels." The authenticity (earnest desire) of the landowner is an essential factor to be considered, but for the conversion and operation process, it is paramount to take a careful stance after examining the negative aspects through the lenses of various plans. Moreover, on a personal level, each individual has their own intentions and desires, and although they are the landowners, there is a risk that the landscape, ecosystem, and aspects of the "region" will be changed. Thus, "regional consensus or agreement" is necessary to achieve a comprehensive and systematic utilization plan.

Author Contributions: Conceptualization, R.K. and S.K.; Methodology, R.K. and S.K.; Validation, R.K. and S.K.; Data Curation, R.K. and S.K.; Writing–Original Draft Preparation, R.K. and S.K.; Writing–Review & Editing, R.K. and S.K.; Supervision, R.K.; Project Administration, R.K. Funding Acquisition, R.K. All authors have read and agreed to the published version of the manuscript.

Funding: This research was funded by JST RISTEX Grant Number JPMJRX20B3; JSPS KAKENHI Grant Numbers JP22H03852; JP21K18456; JP20K12398; JP20K01417 JP17K02105; JST Grant Number JPMJPF2110; Heiwa Nakajima Foundation (2022); and Asahi Group Foundation (2022).

Institutional Review Board Statement: Not applicable.

Informed Consent Statement: Informed consent was obtained from all respondents interviewed in the study.

Data Availability Statement: Not applicable.

Acknowledgments: The authors would like to thank Jay Mar D. Quevedo, Yoshitaka Miyake, and Yuta Uchiyama for their valuable comments and suggestions to improve the manuscript.

Conflicts of Interest: The authors declare no conflict of interest.

References

1. Gava, O.; Bartolini, F.; Venturi, F.; Brunori, G.; Pardossi, A. Improving policy evidence base for agricultural sustainability and food security: A content analysis of life cycle assessment research. *Sustainability* **2020**, *12*, 1033. [CrossRef]
2. Savilaasko, S. Beliefs, facts, and practices: Towards evidence-based decision-making in the forestry sector in Finland. In Proceedings of the 5th European Congress of Conservation Biology, Jyväskylä, Finland, 12–15 June 2018.
3. Delgado, C.; Brooks, K.; Derlagen, C.; Haggblade, S.; Lawyer, K. *Use of Evidence to Inform Agricultural Policy Decision: What Have We Learned from Experience in Africa?* World Bank: Washington, DC, USA, 2019. Available online: https://openknowledge.worldbank.org/handle/10986/34337 (accessed on 25 March 2022).
4. Phillips, P.W.B.; Castle, D.; Smyth, S.J. Evidence-based policy making: Determining what is evidence. *Heliyon* **2020**, *6*, e04519. [CrossRef] [PubMed]
5. Morikawa, M. Evidence of "Evidence-Based Policymaking". (Japanese) EconPapers. 2017. Available online: https://www.rieti.go.jp/jp/publications/pdp/17p008.pdf (accessed on 25 March 2022).
6. Sugitani, K. Can evidence-based policy contribute to correcting overreaction?—Considering democracy and rationalized policy making. *Int. Relat. Dipl.* **2018**, *6*, 306–318.
7. JST RISTEX–Ryo KOHSAKA 2020. Building and Implementing a Mapping Consensus-Building System with Scientific Evidence to Balance Environmental Conservation and Productive Activities in the Agriculture and Forestry Sector. Available online: https://www.jst.go.jp/ristex/stipolicy/project/project40.html (accessed on 15 March 2022).
8. Martinez-Fernandez, C.; Weyman, T.; Fol, S.; Audirac, I.; Cunningham-Sabot, E.; Wiechmann, T.; Yahagi, H. Shrinking cities in Australia, Japan, Europe and the USA: From a global process to local policy responses. *Prog. Plan.* **2016**, *105*, 1–48. [CrossRef]
9. Rieniets, T. Shrinking cities: Causes and effects of urban population losses in the twentieth century. *Nat. Cult.* **2009**, *4*, 231–254. [CrossRef]
10. Uchiyama, Y.; Takatori, C.; Kohsaka, R. Designing participatory green area management and biodiversity conservation strategies in the era of population shrinkage: Empirical analysis of multi-generational perceptions on Satoyama rare species in central Japan. *Landsc. Ecol. Eng.* **2022**. [CrossRef]
11. Iwasaki, Y. Shrinkage of regional cities in Japan: Analysis of changes in densely inhabited districts. *Cities* **2021**, *113*, 103168. [CrossRef]
12. Kobayashi, Y.; Higa, M.; Higashiyama, K.; Nakamura, F. Drivers of land-use changes in societies with decreasing populations: A comparison of the factors affecting farmland abandonment in a food production area in Japan. *PLoS ONE* **2020**, *15*, e0235846. [CrossRef] [PubMed]
13. Shi, Y.; Shi, Y. Spatio-temporal variation characteristics and driving forces of farmland shrinkage in four metropolises in East Asia. *Sustainability* **2020**, *12*, 754. [CrossRef]
14. Kajima, S.; Uchiyama, Y.; Kohsaka, R. Private forest landowners' awareness of forest boundaries: Case study in Japan. *J. For. Res.* **2020**, *25*, 299–307. [CrossRef]
15. Ministry of Agriculture, Forestry and Fisheries. Available online: https://www.maff.go.jp/j/press/kanbo/kihyo01/190830_15.html?msclkid=9fcbf5bcabc711ec92c6eacb29074bf1 (accessed on 15 March 2022).
16. Yoshida, Y. Potential of agrivoltaic systems: The integration of agricultural development and renewable energy promotion. *Energy Resour.* **2019**, *40*, 94–97.
17. The Japan Agricultural News. Area Division into People/Cropland Plan "Agricultural Use" and "Conservation" Ministry of Agriculture, Forestry and Fisheries policy (1 March 2022). 2022. Available online: https://www.agrinews.co.jp/news/index/60476 (accessed on 15 March 2022).
18. Cropland Act (Act No. 249 of 1952). Available online: http://www.japaneselawtranslation.go.jp/law/detail/?ft=2&re=2&dn=1&yo=%E8%BE%B2%E5%9C%B0%E6%B3%95&x=73&y=18&ia=03&ja=04&ph=&ky=&page=1 (accessed on 15 March 2022).
19. Ministry of Agriculture, Forestry and Fisheries. Explanation of Technical Terms. Available online: https://www.maff.go.jp/j/wpaper/w_maff/h18_h/trend/1/terminology.html (accessed on 15 March 2022).
20. Nihon-Keizai Newspaper. Abandoned Cultivated Land, as Large as Shiga Prefecture (Today's Words) (11 August 2013). Available online: https://www.nikkei.com/article/DGXNASGF1000V_Q3A810C1NN1000/ (accessed on 15 March 2022).
21. Ministry of Agriculture, Forestry and Fisheries. About Devasted Cropland in 2020 (Reiwa 2). Available online: https://www.maff.go.jp/j/press/nousin/nihon/211111.html (accessed on 15 March 2022).
22. Ministry of Agriculture, Forestry and Fisheries. About Agricultural Power Generation Facilities, Solar Sharing. Available online: https://www.maff.go.jp/j/nousin/noukei/totiriyo/attach/pdf/einogata-39.pdf (accessed on 15 March 2022).

23. Ministry of Agriculture, Forestry and Fisheries. About Agricultural Power Generation Facilities, Solar Sharing. Available online: https://www.maff.go.jp/j/shokusan/renewable/energy/einou.html (accessed on 15 March 2022).

24. Civil Code (Act No 89 of 1896). Available online: http://www.japaneselawtranslation.go.jp/law/detail/?ft=1&re=2&dn=1&x=36&y=17&co=01&ia=03&ja=04&ky=%E6%B0%91%E6%B3%95&page=91 (accessed on 15 March 2022).

25. Rural Renewable Energy Act (Act on Promoting Sound Development of Agriculture, Forestry and Fisheries and Harmonious Power Generation of Renewable Energy Electricity: Act No, 81 of 2013). Available online: https://www.maff.go.jp/j/nousin/noukei/totiriyo/einogata.html (accessed on 15 March 2022).

26. Ministry of Agriculture, Forestry and Fisheries. About Agricultural Power Generation Facilities, Solar Sharing Situation (as of the End of 2020 (Reiwa 1). Available online: https://www.maff.go.jp/j/nousin/noukei/totiriyo/attach/pdf/einogata-49.pdf (accessed on 15 March 2022).

27. The Japan Agricultural News. 10% of Farm-Type Solar Power Generation "Yield Not Reached" Businesses with Insufficient Knowledge also Surveyed by the Ministry of Agriculture, Forestry and Fisheries. Available online: https://www.agrinews.co.jp/news/index/62608?fbclid=IwAR2A-UboCSh3lDE_rUBXtFS8grfHkFb0IakkTIjlQ0RFOmYwObgKS_B0QCA (accessed on 15 March 2022).

28. Ministry of Agriculture, Forestry and Fisheries. Regarding the Handling under the Farmland Conversion Permit System for Solar Power Generation Facilities, etc. That Continue Farming with Columns. Available online: https://www.maff.go.jp/j/nousin/noukei/totiriyo/attach/pdf/einogata-42.pdf (accessed on 15 March 2022).

29. Ministry of Agriculture, Forestry and Fisheries. Regarding the Handling of Permission under Article 3 (1) of the Cropland Act Regarding the Installation of Farm-Type Power Generation Facilities. Available online: https://www.maff.go.jp/j/nousin/noukei/totiriyo/attach/pdf/einogata-46.pdf (accessed on 15 March 2022).

30. Nihon-Nogyo Newspaper 2022. Collective Transfer of Cropland Rights. To Prevent Devastation and Make Smooth Use. Amendments of Agriculture, Forestry and Fisheries Vitalization Act. (accessed on 14 January 2022). Law Amendments. Available online: https://www.maff.go.jp/j/kasseika/k_law/ (accessed on 21 April 2022).

31. Ministry of the Environment. 2050 Municipalities Expressing Virtually Zero Carbon Dioxide (as of 28 February 2022). Available online: https://www.env.go.jp/policy/zero_carbon_city/01_ponti_20220228.pdf (accessed on 15 March 2022).

32. Act on Promotion of Global Warming Counter Measures (Act No. 117 of 1998). Available online: http://www.japaneselawtranslation.go.jp/law/detail/?ft=2&re=2&dn=1&yo=Act+on+Promotion+of+Global+Warming+Counter+measure&x=67&y=15&ia=03&ja=04&ph=&ky=&page=1 (accessed on 15 March 2022).

33. Ministry of the Environment. Basic Concept Regarding Promotion of Regional Decarbonization Promotion Business (September 2021). Available online: https://www.env.go.jp/policy/council/51ontai-sekou/y510-02b/mat04_1-1-1.pdf (accessed on 15 March 2022).

34. Japan Agricultural Communications 2022. Ministry of Agriculture, Forestry and Fisheries to Prevent the Devastation of Cropland "Tree Planting" Added to the Support Target from 2022. Available online: https://www.jacom.or.jp/nousei/news/2022/02/220224-57098.php (accessed on 15 March 2022).

35. Ministry of Agriculture, Forestry and Fisheries. Agricultural Optimal Land Use Measures (Agricultural, Mountain and Fishing Village Promotion Grant). Available online: https://www.maff.go.jp/j/nousin/tikei/houkiti/saitekitochiriyo.html (accessed on 15 March 2022).

36. Ministry of Agriculture, Forestry and Fisheries. Agricultural Optimal Land Use Measures (Agricultural, Mountain and Fishing Village Promotion Grant) (February 2022). Available online: https://www.maff.go.jp/j/nousin/tikei/houkiti/attach/pdf/saitekitochiriyo-9.pdf (accessed on 15 March 2022).

37. Ministry of Agriculture, Forestry and Fisheries. Outline of 2021 (Reiwa 3) Application District. Available online: https://www.maff.go.jp/j/nousin/tikei/houkiti/attach/pdf/saitekitochiriyo-7.pdf (accessed on 15 March 2022).

38. River Act. (Act No, 167 of 1964). Available online: https://elaws.e-gov.go.jp/document?lawid=339AC0000000167 (accessed on 15 March 2022).

39. Ministry of the Environment. Benefits to the Local Economy by Decarbonizing (Introducing Renewable Energy). Available online: https://www5.cao.go.jp/keizai-shimon/kaigi/special/reform/wg6/20210423/pdf/shiryou1-2-11.pdf (accessed on 15 March 2022).

40. Ministry of the Environment. About the Revised Act on Promotion of Global Warming Countermeasures (October 2021). Available online: https://www.enecho.meti.go.jp/category/saving_and_new/saiene/community/dl/05_07.pdf?msclkid=c2e552fcaa9b11eca9fc0300bb5083c6 (accessed on 23 March 2022).

MDPI
St. Alban-Anlage 66
4052 Basel
Switzerland
Tel. +41 61 683 77 34
Fax +41 61 302 89 18
www.mdpi.com

Land Editorial Office
E mail: land@mdpi.com
www.mdpi.com/journal/land